Planning for Climate Proof Cities

Series Editor

Francesco Musco, Dipartimento di Progettazione e Pianificazione, Università Iuav di Venezia, Venezia, Italy

International commitments to design resilient and adaptive cities, as consequence of the Global Climate Change, can represent the most relevant issue (both political and scientific) of the twenty-first century, requiring an explicit revision of the processes of planning and managing cities.

It appears evident that urban contexts are mostly artificial, therefore characterized by low resilience, so their adaptation capacity is related to a punctual action that should be supported with the definition of a "new culture" of spatial planning and urban design specifically oriented to resilience.

The series will contribute to an international and oriented debate on the role of cities and territories to counteract climate change, also in connection with metabolism improvement, resilience enforcing, risk reduction due to extreme climate local events.

The series will investigate the definition of a "new toolkit" for urban scholars, policy makers and city/urban planners finalized to include and support adaptive and resilient approach in planning, based on recent international researches and practices promoted by universities, cities and public bodies in Europe, US and in the rest of the world.

The series will be strongly interdisciplinary, connecting disciplines that can contribute to design and plan "climate proof cities" of the next decades.

Massimiliano Granceri Bradaschia ·
Filippo Magni · Francesco Musco
Editors

Climate Change Adaptation, Flood Risk, and Beyond

State of Play in the Science-Policy-Action Nexus

Springer

Editors
Massimiliano Granceri Bradaschia
IUAV University of Venice
Venice, Italy

Filippo Magni
IUAV University of Venice
Venice, Italy

Francesco Musco
IUAV University of Venice
Venice, Italy

ISSN 2523-8329 ISSN 2523-8337 (electronic)
Planning for Climate Proof Cities
ISBN 978-3-031-65462-6 ISBN 978-3-031-65463-3 (eBook)
https://doi.org/10.1007/978-3-031-65463-3

This Springer imprint is published by the registered company Springer Nature Switzerland AG
The registered company address is: Gewerbestrasse 11, 6330 Cham, Switzerland

Foreword

Adaptation to climate change is a societal and ecological challenge that must be responded to from a multi-lens point of view. The information and knowledge that we use to adapt to climate change will depend on what types of data we consider important and relevant and also what types of sources of data are considered. We live in a world of multiple knowledge and multiple stakes and failing to take this into account when preparing the ground to adapt will end in simplistic solutions and ways of doing that do not sufficiently reflect the needs and realities of societies and ecosystems. We have experienced almost three decades of science and policy development to prepare for the (now current and future) impacts of climate change, but still important challenges persist when it comes to understanding the challenges ahead. For one, when planning to adapt to climate change one must realise that it is not something that can be addressed in isolation. We live in a reality of compound crises and risks and climate change is just one of them. Assessments of risks and vulnerabilities are therefore required to adopt a complex and diverse understanding of needs, resources, and capacities. In the same way, for actions on the ground to be successful, they need to address not only climate change impacts but also existing and compounded ecological and societal structural vulnerabilities. However, for this to happen, the way we think of adaptation should also change. Processes of planning and decision-making should embrace more interdisciplinary and transdisciplinary approaches where scientific, policy and local expert knowledge can co-exist to provide the best available information to adapt to climate change impacts in context-specific settings. Finally, there is a need for society to understand that adaptation is not an end but a culture of doing and living in this world subject to changing climates and unstable socio-political systems. We must pursue approaches to policy and action that are grounded on the continuous re-evaluation of impacts and processes on the ground to learn and adapt to those changing socio-ecological conditions. This book delves into the different layers of complexity that characterise climate change adaptation from problem understanding to planning and action, demonstrating that it is not a wicked problem, but rather a challenge that requires a different set of socio-political and scientific skills and mindsets.

Marta Olazabal
Adaptation Research Group
Basque Centre for Climate Change
Bilbao, Euskadi, Spain

Acknowledgments

The Editors would like to express their sincere gratitude to the Authors, who wrote the chapters of this book.

The Editors would also like to thank the Reviewers, who contributed with their valuable time and knowledge on providing their detailed reviews.

Finally, the Editors would like to acknowledge the support from the European Commission (EC), which has financed STREAM project under the Interreg Italy-Croatia 2014–2020 programme period. This book is the output of a two-day seminar, under the STREAM project, on climate change adaptation and flood risks held in Venezia on 29th and 30th of May 2024.

The Editors confirm that the content of this book is the sole responsibility of its Authors and Editors and does not necessarily reflect the views of the EC.

Venice, Italy

Dr. Massimiliano Granceri Bradaschia
Dr. Filippo Magni
Prof. Francesco Musco

Contents

About the Editors

Massimiliano Granceri Bradaschia is a post-doc researcher at IUAV University of Venice and an independent consultant with more than 10 years of work experience in Urban and Environmental Planning, Climate Change Adaptation, Monitoring and Evaluation, Public policy analysis. He is a professional with diverse and multicultural experiences as both scholar and practitioner in Brazil, France, Spain, Italy, and in the lusophone African countries. He holds a Ph.D. in Urban and Regional Development (2020) and a M.Sc. in Urban and Environmental Planning (2014).

He is officially chartered to the Italian Urban Planning National Register (2014) and to the Turin Architects and Urbanists register (2020).

He's been working at the services of: (i) governments, (ii) private companies and consortia, (iii) learning and research institutions, (iv) donors and funding organizations, supporting the improvement and update of their operative frameworks with the aim to mainstream Sustainability and Climate Change Adaptation.

Filippo Magni urban planner, holds a Ph.D. in planning and public policies for the territory at Università Iuav di Venezia, a master's degree in planning at Iuav with a double title from the International Master in Estudios Territorials i de la Població issued by the Universitat Autonoma de Barcelona (UAB). He is currently a researcher in urban and regional planning.

He has carried out research and teaching abroad at Universitat Autonoma de Barcelona, Universidad de Sevilla, and the Drexel University of Philadelphia, where, through a multidisciplinary approach, he focused on the need to redesign urban planning tools, recognised as the main vectors capable of increasing greater resilience to climate change.

He is the scientific co-coordinator of the Interreg ITA-CRO project "ADRIADAPT and of the project" Venice 2021-Line 5.3.1. He is the coordinator of the technical-scientific consultancy unit for the implementation of the climate strategies of the municipalities of the Lombardy region, activated by Iuav and Cariplo Foundation within the "F2C-Fondazione Cariplo per il clima" project. Previously he coordinated the technical support unit for the Interreg MED 2014-2020 projects "COEVOLVE and LIFE MASTER ADAPT".

Francesco Musco urban and environmental planner, Ph.D. in Analysis and Governance of Sustainable Development (Ca' Foscari, Venice), Full Professor in Urban and Environmental Planning at the Department of Architecture and Arts—University Iuav of Venice. Currently teaches "Environmental planning", "Spatial Planning for Climate Change", "Urban design".

Currently is Director of Research at Iuav (2021–2024) and Director of the MS in Urban Planning for Transition and professor of Ph.D. program in "Spatial planning and public policies for the territory" at Iuav.

Scientific coordinator of numerous EU-funded projects and research and he obtained more than 8 million euros of competitive research funds in the period 2011–20. Among his publications: with M. van Staden (2010), Local Government for Climate Change, Springer, NY; (2016); (ed) Counteracting Urban Heat Island Effects in a Global Climate Scenario, Springer NY.

Chapter 1
Introduction: How Climate Change Adaptation is Addressed Through Science-Policy-Action Nexus in This Book

Massimiliano Granceri Bradaschia and Filippo Magni

Abstract In a historical period characterised by the increasingly serious impacts of climate change, the need to develop adaptive strategies to address complex and interconnected challenges has become an imperative that can no longer be postponed. The need to address this phenomenon calls for the adoption of a comprehensive, multi-risk approach, requiring the integration of cutting-edge scientific knowledge and practical experience. This approach includes collaborative public policy formulation, leveraging collective stakeholder understanding and facilitating pervasive and strategic actions on the ground. Precisely the Science-Policy interface emerges as a crucial dimension in this scenario of change, requiring the transfer and co-production of knowledge between climate experts, spatial planners, and policy makers. Also, the Policy-Action interface is central, requiring a re-evaluation of spatial policy frameworks and the development of new operational tools and processes. Starting from these premises, the book emphasises the collaborative efforts of the main actors that characterise this triple science-policy-action relationship: scholars, policy makers, civil servants, and practitioners. This publication unfolds in three distinct sections, each meticulously addressing a critical dimension of the overall challenge. The narrative is structured to traverse the intricate landscape of scientific inquiry, policy formulation, and the imperative for transformative action. As this scientific exploration proceeds, the integration of different disciplines and the imperative for collaborative efforts underscore the need for a holistic approach in addressing the multiple impacts of climate change.

Keywords Science-policy · Policy-action · Holistic approach · Climate change adaptation

In a historical period characterised by the increasingly serious impacts of climate change, the need to develop adaptive strategies to address complex and interconnected

M. Granceri Bradaschia (✉) · F. Magni
Department of Architecture and Arts, Iuav University of Venice, Venice, Italy
e-mail: mgranceri@iuav.it

M. Granceri Bradaschia et al. (eds.), *Climate Change Adaptation, Flood Risk, and Beyond*, Planning for Climate Proof Cities,
https://doi.org/10.1007/978-3-031-65463-3_1

challenges has become an imperative that can no longer be postponed. This volume aims to explore the intricate landscape of climate change adaptation processes, characterising it as a multi-disciplinary, multi-scalar and multi-actor challenge, requiring a holistic and proactive response.

The need to address this phenomenon calls for the adoption of a comprehensive, multi-risk approach, requiring the integration of cutting-edge scientific knowledge and practical experience. This approach includes collaborative public policy formulation, leveraging collective stakeholder understanding and facilitating pervasive and strategic actions on the ground. The imperative is underscored by the intricate nature of climate change impacts, which require a pervasive and adaptive strategy that considers multiple risks and integrates knowledge from multiple disciplines. Consequently, a robust response requires the synthesis of scientific expertise, social sharing in policy development, and sustained efforts by actors on the ground to effectively mitigate and adapt to the multifaceted challenges posed by climate change.

Precisely the Science-Policy interface emerges as a crucial dimension in this scenario of change, requiring the transfer and co-production of knowledge between climate experts, spatial planners, and policy makers. Within the framework presented, the analysis of ecosystem services takes on a new and increasingly central role as a key component in climate risk assessments. This analytical approach clarifies the intricate interdependencies between natural systems and human vulnerabilities, providing a broad understanding of ecosystem capacity to mitigate or exacerbate climate-related risks. The use of advanced methodologies in the analysis of ecosystem services enables a comprehensive assessment of the multiple contributions of ecosystems to human well-being, emphasising their role as critical components in climate adaptation and resilience strategies. This integrative assessment further emphasises the importance of recognizing and harnessing the ecosystem services provided by natural systems to inform evidence-based decision making in the face of complex climate challenges.

In the current scientific arena, extensive investigations have been conducted to clarify the cause-effect and effect-response relationships associated with climate change. However, there has been a significant shift in focus, bringing to the forefront the hitherto underestimated interface between policy dynamics and implementable strategies. This paradigmatic shift underscores the critical role of the policy dimension in contemporary climate adaptation processes. It underscores the imperative to transcend theoretical understanding and effectively translate climate knowledge into tangible actions. Today, the Policy-Action interface is central, requiring a re-evaluation of spatial policy frameworks and the development of new operational tools and processes. This reconceptualization accentuates the urgency of implementing policy measures oriented toward mitigation and adaptation to climate impacts, thus fostering a synergistic integration of scientific knowledge with site-specific and target-oriented policy initiatives.

Nevertheless, one key obstacle remains evident in the area of climate change adaptation: the absence of a well-defined and effective monitoring and evaluation (M&E) framework to assess this complex process. Despite the abundant availability of data on various climate phenomena such as floods, heat waves, and droughts, there is a lack of a comprehensive and unambiguous M&E system that facilitates the assessment of adaptation efforts and related spatial impacts. This deficiency in the M&E apparatus hinders processes for assessing the effectiveness of implemented adaptation strategies and inhibits the development of a holistic understanding of their spatial implications. A robust M&E framework is essential to promote scientific (and economic) evidence-based decision-making processes, enabling policymakers to assess the success of adaptation interventions, identify areas for improvement, and refine strategies in response to changing climate dynamics. Consequently, the development of an effective M&E system emerges as an imperative step to improve the effectiveness and resilience of climate adaptation initiatives.

Starting from these premises, the book emphasises the collaborative efforts of the main actors that characterise this triple science-policy-action relationship: scholars, policy makers, civil servants, and practitioners. Structured in three distinct parts, the text outlines the complexities of climate change adaptation.

This publication unfolds in **three distinct sections**, each meticulously addressing a critical dimension of the overall challenge. The narrative is structured to traverse the intricate landscape of scientific inquiry, policy formulation, and the imperative for transformative action. As this scientific exploration proceeds, the integration of different disciplines and the imperative for collaborative efforts underscore the need for a holistic approach in addressing the multiple impacts of climate change.

The first part of the volume delves into the interface between science and policy, focusing mainly on the Adriatic region. Here the emphasis is on the development and monitoring of multi-risk assessments. This scientific effort aims to serve as a conduit, effectively bridging the gap between the reservoir of scientific knowledge and the construction of public policy. The ultimate goal of this first part is to equip decision-making processes with a solid foundation, ensuring that policies are not only aware of scientific insights but are also adaptive to the dynamic challenges posed by climate change.

With a seamless transition, **the second part** focuses on the intricate interplay between policy and action. Within this segment, how theoretical frameworks are translated into practical tools is explored, touching on areas ranging from land-use planning to climate finance. The focus extends to the design of tailored operational plans for both coastal and river flooding, clarifying the complex issues involved in mitigating the impacts of climate change at the ground level. In addition, this section addresses the increasingly pressing challenge of M&E for flood-related projects, offering innovative methodologies to bridge the gap between policy intentions and tangible results achieved by these projects.

The third part of this scientific exploration, titled "From Action to Policy," shifts the focus within Brazil's experiential landscape. Here, the narrative is polarised toward water-sensitive and art-driven approaches, recounting ongoing experiences that redefine conventional perspectives on urban water and secondary river systems. Through an analytical lens, these experiences are presented as a model for both local communities and policymakers, highlighting the transformative potential of creativity and cultural elements in the field of climate change adaptation.

In this book, the multi-faceted trajectories of the several scientific explorations led to the statement of that necessary holistic condition recalled in the opening. Adaptation to climate change is presented no longer as a compartmentalised effort; instead, it becomes a dynamic and interconnected challenge to be responded to with collaboration and integrated methodologies.

By addressing existing gaps in monitoring and evaluation, ecosystem services analysis, and drawing inspiration from bottom-up experiences, this publication aims to establish a solid foundation for effective and sustainable adaptation to climate change. The collaborative engagement of key stakeholders and contributors along the science-policy-action nexus emerges as the force that can drive new land management and flood risk reduction processes, accompanying through the intricate complexities of climate change toward a future characterised by resilience and adaptability.

Part I
From Science to Policy

Chapter 2
Assessing Hydrogeological Vulnerability Within Northern Apennines: An Integrated Spatial Analysis in Emilia-Romagna Region (Italy)

Gianfranco Pozzer

Abstract The steady increase in climate change-related disaster phenomena in Italy forces territorial (and local) governments to meet a growing demand to revise mitigation and adaptation protocols. Housing solutions and models are called upon to update resilience strategies *vis-à-vis* phenomena such as the spatial convergence of various impacts and vulnerabilities. Complex vulnerability domains, often unpredictable, require new computational technologies and a clear definition of "spatial knowledge" according to critical issues, emerging challenges, and settlement perspectives. This paper provides an opportunity to test this hypothesis using spatial modelling and remote sensing techniques. The objective is to ease the identification of physical-environmental correlations/associations among endogenous and exogenous factors, namely between morphological states and climatic stressors "captured" through satellite images. The contribution is part of the European project AdriAdapt Interreg Italy-Croatia coordinated by CMCC (Euro-Mediterranean Center on Climate Change). It develops a test of spatial interaction between flooding, landslides, and drought. The test is conducted within the sample territory of Unione dei Comuni Valle del Savio (Emilia-Romagna Region), an area frequently affected by flooding events and hydrogeological instability. The results show a significant overlap between hydraulic performance and spatial texture, providing a new vulnerability vision in a multi-system perspective.

Keywords Climate proofing · Anthropogenic activity · Runoff modelling · Remote sensing · Multi-vulnerability · Spatial planning · Uncertainty

G. Pozzer (✉)
Università Iuav di Venezia, Venice, Italy
e-mail: gpozzer@iuav.it

M. Granceri Bradaschia et al. (eds.), *Climate Change Adaptation, Flood Risk, and Beyond*, Planning for Climate Proof Cities,
https://doi.org/10.1007/978-3-031-65463-3_2

2.1 Introduction[1]

The Italian peninsula is a territory at high hydraulic and hydrogeological risk due to morphological reasons, a fast and dense urbanization and inefficient land use.[2] According to estimates from the Institute for Environmental Protection and Research (ISPRA) for 2021 (Trigila et al., 2021), over 90% of Italian municipalities are affected by hydrogeological criticalities or coastal erosion phenomena. Additionally, about 7 million people live in areas with high flood risk. In this context, current climate anomalies will have a significant impact, posing new challenges to land planning, management, and regulation devices. According to the Intergovernmental Panel on Climate Change (IPCC) predictions, climate change-related phenomena will intensify in the coming decades, and extreme events will increasingly increase rise risk to environmental protection and territorial safety (Espon Climate, 2022; IPCC, 2018; Iturbide et al., 2021).

Recent flooding events (Spring of 2023) in various provinces of Emilia-Romagna are unquestionable evidence. They demonstrate how the presence of meteorological phenomena of particular intensity and duration, combined with morphological fragility, partly related to climate change and partly to anthropic factors,[3] can worsen territorial instability and environmental degradation (Raikes et al., 2019). These

[1] This contribution provides a brief overview of the results of a scientific research project on climate change (2019–2021), initially conducted within the Planning and Climate Change Lab of the University Iuav of Venice (coordinated by the scientific supervisor Francesco Musco, with the technical and methodological support of Filippo Magni, Denis Maragno, Gianfranco Pozzer, and Giovanni Carraretto). The research results have recently been incorporated into IR.IDE (a platform for scientific innovation established as part of the Department of Excellence Iuav project, 2018–2022), thanks to the collaboration between the VAULT study centre of IR.IDE (Value Activators in Urban Landscape and Territory) and the Planning and Climate Change Lab of Iuav. The contribution presents the methodological results of the AdriAdapt project (experimental research funded under the European Interreg Italy-Croatia program). The contents partly refer to a previous paper presented at the XXIII National Conference of SIU—Italian Society of Urban Planners (Pozzer et al., 2021).

[2] For many years, Italy has been grappling with the consequences of climate change due to its unique combination of morpho-climatic, morphological, topographic, and geological characteristics. The storm "Ciaran" on November 2nd and 3rd, 2023, generated more than 73 extreme meteorological events (wind gusts and heavy rainfall), hitting many regions hard, with Lombardy and Tuscany being among the most affected, resulting in casualties, missing persons, and significant damage (source: Dire—National Press Agency, November 4, 2023). In Italy, flood risk management requires a high level of interaction between hydraulic engineering and territorial planning, and policies must now work synergistically and systemically more than ever to counter the increasing territorial vulnerabilities (Landauer et al., 2019). These policies should promote sustainable land use, infrastructure, and mobility models adapted to the changing climate, as Mims (2021) highlighted. Regarding the possibilities of reducing risk, the recommendations of the De Marchi Commission, proposed immediately after the floods of 1966, remain relevant. In addition to structural interventions, non-structural works should also be considered, enabling the prediction of when and with what intensity a flood event could occur. This can facilitate the creation of new communication channels for managing emergency phases, alerting the population and ensuring safety where needed.

[3] For instance, sea-level rise, drought, erosion, deforestation, intensive crop cultivation, soil consumption, etc.

phenomena generate violent and extensive flooding, often exacerbated by hydraulic inefficiencies in drainage systems (Galimberti & Balbo, 2023),[4] soil compaction processes, or a significant increase of sealed surfaces.[5]

In Emilia-Romagna, soil artificialization reduces the resilience of local ecosystems with adverse effects on vulnerability and risk components (Munafò, 2022). There is a noticeable reduction in the surface needed for natural rainwater infiltration, which flows into highly urbanized environments much more rapidly than in the past (Avand & Moradi, 2021). This process can be monitored through specific "inflow-outflow" dynamic simulation models with logical criteria on flooding based on a spatial association between land use, runoff and terrain morphologies (Pistocchi, 2018).

Such a simulation helps to draw potential risk scenarios, assessing the effectiveness of mitigation and adaptation actions during planning and programming activities (European Commission, 2021). In this perspective, the present work illustrates an experimental approach for assessing vulnerability to landslides and floods, focusing on hydraulic and hydrogeological features in spatial analysis. The European project AdriAdapt—Resilience information platform results for Adriatic cities and towns (PAP/RAC, 2021) are used here. They were developed at the Department of Culture of the Project at the Iuav University of Venice in collaboration with the CMCC (Euro-Mediterranean Center on Climate Change). The study area is located in the Valle Savio (Emilia-Romagna, Italy), and an analysis impacts model on flooding, drought, and landslides is applied.[6] The factors considered vary with data availability as parameters and physical properties are acquired through remote sensing. The simultaneous consideration of endogenous and exogenous factors facilitates the construction of an integrated geo-dataset capable of catching spatial correlations between phenomena considered by climate-proof planning tools (Magni, 2019). The result helps define a planning dimension capable of recognizing territorial vulnerability

[4] In addition to these conditions, extensive storm surges may prevent the natural drainage of river waters into the sea due to the temporary rise in sea level.

[5] In many Italian territories, the causes contributing to soil consumption are partly attributed to the correlation between the building and construction cycle, infrastructural policies (Patassini & Pozzer, 2018), and the overall regional economic cycle, where the role of the financial component is much more prominent than in the past (Fregolent & Savino, 2014). The dynamics of soil sealing activate specific consumption levels related to settlement morphologies, construction processes, and coverage patterns. Settlement morphologies represent the model of urbanization as an organizational component aimed at forms of consumption uses, and functions. Generally, these phenomena interact, generating synergistic morpho-typological effects, compensatory or decidedly conflicting. The effects of consumption tend to change depending on settlement contexts and the planning cultures, giving rise to territorial fragilities and vulnerabilities that differ in intensity, territorial size, morphological composition, and settlement patterns. Changes in land use intensity can also be traced back to territorial dynamics already activated by administrative fragmentation, which, combined with other economic and demographic factors (demographic balances), tend to define forms of consumption with a contextual peculiarity (Patassini & Pozzer, 2018).

[6] Related impacts occur as first-order, overlapping or contiguous "joint evidence", while cumulative impacts result from additive or multiplicative effect interactions. Cumulative impact is also understood as "second-order evidence" with a nonlinear or complex effect (conversation with Patassini, 2023).

as an outcome of cumulative impacts generated (in part) by the effects of climate change (de Sherbinin et al., 2019; IPCC, 2022). The concept of spatial association (semantically contiguous to correlation but different in metric terms) emerges from structural and contextual informative layers (Fritzsche et al., 2014; O'Brien et al., 2007; Tapia et al., 2015).

In summary, the contribution demonstrates how processing new spatial information generated by modelling cumulative impacts combined with remote sensing analysis enriches understanding of spatial vulnerability assessment in a multi-systemic learning perspective (see, for example, Maragno et al., 2023).[7] The choice of study area enhances the convergence of these new data, comparing them with informative sources, theoretical aspects, and procedural elements validated by various scientific contributions (Kamel Boulos & Wilson, 2023; Maragno et al., 2021).

The article is divided into four parts. The first part (Sect. 2.2) presents the scope of work, research design, and the study area. The second part (Sect. 2.3) introduces an integrated modelling approach. Techniques and data analysis procedures for studying impacts illustrate the logic and operational steps, from constructing individual vulnerability indices to recognizing an aggregated mapping. The third part (Sect. 2.4) discusses the results of empirical research. They reflect the potential of a working methodology capable of recognizing the existing correlation between morphological states and incremental surface runoff. This involves estimating the ratio between impact indicator values and using topological overlapping.[8] The fourth and final part (Sect. 2.5) discusses the procedure's usefulness and the reliability of its results. The method demonstrates how land use and impacts are partly connected to the morphologies of the territory and, in part, directly comparable and assessable in a meta-evaluation process.

The article concludes with recommendations and suggestions for future practice and research.

[7] Here, "systemic vulnerability" represents (and measures) an area's propensity to manifest multidimensional issues, considering the combined reading of three impact conditions: floods, droughts, and landslides. The systemic (or multi-systemic) approach favors integrated assessment cues, facilitating the process of selecting adaptation strategies (and priority areas for intervention) through Decision support systems (Dss) or multicriteria analysis techniques.

[8] It is important to remember that the resurgence of geological phenomena (such as landslides) is a very complex dynamic, where associations like drought landslide and runoff landslide require appropriate interpretative contextualization. This work assumes a strategic-knowledge connotation to support the definition of planning practices oriented towards climate-proofing.

2.2 Scope of Research: References, Design, and Case Study

2.2.1 *Theoretical Background*

This contribution is part of the European program AdriAdapt Interreg Italy-Croatia,[9] coordinated by the CMCC. Its main aim is to increase the capacity of the Upper Adriatic to adapt urban and coastal areas to climate change. The article focuses on the study's progress related to three particularly adverse classes of impact within the territories of Valle Savio[10]: flooding, landslides and drought.[11]

Innovative elements lead to spatial analyses that estimate the differential hydraulic risk linked to morphologies and land use changes (Ungaro et al., 2014). The work delves into a research field recognized at institutional,[12] academic, and operational levels in urban and territorial management practices (Musco, 2016; Pietrapertosa et al., 2019).[13] In territorial governance, assessing hydraulic danger (and flooding) is increasingly recognized with attention in mitigation and adaptation practices (Kourtis & Tsihrintzis, 2021; Wilby & Keenan, 2012). Adaptation is a complex[14] process that relies primarily on the geographic, geomorphological, and climatic specificities of the location (Wamsler et al., 2013), as well as on the "culture" of local communities (Patassini, 2022; Romero Lankao & Zwickel, 2015): it activates

[9] The project analyses impact types that interact with the vulnerability dynamics of the following territories: the Municipality of Udine, the Municipality of Cervia (RA), the Unione Valle Savio, the Municipality of Vodice (Croatia), and the County of Šibenik (Croatia). The following impacts are recognized analytically and cartographically: Urban Heat Islands, Urban flooding, Wildfires, Drought, Landslides, Sea-level rise, and Salt intrusion. The project aims to provide local authorities with the territorial knowledge necessary to prepare procedures and planning models capable of interpreting and raising the resilience levels of settlement environments, preparing them to withstand the mechanical stresses induced by increasingly severe climate variations.

[10] The Union of Municipalities Valle del Savio includes the following territorial units: Bagno di Romagna, Cesena, Mercato Saraceno, Montiano, Sarsina, and Verghereto.

[11] Drought results from local transformations (agricultural changes, urbanization, hydraulic management, etc.) and general climatic variations.

[12] Please refer to Italian legal conditions concerning the imposition and management of hydrogeological constraints and the formation of hydrogeological planning plans (Italian acronym: PAI).

[13] In Italy, many investigations and research on Adaptation to climate change are commissioned and published by the *Fondazione per lo Sviluppo Sostenibile* (see www.fondazionesvilupposostenibile.org).

[14] Adaptation to climate variations has characterized human life and migrations from ancient times, as evidenced by some historical reconstructions. Forms of Adaptation had characterized periods when humans lived by hunting and gathering, when they began to develop agricultural and breeding technologies, with the foundation of the first cities and their subsequent transition in industrial and post-industrial revolutions, as well as in more recent analogue and digital phases. Adaptation is a "vital" practice that produces culture, and its exceptionality is generally experienced in actual or threatened catastrophes and transitional phases (conversation with Patassini, 2023).

economies, infrastructures, capacities, and flows, responding to how risk components arise (Füssel, 2010). When looking at the settlement environment, it is strategically relevant to recognize the domains of vulnerability (IPCC, 2014). This requires an in-depth analysis of morphological structures, physical-functional components, eco-systemic characteristics, and, more generally, dynamic equilibrium conditions (Bridle, 2022; Stewart et al., 2014). Such an operational need is partially met by the availability of new sources and actual interoperability of spatial information, which is assumed to be updatable based on multi-temporal and multi-scale devices.[15]

2.2.2 Research Design

The study addresses the spatial relationship between two constitutive components of hydraulic and hydrogeological systems: an endogenous and an exogenous component. The first component considers how land use and runoff coefficients contribute to altering the surface hydraulic regime of the territory, quantifying and spatializing the hydraulic impact at the scale of the hydrogeological basin (runoff or surface runoff). Hydraulic performance is estimated using a specific spatial simulation model that recognizes the correlation between territorial land use patterns, topography, and terrain morphologies (specifically, slopes, depressions, orientations, and valleys). The second component focuses on the dynamics of drought phenomenon linked to climatic-environmental effects studied with parameters and physical properties acquired through remote sensing.

Recognizing these two analytical dimensions allows a topological overlay between runoff, landslide phenomena, and drought environments.

The research follows an exploratory design (*impact analysis and modelling*) organized in two quantitative phases followed by a qualitative synthesis (Phase 3).

Phase 1: Remote Sensing Analysis. This phase identifies descriptors (variables) helpful for spatially detecting the consequences of climate change in terms of heat waves and drought. The drought study uses Landsat-8 satellite data, considering the impact of vegetation and thermal parameters capable of describing the structure of morphological spatial relationships and the varying degrees of vulnerability of anthropic and natural elements.

Phase 2: Morpho-Dynamic Analysis. The second phase develops a morpho-dynamic model of spatial association between runoff coefficients and a digital terrain model (Dtm). By applying direction and accumulation functions (GIS-based hydrologic modelling), it is possible to study the dynamic behaviour of surface runoff and quantify the impacts of land-use change on the evolution of flood vulnerability.

[15] Spatial analysis and the latest approaches in quantitative geography highlight a semantic correlation, before a metric one, between space and time. The interpretation (represented or experienced) of space is spatialization, positioning in a space broader than the lines of time. Thus, space–time positioning becomes a provisional and oscillating pointer within spatial and temporal scales. This operation creates space and time (conversation with Patassini, 2023).

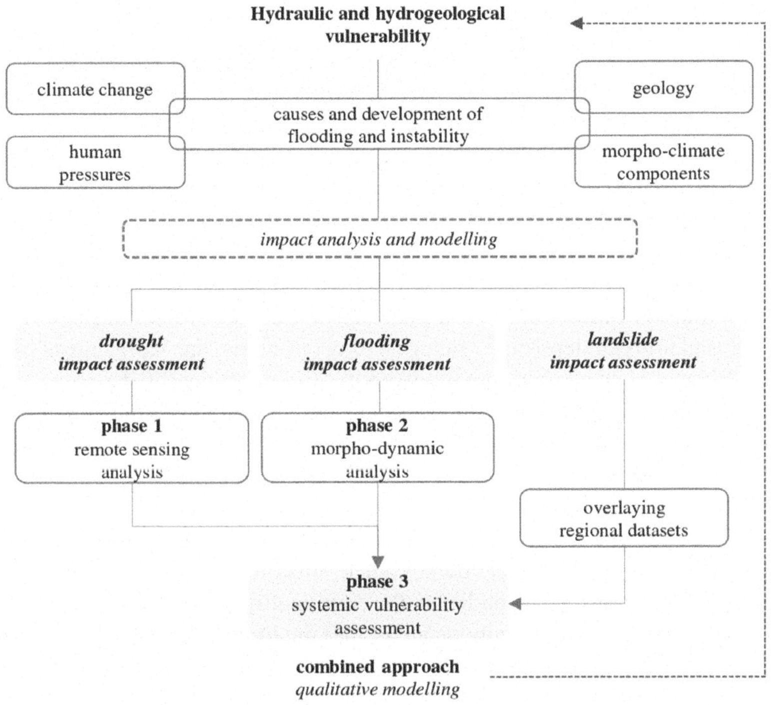

Fig. 2.1 Modelling systemic vulnerability: a logical path

Phase 3: Systemic Vulnerability Assessment. The third phase resumes the procedure for evaluating territorial vulnerability based on a combination of two environmental determinants: sensitivity and adaptive capacity.[16] It, therefore, proposes an evaluative overview based on a logical-spatial integration (at different scales) of ancillary and multi-spectral indices. This leads to recognising a systemic and integrated vulnerability model to landslides and floods (integrated reading of voluntary and mandatory information levels).

This phase generates a qualitative mapping of the most unstable areas and those susceptible to failure, aiming to highlight a potential spatial and morphological association between runoff and active landslides. This final step allows for an initial validation of the entire methodological framework.

The logical path of the research is outlined below (Fig. 2.1).

[16] According to the IPCC approach, sensitivity represents the degree to which a given exposure negatively influences a system. It depends on the specific properties of the system under consideration. Adaptive capacity is the ability of a natural or built system to adapt to climate change in order to moderate potential impacts or damages.

2.2.3 *Study Area Overview*

The study area covers the Valle del Savio Municipal Union. It is a geographic-administrative area located along the Savio River, covering 810 km^2, almost entirely in the Province of Forlì-Cesena, with a very short stretch in the Provinces of Rimini and Ravenna. The Savio basin is between Valle del Bidente (Province of Forlì-Cesena) and Valle del Marenchia (Provinces of Arezzo, Pesaro-Urbino, and Rimini). The morphological profile of the area is characterized by an altitude ranging from 5 to 1520 m above sea level (mountain ranges of Tuscan-Romagnolo Apennines). The most common lithotypes from a pedological perspective are argillites, marly-sandstone, and alluvial deposits.[17] In terms of ecosystems, the Valley hosts multiple natural habitats and rich landscape biodiversity. Bio-climatically, the area is mainly belonging to the temperate region. Besides, the flat areas of the hinterland are moderately urbanized, while those near coastal shores are densely built-up (see, for evidence, urban areas of Cesena).

Only a portion of the Valle's territories falls under hydraulic constraint (Fig. 2.2).[18] In these areas, hydrogeological risk is widespread (Fig. 2.3), and it is primarily associated with slope instability or river courses due to specific environmental, meteorological, and climatic conditions. In addition, the combination of intense urbanization, the gradual abandonment of mountainous areas, and the increased frequency of extreme rainfall poses a severe environmental risk to settlements and human activities.

The analysis focuses on a sample area in the municipality of Bagno di Romagna (FC), providing a combined (correlated/associated) assessment of impacts.

The following factors justify the attention given to Valle Savio:

- presence of urban sprawl along the E45 European road, which at times runs parallel to the Savio River;

---

[17] Source: Soil Cartography of the Emilia-Romagna Region.

[18] Focusing on the assessment framework of hazards and risks associated with floods—updated by the District Basin Authorities in December 2020 (see Article 6 of the European Floods Directive 2007/60/CE, second cycle of Management)—a context emerges in which Emilia-Romagna stands out as a region characterized by percentages of territory potentially subject to flooding significantly higher than the national average (Trigila et al., 2021). This situation increases the population's exposure to flood risk, considering three scenarios of hazard and probability (medium, low, high). As reported in the technical document by ISPRA (ibidem), the extensive extension of potentially floodable areas, especially starting from the medium hazard scenario in Emilia-Romagna, can be attributed to the complex and extensive network of drainage collectors and minor watercourses that develop over vast areas characterized by depressed morphology. These areas often host narrow riverbeds and raised embankments and are subject to a series of regulation and rectification interventions, especially in the plain regions. However, for events with return periods longer than those expected for a high-hazard scenario, it has emerged that the drainage system (mechanically or hybrid-managed) needs to be improved. This deficiency promotes widespread flooding over large portions of the regional territory, increasing flooding vulnerability. The provinces with the highest percentages of susceptible territory to floods are Ravenna and Ferrara, standing at 80% and almost 100%, respectively, considering a simulation related to a medium hazard scenario. The percentage of flood-prone areas in a medium-hazard scenario is approximately 18% for the Valle del Savio area.

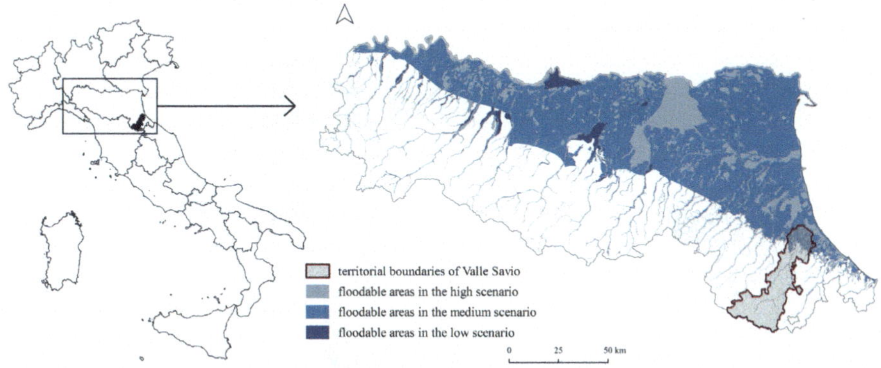

Fig. 2.2 Valle Savio in Emilia-Romagna Region. Floodable areas according to three different scenarios: high, medium, and low probability. *Source* Author's elaboration on the basis of vector maps of floodable created by ISPRA (database 2020)

- urban areas affected by periodic flooding;
- landslide movements related to slope instability, exacerbated by May 16 and 17, 2023 weather events.

2.3 Materials and Methodology

The research tests an exploratory model capable of analyzing geographic domains adversely influenced by a spatial convergence of impacts of different natures, namely drought, runoff, and landslides.

The approach can be considered experimental and involves:

- collection of data from interoperable sources (Sect. 3.1);
- application of algorithms for processing morphological information of impacts (Mapping: Sects. 3.2.1. and 3.2.2);
- cross-evaluation of impacts for a chromatic representation of vulnerability according to spatial coordinates (Modelling: Sect. 3.2.3).[19]

2.3.1 Data Acquisition

The study was conducted using various sources of information (Table 2.1). Some data refer to spatial and alphanumeric information already available at municipal levels (primary and general thematic maps); others come from diachronic readings or modelling states and dynamics through spatial analysis and remote sensing. The methodology also relies on a spatial correlation between satellite data from the

[19] Landslide movements are incorporated into the assessment using the overlay technique.

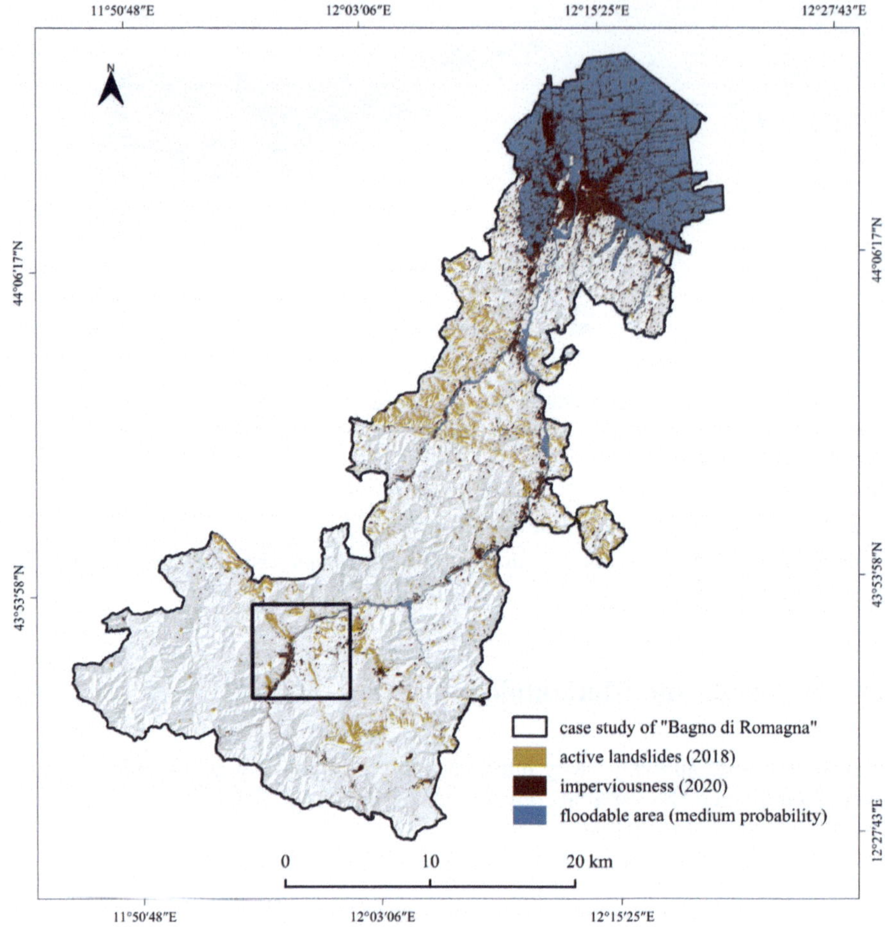

Fig. 2.3 Case study and hydrogeological risk profiles of Valle Savio. *Source* Author's elaboration based on vector maps of floodable areas and active landslide provided by ISPRA (database 2020) and Emilia Romagna Region (database 2018)

European Copernicus program and those recorded by national-scale monitoring and control systems (see, for example, ISPRA Institute).

Remote sensing and Geographic Information Systems (GIS) have been jointly utilized. Remote sensing analysis enables the estimated drought index linked to biophysical parameters derived from processing satellite images. The processing includes a hydro-morphological modelling algorithm for the dynamic simulation of

Table 2.1 Spatial information: category, layer, format and source

Category	Layer	Formats	Edition	Source
Themes of main geo-data	Hydrographic network of the Emilia-Romagna region	Vector	2020	Knowledge Framework of the Emilia-Romagna region
	Digital terrain model (Dtm) 5-m pitch of the Emilia-Romagna region	GeoTiff	2020	
Specific geo-data	Potentially floodable areas in the high, medium and low probability scenario—Emilia-Romagna region	Vector	2020	ISPRA
	Land use and land cover of the Emilia-Romagna region	Vector and raster	2020	Knowledge framework of the Emilia-Romagna region
	Emilia-Romagna Region landslide inventory map: active landslide deposition level[20]	Vector	2018	Geological databank of the Emilia-Romagna region
Remote sensing	LC08_L1TP_192029_20190820_20190903_01_T1	GeoTiff	2019	Landsat 8, USGS
	LC08_L1TP_192029_20181004_20181010_01_T1	GeoTiff	2018	
	LC08_L1TP_192029_20170729_20170811_01_T1	GeoTiff	2017	
	LC08_L1TP_192029_20160827_20170321_01_T1	GeoTiff	2016	
	LC08_L1TP_192029_20150708_20170407_01_T1	GeoTiff	2015	

surface runoff.[21] These techniques facilitate the definition of a guiding model for the multi-systemic and multi-dimensional assessment of hydrogeological vulnerability.

[20] An active landslide refers to a deposit showing evidence of movement in the last seasonal cycle, regardless of the magnitude and speed (Source: Emilia-Romagna Region landslide inventory map).

[21] The algorithm developed is based on an adjustment of the method for hydraulic invariance by the Autorità dei Bacini regionali romagnoli (AdBRR—Emilia-Romagna, Italia). The operational concept of hydraulic invariance by AdBRR allows measuring the minimum reservoir volume to be allocated to areas transforming (land consumption, impermeabilization, urbanization) through the use of a planning index expressed in reservoir volume to associate with new impermeabilizations (Pistocchi, 2001). The criterion of hydraulic invariance is associated with the positive effects of urban drainage networks and the natural attenuation of floods due to the altimetric variations of the terrain (slopes and depressions).

2.3.2 *Mapping and Modelling*

2.3.2.1 Vegetation Health Index

The drought phenomenon is assessed through the Vegetation Health Index (Vhi).[22] This index sketches the intensity of drought and its spatial extent (Bento et al., 2018; Tripathi et al., 2013).[23] Landsat 8 satellite images are used to calculate the index. The estimation is indirect in that it is based on the forest and agricultural response of the vegetation. It captures thermal nature stresses or soil moisture changes.[24] Vhi is based on the ratio of two derived satellite indices: the Temperature Condition Index (Tci) and the Vegetation Condition Index (Vci). The Tci calculation uses thermal data obtained from Land Surface Temperature (Lst),[25] while Vci is based on vegetation data, Normalized Difference Vegetation Index (Ndvi),[26] capable of reflecting soil moisture conditions. Vci (2.1) is expressed with linearly standardized values (in %) that reflect vegetative stress associated with a low moisture rate (Kogan, 1995).

$$\mathrm{Vci} = (\mathrm{Ndvi}_\mathrm{J} - \mathrm{Ndvi}_{\mathrm{min})})/(\mathrm{Ndvi}_\mathrm{max} - \mathrm{Ndvi}_\mathrm{min}) * 100 \tag{2.1}$$

TCI (2.2) is expressed with standardized values (in %) related to vegetative stress associated with high temperatures (Kogan, 1995).

---

22 The measurement of drought can be done according to different indices. Among the most widely used and internationally recognized ones is the Spi (Standardized Precipitation Index). This is a standardized indicator for detecting and assessing precipitation deficit (drought) at different time scales. The Spi index makes it possible to quantify the water surplus or deficit concerning the climatology of the area under consideration (see: https://www.isprambiente.gov.it/pre_meteo/siccitas/index.html). For a more in-depth discussion of the topic, see Vergni and Todisco (2010, 2011).

23 The Vhi index can be associated with the strong inverse correlation between Ndvi and Lst under the hypothesis that a constant and intense increase in soil temperature negatively impacts vegetation vigour. This may contribute to the increased propensity of different tree species to be more or less easily attacked by fire. Against this background, the European Commission recently presented a draft regulation to track threats like fires fueled by climate change and illegal logging.

The data, produced mainly by satellite platforms, are already used at the European level and by member states to provide forest inventories continuously, identify land use (change) and monitor tree health. Information from satellite imagery will help identify illegal logging and mismanagement practices. The monitoring will allow shared forest data to be collected on the ground and via satellite. The goal is to have reliable data available to make faster and more timely decisions. These examples guide forest management in the digital age.

24 To make the best use of Landsat-8 satellite imagery, choosing satellite data by concurrently evaluating temperatures during the hottest periods is essential. The selection of satellite images is based on four criteria: (i) acquisition year, (ii) acquisition month, (iii) daily average temperature, and (iv) absence of significant cloud cover. The selection considers orbital acquisition times with minimal atmospheric cloud presence.

25 Lst is processed using a specific algorithm for extracting the land surface temperature based on the processing of thermal data acquired by Landsat 8 (Guha et al., 2018).

26 Ndvi is calculated by measuring the spectral reflectance of vegetation, water, and bare soil in the visible, near-infrared, and red spectral bands (ibidem).

$$Tci = (Lst_{max} - Lst_{j)})/(Lst_{max} - Lst_{min}) * 100 \quad (2.2)$$

Vhi (2.3) is a proxy for the vegetation's health estimated based on a ratio between humidity values and more stressful thermal conditions (Kogan, 1995),

$$Vhi = a * Vci + b * Tci \quad (2.3)$$

where "a" and "b" are coefficients quantifying the contribution of Vci and Tci to vegetation's response (with values ranging from 0 to 1). Low Vhi values identify areas more affected by drought.

Long-term Vhi analyses (conducted on various temporal scales) can represent geomorphologically, vegetation-wise, and climatically complex territorial realities. Notably, in conditions of intense and prolonged extreme events, Vci values can identify areas more affected by climatic phenomena during periods of drought (forest areas), average conditions, or in the presence of a rainfall surplus. The spatialization of Vhi can help identify different gradients of drought stress that, if adequately correlated with specific contextual information (endogenous and exogenous), can express the potential propensity of a specific forest or tree type to be affected by fires.[27]

2.3.2.2 Surface Runoff Estimation

The estimating hydraulic spatial performances is often entrusted to specific models simulating the "inflow-outflow" dynamics (Pistocchi, 2001, p. 5). The surface runoff is modelled based on a spatial association of land uses with terrain morphologies. An *ad-hoc* statistical model—developed in a GIS environment— clusters surface runoff dynamics into different land use categories (agricultural, urban, residential and industrial; woodland; wetland and semi-natural) using flow direction (FlowDir)[28] and accumulation (FlowAcc)[29] functions calculated at the basin scale.

---

[27] Reading the Vhi satellite index in forestry can indicate moderate, severe or extreme water and heat stress conditions. Forests with low moisture levels can promote the rapid spread of fires, even on a large scale.

[28] The "Flow Direction" function processes a raster flow grid, assigning a numerical value (D8) to each pixel (spatial unit) based on the maximum slope. The D8 method recognizes eight possible drainage directions counterclockwise: 1—East; 2—Northeast; 4—North; 8—Northwest; 16—West; 32—Southeast; 64—South; 128—Southwest. The first pixel contains spatial information indicating towards which of the eight adjacent pixels the pixel itself is draining.

[29] The "Flow Accumulation" function makes preferential water flow paths visible. Where pixel values are very high, water converges massively (values will be higher closer to the valley). This results in a raster grid that assigns a numerical value to each cell, referring to the number of cells connected to that cell through the path of maximum slope. Cells corresponding to the watershed will have a value of 1, and the cell at the closing section will have a value equal to the sum of all cells in the basin. It should be noted that in Flow Accumulation, the actual water network does not emerge. It depends not only on topographic heights (Digital Elevation Model—Dem) but also on

Hydrological functions define the relationship (2.4), allowing the estimation of hydraulic impacts of land use changes. The necessary data for hydrological modelling are as follows:

- Digital Terrain Model (Dtm).
- Land use (land use and cover in the Emilia-Romagna Region in 2020).
- Hydrographic network and delimitations of reclamation consortia.

The function is as follows (Pozzer, 2015):

$$\varphi_i = \{\frac{[p \cdot F_U + p^\circ \cdot (F - F_U)]}{F}\}_i \tag{2.4}$$

where,

p is the runoff coefficient associated with impermeable areas,
p° is the runoff coefficient associated with permeable areas,
F is flow accumulation calculated on Dtm,
F_U is flow accumulation correlated with land cover,
U is the land use in "i" (spatial unit or pixel).

The function allows assigning the value p to the flow accumulation F, giving hydraulic impacts "i". This correlation is expressed as the percentage of rainfall that transforms into surface runoff (range from 0.2 to 0.9).[30] The procedure, calibrated at the hydrogeological basin scale, is cumulative.[31]

2.3.2.3 Vulnerability Assessment

Vulnerability scenarios are related to strategies of territorial governance. Each scenario is understood not as a combinatorial factor or "abstract possibility" but as a plausible governance outcome.

Two components of opposite signs are used for estimating vulnerability (V): sensitivity (S: learning of a system to suffer an impact) and adaptive capacity (AC: intrinsic capability of a system to mitigate impacts). The two components can be used for a replicable vulnerability study at both urban and territorial scales based on the following additive function[32]:

---

soil type (pedology, geological substrate, etc.) and the artificialization of the territory (creation of embankments, diversion of rivers and streams, artificial channelling, etc.).

30 The choice of runoff coefficients by land use type adheres to the guidelines of civil engineering and hydraulic design manuals. Since runoff coefficients are characterized by margins of error, stochastic or probabilistic formulations are often used.

31 One of the main opportunities for improvement in the study is the need to recalibrate runoff results within an acceptable range based on further investigations, including geological and climatic considerations (indicators of local micro-climate and simulations).

32 A multiplicative version of the function of the type $V = S * AC$ is currently in the testing phase, starting from the hypothesis that there is an interaction between the two propensities. A system may have a high propensity to undergo an impact because it already has (or could have) a propensity to respond. However, the interaction can take different forms.

$$V = S - AC \tag{2.5}$$

In this context, vulnerability (2.5) is a composite concept derived from the so-called "disaster equation" (Pasini, 2020) formalized as follows (2.6):

$$\text{Risk} = f(\text{hazard, exposure, vulnerability}) \tag{2.6}$$

A hazard is a potentially harmful physical event (natural or anthropogenic) capable of causing loss of human lives or injuries, material damage, social and economic disruptions, or environmental degradation. Exposure is defined as a system of people, properties, and populations located in hazardous areas and thus subject to potential losses. Although both components can enhance vulnerability assessment methods, vulnerability is currently used to recognize and understand territorial fragilities that could become threats.

Equation (2.5) provides a simple morphological response to an impact "x" localized through a subtraction operation between sensitivity and adaptive capacity. The method allows for deriving indicators of territorial vulnerability based on parameters describing ecosystem functionalities and territorial fragilities. The outcomes can be sketched by vulnerability maps for each type of impact (Table 2.2), highlighting the most vulnerable areas of the territory.

The multidimensional approach enables querying and cross-referencing information that is otherwise difficult to manage in mainstream land management and regulation models. The method responds to physical-morphological research methods, experimenting with an integrated and systemic vulnerability reading (combined approach, Sect. 4.3).

2.4 Results

The techniques for estimating the Vhi index and runoff described briefly in the third paragraph enable the identification of main areas vulnerable to long-standing drought and intense surface runoff events. The methodology allows for spatial contextualization of hydraulic criticality levels, highlighting how and to what extent land uses influence hydrogeological performances and the consolidation process of landslide-prone areas. Although the impacts under consideration have different physical and mechanical features, all three contribute to a cumulative definition of territorial vulnerability easily derived using topological overlapping.[33]

---

[33] In spatial analysis, topological overlapping allows for categorizing and classifying spaces with their structure or components that can be traced back to a structure. Combinatorial topology highlights how a space can result from combining more straightforward or simplified spaces. Algebraic topology allows for the interpretation and representation of spaces through algebraic invariants capable of recognizing homeomorphism relations.

Table 2.2 Elaborations for territorial vulnerability assessment

Impact	Statistical unit	Data, indicators, and indexes	Vulnerability features		Vulnerability calculation $V = S - AC$
			S	AC	
Drought	Pixel 30 × 30 metri	Vegetation health index (Vhi)	Temperature condition index Tci	Vegetation condition index Vci	"$V = S - AC$" is measured as Vhi
Surface runoff	Pixel 5 × 5 metri	– Dtm – land uses – water flow coefficient	0,9 impermeable areas	0,2 permeable areas	The "$V = S - AC$" relationship is processed in the GIS environment using the hydrology function. The function returns a raster of the qualitative/quantitative behaviour of the territory's hydraulic system The relationship allows the runoff coefficients to be assigned the "Flow_Acc" flow accumulation by returning the "hydraulic impacts" φi (or $\varphi_{\text{cell raster}}$). This correlation is expressed as % of rainfall becoming surface runoff (ranging from 0.2 to 0.9). The procedure, calibrated to the basin scale, is cumulative
Rock fall/landslide	Vector-based analysis and parametrization	Classification of landslides based on the combination of activity state and movement type	The landslide phenomenon is mapped and studied in association with the runoff study		

2.4.1 Drought

For a more in-depth understanding of drought phenomena, the Vhi was calculated for the following periods: July 2015, August 2016, July 2017, October 2018, and August 2019. Using the arithmetic averages calculated on the pixel values of the 5 Vhi, it was possible to identify forested areas with a greater tendency towards drought during the summer/fall months of 2015–2019 (Fig. 2.4).[34] The study on Vhi demonstrates how the temperature increase negatively affects vegetation health.

The prevalent vegetation types are oaks, hornbeams, and chestnut groves, covering an area of approximately 239 square kilometres (Table 2.3). Beeches follow with 72 square kilometres, and mixed forests of conifers and broadleaves cover around 44 square kilometres. Oaks, hornbeams, chestnut groves, willows, and poplars have an average Vhi of 57% (moderate drought), while oaks and ashes show a higher susceptibility to drought with an average Vhi of about 48%.

This level of information helps study landslides triggered by structural and textural issues. The research uses Vhi information to enhance the study of spatial convergence of land uses, land cover quality, and erosive capacity of surface water runoff.

2.4.2 Surface Runoff

The runoff assessment provides a spatial index of hydraulic critical issues. Critical thresholds are calculated and weighted based on soil morphology and hydraulic capacity of land uses to absorb meteoric waters (Fig. 2.5). Inefficient soil consumption and increasing impermeability close meteoric water outflow routes, increasing flood waves and the water charge in receiving bodies.

The modelling procedure enables estimating runoff coefficients at the watershed scale, assessing which land uses contribute the most to modifying the hydraulic safety of a territory, affecting exposure and vulnerability. The results highlight how the ratio between water volumes generated by Dtm hydrological modelling significantly increases runoff coefficients within highly urbanized environments (Fig. 2.6).[35] The map dynamically illustrates hydraulic stresses by corresponding watersheds.

In urbanized areas, the runoff coefficient fluctuates between 0.70 and 0.82, while in rural or less urbanized areas, there is a reduction in surface runoff with values decreasing towards 0.5/0.6. In both contexts, critical environmental and hydraulic conditions remain high. The runoff tends to decrease significantly in natural areas, with values ranging from 0.2 to 0.4 depending on the physical and chemical soil features. The index tends to increase in mountainous and hilly areas, yielding values

[34] Source of forest cover: land use and land cover in the Emilia-Romagna Region as of 2020.

[35] Considering these areas only as geographical-territorial partitions with a particular vulnerability to flooding is appropriate. This vulnerability varies with the intensity of the rainfall event and its spatial distribution.

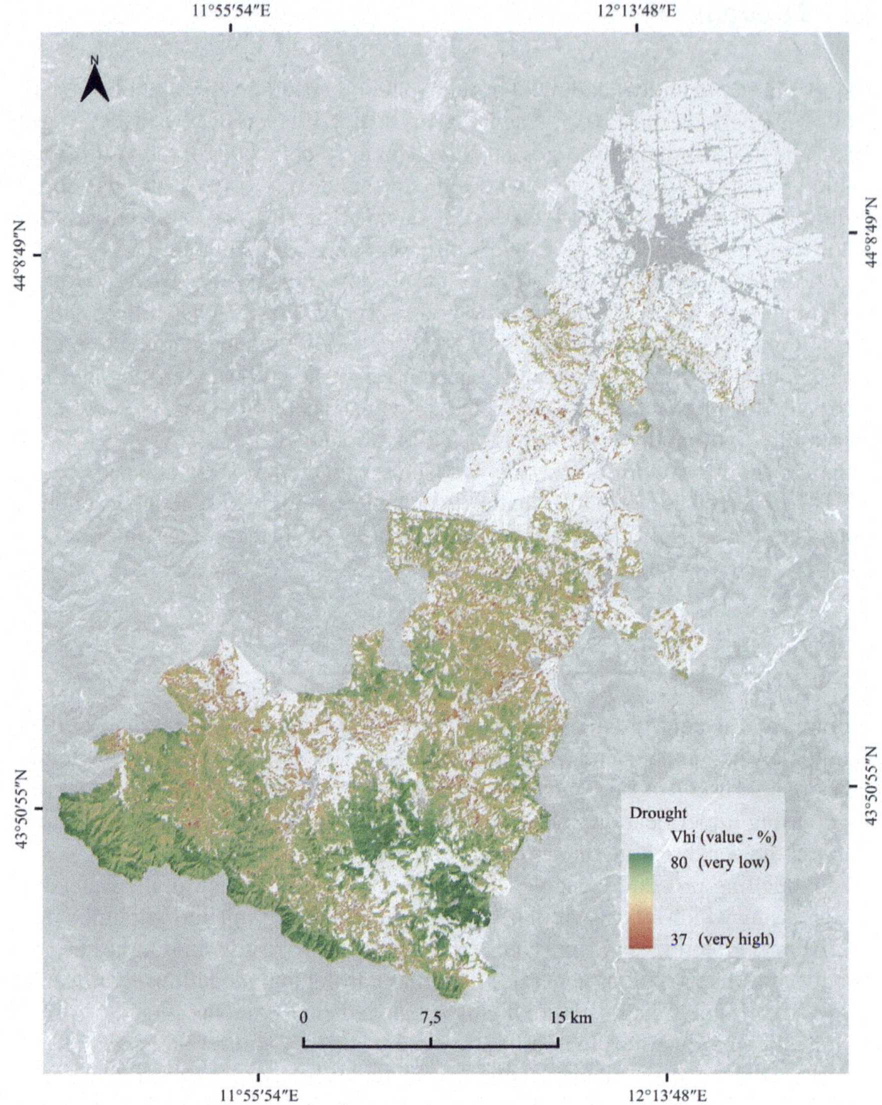

Fig. 2.4 Forest health: vegetation health index (arithmetic mean 2015–2019). *Source* Author's elaboration based on Landsat 8

within a range of 0.6–0.65. This is due to slopes and the presence of gullies related to drought phenomena, as well as erosive actions of runoff water.

Table 2.3 Vhi variations per vegetation class

Typology	Territorial surface (km^2)	Vhi minimum value recorded	Vhi average value	Vhi maximum values recorded
Beech-dominated forests	72.35	55.40	66.35	75.91
Forests with a predominance of oaks, hornbeams and chestnut trees	229.85	49.23	57.43	72.55
Groves with a predominance of willows and poplars	1.37	46.69	56.34	61.60
Coniferous forests	13.84	42.42	61.51	72.70
Mixed coniferous and deciduous forests	43.72	47.68	62.26	76.66
Lowland forests with a predominance of fairness and ash trees	0.06	44.96	48.20	52.08

2.4.3 Combined Approach

The analysis conducted in the municipality of Bagno di Romagna addresses the issue of systemic vulnerability in terms of hydrogeological instability and changes in soil mechanical properties. The assessment is based on a spatial combination of geomorphological and eco-systemic links between hydrodynamic conditions (runoff) and soil thermal regimes (Vhi). The generated value can range from 0 (better condition) to 100 (worse condition) and corresponds to five degrees of vulnerability: very low, low, moderate, high, and very high (Fig. 2.7).

The results, further enriched by landslide overlay, highlight a significant physical and morphological association between landslide phenomena, drought, and surface runoff. The analysis recognizes how geomorphological characteristics and climate change can affect slope stability, leading to a regression in the hydrogeological dynamics of territory and soil quality (performance). Due to prolonged drought periods, the resurgence of hydrogeological risk can modify the territorial structure, rainfall intensity, or soil chemical-structural features. Consequently, flows, soil lithologies, slopes, and land uses can significantly influence landslide phenomena regarding predisposition and prediction (Moreno et al., 2013). In particular, active landslides in Bagno di Romagna significantly and generally lean toward systemic vulnerability. Their runoff surfaces fall into high and very high vulnerability classes for 52.03%, in moderate class for 24.06%, and in low and very low classes for 23.94%.

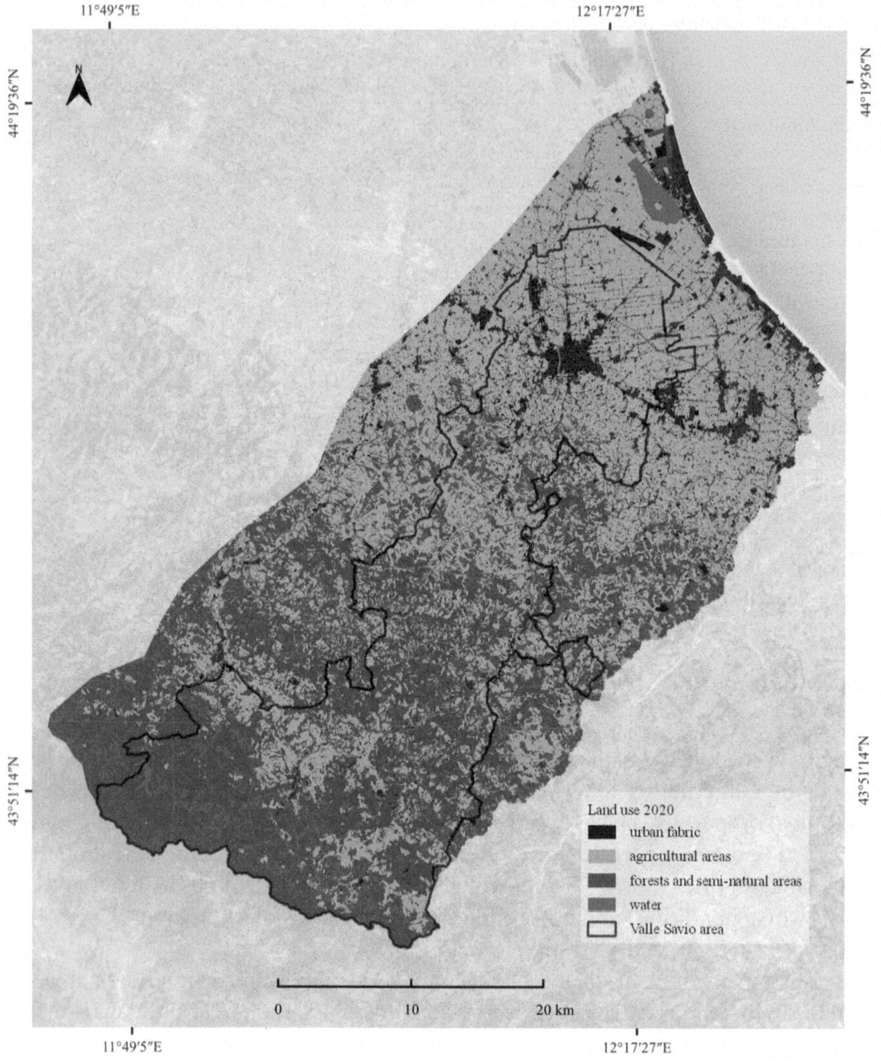

Fig. 2.5 Land use/land cover map: partition of Savio, Rubicone and Fiumi Uniti catchment basin. *Source* Author's processing based on land use/land cover of Emilia-Romagna (database 2020)

2.5 Discussion and Conclusion

A systemic vulnerability model helps to identify spaces and places that may experience a great loss of their ecosystem functionality when exposed to extreme meteorological events. In this regard, the proposed method addresses the links among factors that influence the spatial vulnerability and likelihood of catastrophic events. Such an

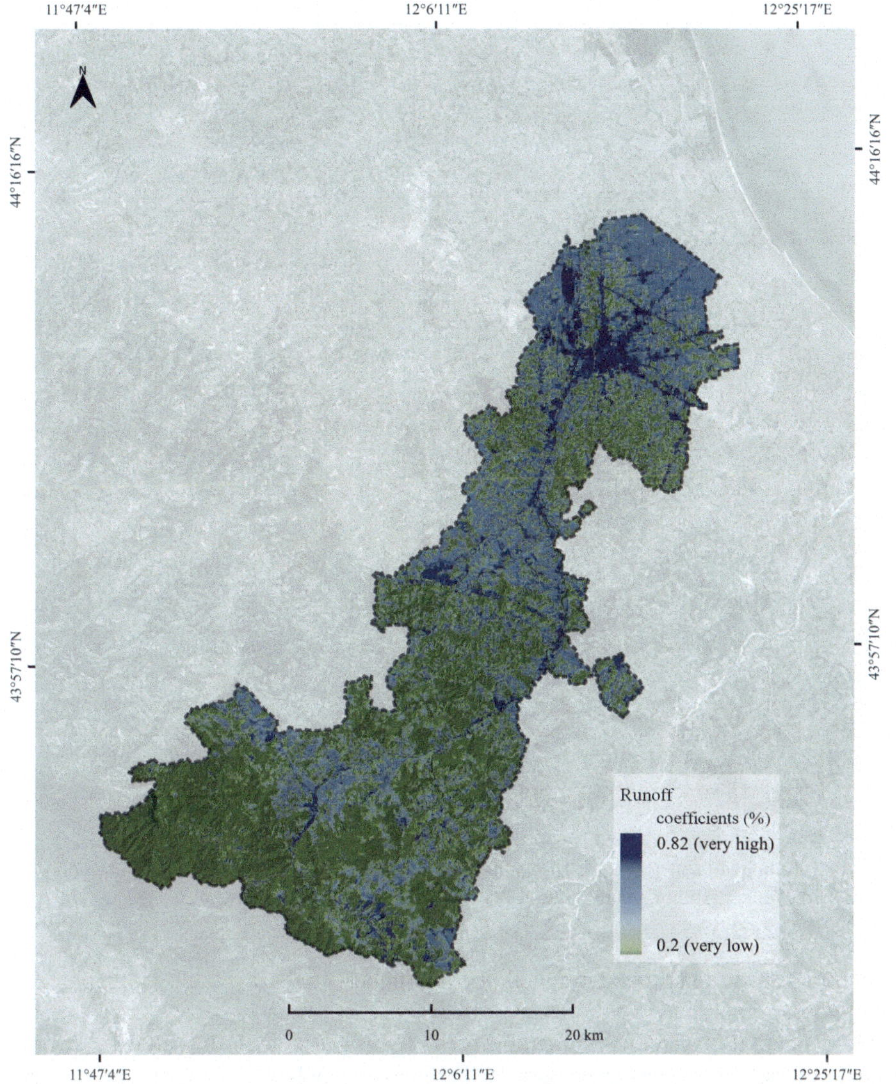

Fig. 2.6 Impact of land use change on hydrological systems (2020). *Source* Author's processing based on land use/land cover and digital terrain model (Dtm)

interaction plays a fundamental and strategic role in detecting climatic impacts and effectiveness of spatial planning and management.

For this reason, an approach of a qualitative-quantitative nature has been developed to reconstruct relationships of complementarity among different levels of information (systemic logic of impacts). The research on territorial complexity underlines that any territory exhibits systemic vulnerability, whereby the consequences of a

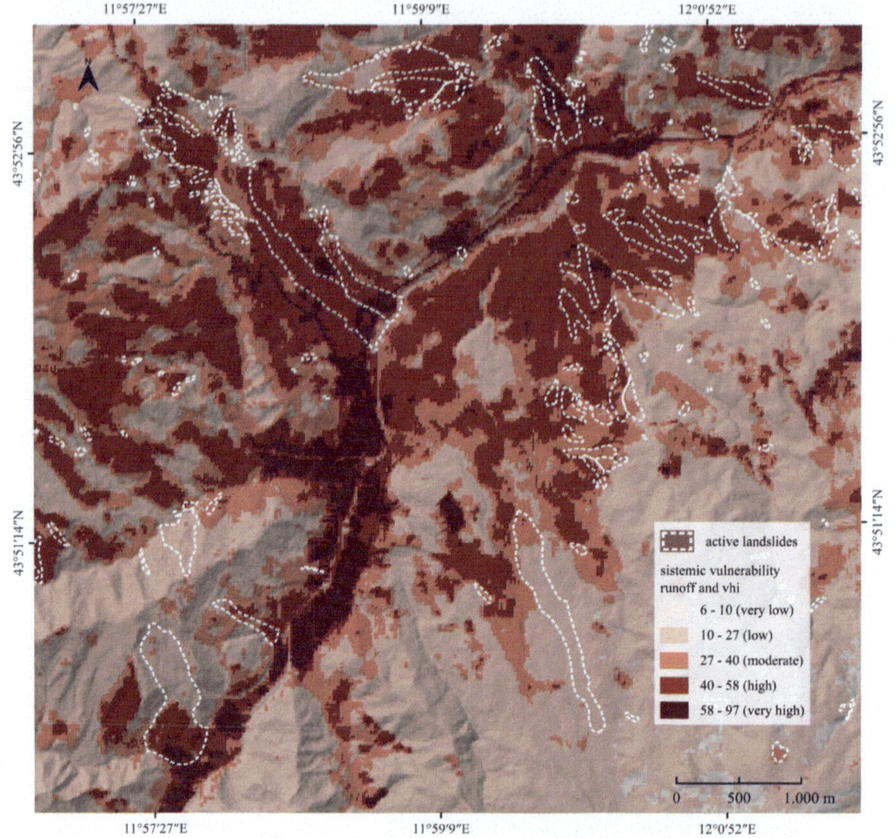

Fig. 2.7 Municipality of Bagno di Romagna: systemic vulnerability to hydraulic and hydrogeological risk. *Source* Author's elaboration

calamitous event tend to extend well beyond the affected area, both geographically and temporally.

A concept of systemic vulnerability to hydrogeological risk stresses probabilistic models in assessing hazard. Probabilistic models work on a temporal (and linear) sequence of events, producing useful projections for medium and long terms. However, the systemic vulnerability discussed here can be linked to interactions between exposed elements influenced by anthropogenic and climatic effects. Consequently, the approach tends to include factors that are generally not considered by traditional planning evaluation methods despite local responses to calamitous events.

In this perspective, the study of systemic vulnerability appears to be a useful analytical and managerial tool to test hypotheses of spatial correlation between land use patterns, biophysical states, and soil impermeability. This approach provides some recommendations for policymakers at the local level and along the Adriatic coasts. It can be helpful in promoting greater awareness and reflexivity in mapping vulnerability and expected risks. It encourages a re-assessment of assumed certainties and needs that often guide spatial actions, opening up to possible experiments and recommendations.

The recommendations focus on three pillars of climate-proof planning: mapping, strategic design, and governance. Regarding mapping, it is advisable to evaluate the eco-systemic performance of a territory using the latest numerical modelling systems and environmental monitoring. New applications of georeferenced information related to satellite remote sensing research programs can be helpful. Recommendations regarding strategic design suggest updating knowledge frameworks oriented to experimenting with mitigation and adaptation policies. Finally, recommendations to governance require local and supralocal policies to favour cooperation on a large scale (in the AdriAdapt project, cooperation assumed a transnational character). This may promote understanding and placement of concepts of territorial protection and quality within a broader process of urban and regional policy renewal (Hincks et al., 2023).

The research development might be considered an additional opportunity to support these general recommendations. The possibility of transferring an integrated methodology tested on territorial security analysis to the planning process will enable using the concept of spatial vulnerability for mapping links between hydraulic conditions and hydrogeological impacts (floods, erosions, landslides). The validity of this hypothesis can already be appreciated by comparing the runoff results in urban–rural areas in Emilia-Romagna affected by the flood event of May 19, 2023.[36] With a focus on the north of Cesena urban centre (Fig. 2.8), the intersection of the two info-layers contributes to the construction of a territorial knowledge that enables locating hydraulic risk according to two rationales: recognition of spatial convergence between possibilities (runoff) and events (flooding); identification of the conflictual relationship between humans (anthropic states and dynamics) and nature (eco-systemic components).

The runoff algorithm helps the planning understand (considering any adjustments) for predictive problems regarding the hydraulic health safety of the territory and its tendency to regress toward less efficient performance. It enriches the literature on flood risk assessment, demonstrating how the produced knowledge can contribute to lead planning, management, and regulation devices with a view to climate change adaptation.

The results acknowledge the effectiveness of the research and encourage further exploration with a second-level hypothesis for estimating significant physical and spatial correlations between water runoff, climatic variables (temperature, precipitation, humidity, wind), and satellite indices.

The new test involves the construction of a richer geo-database based on the spatial integration of morphological-climatic indicators. The exercise will allow us

[36] Flood mapping based on satellite images from the European Copernicus Emergency Management Service (CEMS). The maps produced by Cems depict the extent of the flooded areas in the most affected municipalities (Faenza, Cesena, Forlì, and Conselice), with information on water/water hammer and water volume.

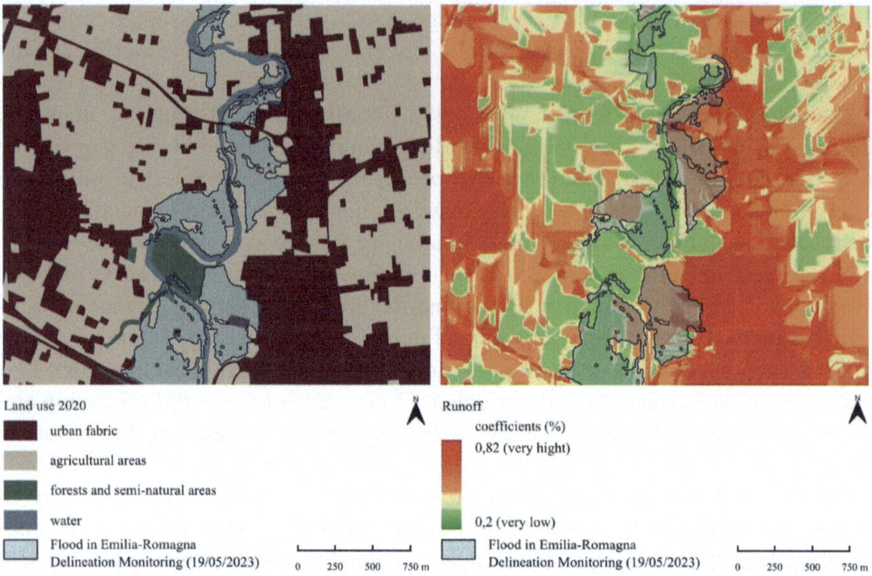

Fig. 2.8 Comparison between surface runoff and flooded areas detected by satellite in the north area of Cesena. *Source* Author's processing based on runoff method and Copernicus Emergency Management Service (May 2023)

to evaluate different hydraulic performances of the territory[37] and relate them to the supply and flow of ecosystem services.[38] Therefore, through multidimensional statistical exploration,[39] it will be possible to recognize the relevance of descriptors and construct a new layer of information (multidisciplinary and inter-scale) capable of supporting spatial planning processes and decision-making. The analysis could significantly update the sensitivity and adaptive capacity parameters for risk assessment, as well as "redefine" the concept of hazard and its empirical significance meaning.

In conclusion, this research should be understood as guidance for those—scientific communities, policymakers, and stakeholders—interested in dynamically assessing

[37] Supported by specific geologic analysis.

[38] The change in ecosystem arrangements, partly accelerated by climate changes and partly driven by the depletion of environmental resources and the progressive reorganization of human activities, plays a fundamental role in defining territorial areas' performance and environmental sustainability. This includes their evolution, conservation, and organizational and functional structure. In the envelopes of the inhabited areas in the Forlì Apennines (but not only), the cultural heritage characterized more than anything else the functional organization of the relationships between the elements of the anthropic-settlement system and the forms of use of the natural space. This space should be understood as an ecosystem in continuous transformation, that is, as a housing model (and way of being) on par with human existence (utilitarian concept vs. holistic concept: see Bridle 2022).

[39] The exploration can use factorization procedures (principal component or correspondence analysis) combined with hierarchical or non-hierarchical clustering.

and governing the overall vulnerability of spatial ecosystems. And that, in relation to plausible and comparable scenarios in climate-proof planning. From this perspective, it is believed that the Valle Savio test can help strengthen decisions and civic awareness regarding urban and territorial resilience issues. The results of this research could enable governments along the Adriatic coast to test new adaptation policies based on developing spatial modelling algorithms and enhancing remote sensing techniques. Implementing of these policies will provide new information to guide exercises in meta-evaluation and monitoring of hydraulic and hydrogeological vulnerability, both at the urban and territorial scales.

References

Avand, M., & Moradi, H. (2021). Using machine learning models, remote sensing, and GIS to investigate the effects of changing climates and land uses on flood probability. *Journal of Hydrology, 595*, 125663.

Bento, V. A., Trigo, I. F., Gouveia, C. M., & DaCamara, C. C. (2018). Contribution of land surface temperature (TCI) to vegetation health index: A comparative study using clear sky and all-weather climate data records. *Remote Sensing, 10*(9), 1324.

Bridle, J. (2022). *Modi di essere*. Rizzoli, Milano.

De Sherbinin, A., Bukvic, A., Rohat, G., Gall, M., McCusker, B., Preston, B., Apotsos, A., & Zhang, S. (2019). Climate vulnerability mapping: A systematic review and future prospects. *Wiley Interdisciplinary Reviews: Climate Change, 10*(5), e600.

ESPON Climate. (2022). *Updating and integrating climate datasets and maps. Final report*. TU Dortmund University, ESPON & IRPUD.

European Commission (EC). (2021). *Forging a climate-resilient Europe—The new EU strategy on adaptation to climate change*. Communication from the Commission to the European Parliament, the Council, the European Economic and Social Committee and the Committee of the Regions, Brussels.

Fregolent, L., & Savino, M. (Eds.). (2014). *Economia, società, territorio. Riflettendo con Francesco Indovina: Riflettendo con Francesco Indovina*. Franco Angeli, Milano.

Fritzsche, K., Schneiderbauer, S., Bubeck, P., Kienberger, S., Buth, M., Zebisch, M., & Kahlenborn, W. (2014). *The vulnerability sourcebook: Concept and guidelines for standardised vulnerability assessments*. Deutsche Gesellschaft für Internationale Zusammenarbeit (GIZ) GmbH.

Füssel, H. M. (2010). *Development and climate change: Review and quantitative analysis of indices of climate change exposure, adaptive capacity, sensitivity, and impacts*. World Development Report Background Papers, World Bank.

Galimberti, G., & Balbo, A. (2023). Cambiamento climatico e drenaggio delle acque: recenti evidenze in Italia, strategie di progettazione e necessità di cambiamento. *Ingegneria dell'Ambiente, 10*(1).

Guha, S., Govil, H., Dey, A., & Gill, N. (2018). Analytical study of land surface temperature with NDVI and NDBI using Landsat 8 OLI and TIRS data in Florence and Naples city, Italy. *European Journal of Remote Sensing, 51*(1), 667–678.

Hincks, S., Carter, J., & Connelly, A. (2023). A new typology of climate change risk for European cities and regions: Principles and applications. *Global Environmental Change, 83*, 102767.

IPCC. (2014). Climate change: Impacts, adaptation, and vulnerability part A: Global and sectoral aspects. In *Contribution of working group II to the fifth assessment report of the intergovernmental panel on climate change*. Cambridge University Press.

IPCC. (2018). Summary for policymakers. In M. Allen, M. Babiker, Y. Chen, H. de Coninck, S. Connors, et al. (Eds.), *Global Warming of 1.5°C: An IPCC special report on the impacts of global warming of 1.5 °C above pre-industrial levels and related global greenhouse gas emission pathways, in the context of strengthening the global response to the threat of climate change, sustainable development, and efforts to eradicate poverty.* World Meteorological Organization.

IPCC. (2022). Climate change 2022: Impacts, adaptation, and vulnerability. In *Contribution of working group II to the sixth assessment report of the intergovernmental panel on climate change.* Cambridge University Press.

Iturbide, M., Fernández, J., Gutiérrez, J.. M., Bedia, J., Cimadevilla, E., Díez-Sierra, J., Manzanas, R., Casanueva, A., Baño-Medina, J., Milovac, J., Herrera, S., Cofiño, A. S., San Martín, D., García-Díez, M., Hauser, M., Huard, D., & Yelekci, O. (2021). Repository supporting the implementation of FAIR principles in the IPCC-WG1 Atlas. *Zenodo.*

Kamel Boulos, M. N., & Wilson, J. P. (2023). Geospatial techniques for monitoring and mitigating climate change and its effects on human health. *International Journal of Health Geographics, 22*(1), 1–7.

Kogan, F. N. (1995). Application of vegetation index and brightness temperature for drought detection. *Advances Space Research, 11*, 91–100.

Kourtis, I. M., & Tsihrintzis, V. A. (2021). Adaptation of urban drainage networks to climate change: A review. *Science of the Total Environment, 771*, 145431.

Landauer, M., Juhola, S., & Klein, J. (2019). The role of scale in integrating climate change adaptation and mitigation in cities. *Journal of Environmental Planning and Management, 62*, 741–765.

Magni, F. (2019). *Climate proof planning: l'adattamento in Italia tra sperimentazioni e innovazioni.* FrancoAngeli.

Maragno, D., dall'Omo, C. F., Pozzer, G., & Musco, F. (2021). Multi-risk climate mapping for the adaptation of the venice metropolitan area. *Sustainability*; *13*(3), 1334.

Maragno, D., Pozzer, G., & dall'Omo, C. F. (2023). Supporting metropolitan Venice coastline climate adaptation. A multi-vulnerability and exposure assessment approach. *Environmental Impact Assessment Review, 100*, 107097.

MIMS. (2021). *Cambiamenti climatici, infrastrutture e mobilità. Soluzioni e strategie per gli investimenti infrastrutturali in un contesto di adattamento ai cambiamenti climatici e di mitigazione delle emissioni di gas-serra.* Rapporto della Commissione cambiamenti climatici, infrastrutture e mobilità sostenibili coordinata da C. Carraro, Roma.

Moreno, J. M., Torres, I., Luna, B., Oechel, W. C., & Keeley, J. E. (2013). Changes in fire intensity have carry-over effects on plant responses after the next fire in southern California chaparral. *Journal of Vegetation Science, 24*(2), 395–404.

Munafò, M. (a cura di). (2022). *Consumo di suolo, dinamiche territoriali e servizi ecosistemici—Edizione 2022.* Report SNPA 32/22, ISPRA, Roma.

Musco, F. (a cura di). (2016). *Counteracting urban heat Island effects in a global climate change scenario.* Springer, AG Switzerland.

O'Brien, K., Eriksen, S., Nygaard, L. P., & Schjolden, A. (2007). Why different interpretations of vulnerability matter in climate change discourses. *Climate Policy, 7*(1), 73–88.

PAP/RAC. (2021). *Guidelines for mainstreaming of climate change adaptation into coastal management along Adriatic coasts.* INTERREG AdriAdapt Project.

Pasini, A. (2020). *L'equazione dei disastri, Cambiamenti climatici su territori fragili.* Codice edizioni.

Patassini, D. (2022), Marginalia. Plausibilità ed efficacia delle teorie dell'adattamento. *Equilibri, 26*(1–2), 128–160.

Patassini, D. (2023). *Cumulative impact as second-order evidence.* Mimeo.

Patassini, D., & Pozzer, G. (2018). Consumo di suolo ad uso non residenziale nei comuni minori del Veneto. Un test sull'effetto "frammentazione amministrativa. In ISPRA (2018) *Consumo di suolo, dinamiche territoriali e servizi ecosistemici—Edizione 2018, Rapporto 288/2018.*

Pietrapertosa, F., Salvia, M., De Gregorio Hurtado, S., D'Alonzo, V., Church, J. M., Geneletti, D., Musco, F., & Reckien, D. (2019). Urban climate change mitigation and adaptation planning: Are Italian cities ready? *Cities, 91*, 93–105.

Pistocchi, A. (2001). La valutazione idrologica dei piani urbanistici: un metodo semplificato per l'invarianza idraulica del piani regolativi generali. *Ingegneria Ambientale, XXX*(7/8), 407–413.

Pistocchi, A. (2018). Hydrological impacts of soil sealing and urban land take. In *Urban expansion, land cover and soil ecosystem services* (pp. 157–168).

Pozzer, G. (2015). Consumo di suolo e gestione del rischio idraulico: test per l'invarianza idraulica nella pianificazione territoriale. In M. Munafò, & M. Marchetti, a cura di (Eds.), *Recuperiamo terreno. Analisi e prospettive per la gestione sostenibile della risorsa suolo* (pp. 168–180). FrancoAngeli.

Pozzer, G., Maragno, D., Magni, F., & Musco, F. (2021). Remote sensing e spatial modelling per strategie di adattamento ai cambiamenti climatici: caso studio Valle Savio. In *Atti della XXIII Conferenza Nazionale SIU, Downscaling, Rightsizing. Contrazione demografica e riorganizzazione spaziale, Torino, 17–18 giugno 2021*. Planum Publisher, Roma-Milano.

Raikes, J., Smith, T. F., Jacobson, C., & Baldwin, C. (2019). Pre-disaster planning and preparedness for floods and droughts: A systematic review. *International Journal of Disaster Risk Reduction, 38*, 101207.

Romero Lankao, P., & Zwickel, T. (2015). A conceptual framework for an urban areas typology to integrate climate change mitigation and adaptation. *Urban Climate,* 14, 116–137.

Stewart, I. D., Oke, T. R., & Krayenhoff, E. S. (2014). Evaluation of the 'local climate zone' scheme using temperature observations and model simulations. *International Journal of Climatology.*

Tapia, C., Guerreiro, S., Mendizabal, M., Kilsby, C., Feliu, E., Glenis, V., Dawson, R., & Eluwa, C. (2015). High level quantified assessment of key vulnerabilities and priority risks for urban areas in the EU. *RAMSES project deliverable D, 3.*

Trigila, A., Iadanza, C., Lastoria, B., Bussettini, M., & Barbano, A. (2021). *Dissesto idrogeologico in Italia: pericolosità e indicatori di rischio—Edizione 2021.* ISPRA, Rapporti 356/2021.

Tripathi, R., Sahoo, R. N., Gupta, V. K., Sehgal, V. K., & Sahoo, P. M. (2013). Developing Vegetation Health Index from biophysical variables derived using mod is satellite data in the trans-gangetic plains of India. *Emirates Journal of Food and Agriculture*, 376–384.

Vergni, L., & Todisco, F. (2010). Valutazione di un indice standardizzato per lo studiodella frequenza e della durata del deficit idrico. In *Atti del XXIII ConvegnoNazionale di Idraulica e Costruzioni Idrauliche* (pp. 14–17). Palermo.

Vergni, L., & Todisco, F. (2011). Spatio-temporal variability of precipitation, temperature and agricultural drought indices in Central Italy. *Agricultural and Forest Meteorology, 151*, 301–313.

Ungaro, F., Calzolari, C., Pistocchi, A., & Malucelli, F. (2014). Modelling the impact of increasing soil sealing on runoff coefficients at regional scale: A hydropedological approach. *Journal of Hydrology and Hydromechanics, 62*(1), 33–42.

Wamsler, C., Brink, E., & Rivera, C. (2013). Planning for climate change in urban areas: From theory to practice. *Journal of Cleaner Production*, 50, 68–81.

Wilby, R. L., & Keenan, R. (2012). Adapting to flood risk under climate change. *Progress in Physical Geography, 36*, 349–379.

Chapter 3
Susceptibility of Flood-Prone Areas to Remain Inundated for Shorter or Longer Periods During a Flooding Event: An Assessment Method Applied in the Friuli Venezia Giulia Region (Italy)

Davide Longato

Abstract River and coastal floods are triggered by meteorological events such as heavy rainfall and storm surges, melting snow in upstream areas, tidal-related influences, or a combination of them. They occur when the river runoff volume exceeds local flow capacities or as a consequence of river bank/coastal defense system breakage/failure. Depending on land use and cover conditions, soil capacity to drain surface water, terrain morphology, and amount of exceeding flood water volume, the vulnerability of flood-prone areas in relation to their susceptibility to remain inundated for longer or shorter periods can be higher or lower even in areas affected by similar conditions of flood hazard. Especially in areas with flat slopes and in deltaic areas the fall of water can be particularly long, possibly exacerbating the flooding impacts and consequent damages. These events are also expected to increase in the future as a consequence of climate change, with Italy being particularly exposed due to its geographical location and geomorphological conditions. However, attempts to assess the temporal dimension of flooding are very rare, despite its recognized importance in influencing flood damages and losses. This work presents a first attempt to assess the vulnerability of flood-prone areas in relation to their susceptibility to remain inundated for shorter or longer periods during a flooding event by proposing a rather simple and quick method based on spatial multi-criteria decision analysis. The assessment is tested in the plain and coastal zone of the Friuli Venezia Giulia region, one of the Italian areas with the largest flood hazard potential. It combines the expected flood water depths during a major flooding event and the main geophysical characteristics that may potentially influence the duration of inundations to spatially determine vulnerability patterns in flood-prone areas. Results can be further refined by expert involvement and used to complement existing flood hazard/risk data to

D. Longato (✉)
Department of Architecture and Arts, University IUAV of Venice, Venice, Italy
e-mail: dlongato@iuav.it

M. Granceri Bradaschia et al. (eds.), *Climate Change Adaptation, Flood Risk, and Beyond*, Planning for Climate Proof Cities,
https://doi.org/10.1007/978-3-031-65463-3_3

support decisions during both the emergency response and the flood risk mitigation planning phases.

Keywords River floods · Coastal floods · Flood hazard · Flood duration · Vulnerability assessment · Spatial multi-criteria decision analysis

3.1 Introduction

Floods are one of the most common hazardous natural events affecting territories and people worldwide. It is a (recurring) disaster that threatens people's lives and homes, leading to damage of properties, crops, infrastructure, and transportation systems, among others, and causing considerable losses. A flood event can be simply described as a usually dry area that (more or less suddenly) gets submerged by water.

Different types of flood exist, including river floods (i.e., river run-off volume exceeding local flow capacities), flash floods (i.e., rapid accumulation and release of run-off waters from upstream mountainous areas), coastal floods (i.e., occurring during storm surges when there are temporary increases in sea levels above the normal tidal range), urban drainage floods (i.e., insufficient capacity of the urban drainage/piped system to drain stormwater), and groundwater floods (i.e., slow move of groundwater to low-laying areas where the groundwater table breaks the groundwater due to extensive period of precipitations) (European Environment Agency, 2012).

In Europe, floods are the natural hazard that causes the greatest economic losses. River and coastal floods are of particular relevance since many people live close to riverine and coastal zones, given the fact that many urban settlements were founded along major rivers and the coast for transport and accessibility reasons (Barredo, 2006; Lugeri et al., 2010).

River and coastal floods are triggered by meteorological events such as heavy rainfall and storm surges, melting snow in upstream areas, tidal-related influences, or a combination of them. In general, they occur when the river runoff volume exceeds local flow capacities, or as a consequence of river bank/coastal defense system breakage/failure.

Flood damage depends on many factors, including flow velocity, duration of inundation, sediment concentration, and the quality of external response in a flood situation, among others (Merz et al., 2010). In particular, flood duration is a key factor since the longer the period an area remains submerged (from a few hours to several weeks), the greater impacts may be expected, increasing the probability of (more severe) damage, especially with concerning indirect losses (Genovese et al., 2007). For example, the longer an area is inundated, the greater may be the water absorbed by and saturation of building structure and contents, the more severe the anoxia of crops, and the higher infrastructure system interruption losses, among others (Dutta et al., 2003; Lazzarin et al., 2022; Merz et al., 2010).

Terrain topography plays a fundamental role in the hydrological dynamics of an area, exerting significant control over the movement and infiltration of surface water and affecting its depth and spatial–temporal distribution at the local scale. In a study by Titley and Colleagues (2021), the duration and magnitude of the flooding were found to be strongly dependent on the characteristics of the catchment area, with gentler slopes having a much greater conditional probability of severe flooding in terms of inundation duration and water depth. In particular, depressed areas are more likely to accumulate water while not being able to evacuate it and, thus are likely to be submerged for more time compared to non- or less-depressed areas (e.g., Zhang et al., 2023). In the case of floodwater, depending on the magnitude of the flooding event, this could lead to having areas that remain submerged for a very long period (up to several weeks), exacerbating flood impacts and consequent damages/losses.

An exemplary case is the recent major river flood event that occurred in May 2023 in the Emilia Romagna region (Italy). Apart from the impressive extent and intensity of the event (i.e., more than 20 rivers flooded and an estimated 58 km^2 of land was submerged by 6 million m^3 of river floodwater causing significant human and economic losses (CIMA Research Foundation, 2023)), one of the key elements characterizing this event was the long-standing floodwater evacuation time in large portions of land, which even after more than a week from the event still were submerged by floodwater. The satellite-derived map of inundated areas created by the Joint Research Center of the European Union (ERCC—Emergency Response Coordination Centre, 2023) shows the temporal evolution of the flood-detected areas, with significant portions of land still inundated ten days after the event (Fig. 3.1).

This event reminds us how fragile are our territories in the light of this natural hazard, with Italy being one of the most exposed countries. It showed that lowland areas, which in the case of Emilia Romagna region mostly correspond to reclaimed land that used to be wetland areas in the past, are likely to suffer from long-standing inundations after a major flooding event which can submerge, for example, crops, buildings, and infrastructures for a prolonged period, exacerbating impacts and extending disaster recovery time.

However, to the author's knowledge, attempts to assess the vulnerability (or susceptibility) of flood-prone areas to remain inundated for longer or shorter periods during a flooding event are lacking, except a very few cases (e.g., Tingsanchali & Promping, 2022; Usman Kaoje et al., 2021). This is confirmed by other authors, stating that the duration of inundation (here intended as the period during which the land remains submerged by floodwater) received limited attention despite its recognized consequences in terms of (indirect) economic losses and health-related issues due to business interruptions and disruption of supply chains (e.g., Haraguchi & Lall, 2015; Koks et al., 2015), or to the negative impact on clean water supply and sanitation (e.g., Dang et al., 2011), for instance.

This study aims to provide a first attempt to assess the vulnerability of flood-prone areas in relation to their susceptibility to remain inundated for longer or shorter periods by proposing a quick assessment method based on a spatial multi-criteria decision analysis. The method is applied to the flood-prone areas of the plain and

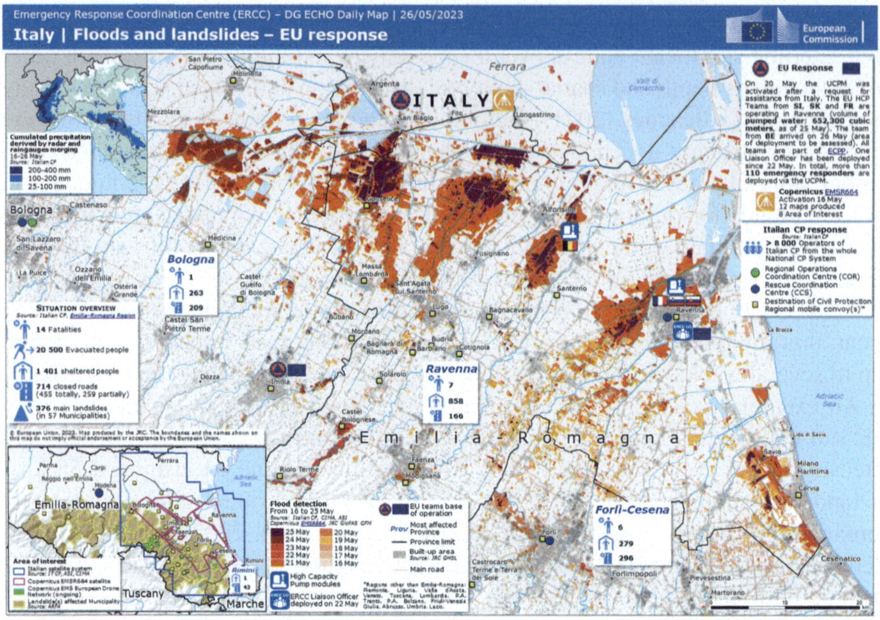

Fig. 3.1 Map showing the spatial and temporal distribution of the flood-detected areas from satellite images produced by the Joint Research Center from various sources (ERCC—Emergency Response Coordination Centre, 2023)

coastal zone of Friuli Venezia Giulia region (Italy), another region characterized by large-scale river and coastal flood hazards.

3.2 Materials and Methods

3.2.1 Case Study

The study area is represented by the central and southern portion of the Friuli Venezia Giulia region (Italy). Most of the area is characterized by a floodplain slowly degrading from north to south towards a coastal lagoon, near which the land is predominantly below the sea level, being the results of hydraulic reclamation works carried out over the past centuries to drain former wetlands and lagoons and create farmlands (Da Lio & Tosi, 2018). To a lesser extent, it also includes a foothill landscape with low-lying hills in the northern part and a hilly karst area with predominant high coast alternating small inlets in the southeastern part. The case study is located within the so-called Eastern Alps River Basin District, which includes several river basins where numerous rivers (both alpine and resurgent rivers) flow towards the North Adriatic Sea. According to the hazard data modeled within the framework of

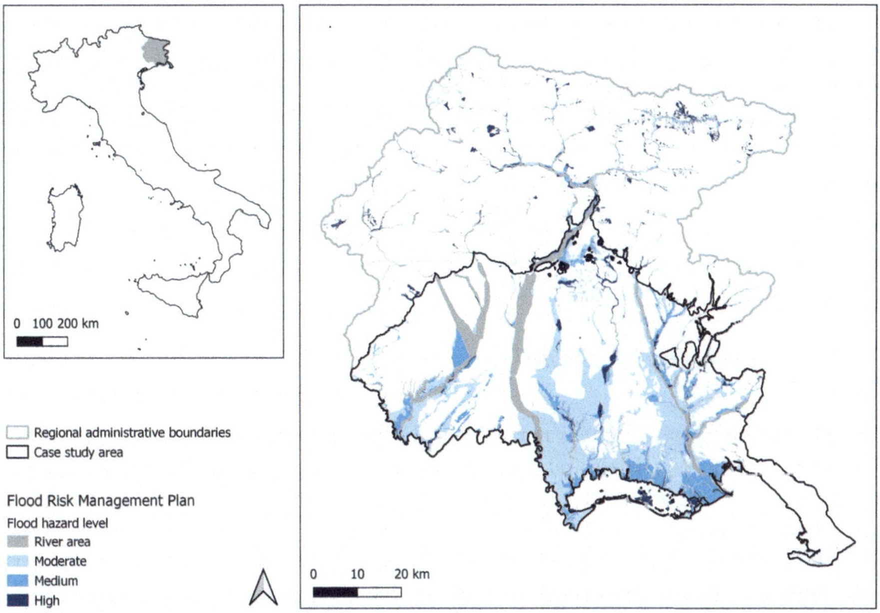

Fig. 3.2 Case study area and flood hazard map in Friuli Venezia Giulia region according to the data modeled within the Flood Risk Management Plan of the Eastern Alps River Basin District (Autorità di bacino distrettuale delle Alpi Orientali, 2021)

the Flood Risk Management Plan of the Eastern Alps River Basin District (Autorità di bacino distrettuale delle Alpi Orientali, 2021), large portions of this area are prone to riverine and coastal flooding (Fig. 3.2). Available statistical elaborations (ISPRA—Istituto Superiore per la Protezione e la Ricerca Ambientale, n.d.) reports that the 21,5%, 14,6%, and 9,6% of the Friuli Venezia Giulia region is characterized by low, medium, and high flooding hazard, respectively, being among the Italian regions with more flood-prone areas in proportion to the regional territory.

3.2.2 *Data*

The spatial assessment of the vulnerability of flood-prone areas in relation to their susceptibility to remain inundated for longer or shorter periods during a flooding event—thus exacerbating the impacts on ecosystems and human activities—is performed by combining available spatial datasets that capture the main intrinsic conditions of the territory that may influence its susceptibility. Each intrinsic condition is represented by one or more spatial variable(s), which in turn can be considered as the vulnerability factors of the territory. The intrinsic conditions considered in the assessment are the amount of excess water covering the territory during a flooding event, the topography of the area that controls the movement and drainage of water, and the surface water infiltration potential into the soil. The latter, in turn, depends on soil infiltration capacity and the rate of impervious cover impeding the surface water to reach and infiltrate into the soil. Table 3.1 shows the intrinsic conditions considered in the assessment and the corresponding vulnerability factors together with the spatial datasets used to represent each one of them.

Figures 3.3, 3.4, 3.5, and 3.6 show the spatial layers of the four vulnerability factors used in the assessment and described in Table 3.1.

3.2.3 Assessment Methodology

The vulnerability assessment is carried out through a spatial multiple-criteria decision analysis (MCDA). First, the criteria, namely the vulnerability factors, are organized in hierarchical levels. Second, the classes/values of each criterion are assigned with vulnerability scores in relation to their influence in increasing or decreasing the susceptibility to remain inundated for longer periods, with scores from 1 (lowest) to 5 (highest). Third, the criteria are progressively spatially combined (through the spatial overlay of datasets by applying a weighted sum of their vulnerability scores) according to the criteria tree hierarchy. During this phase, different weights are assigned to each criterion to attribute more or less importance to their relative influence in determining the overall vulnerability condition. In particular, higher importance is attributed to the vulnerability factor represented by the water depth expected to cover the territory during a flooding event (weight of 0.5), followed by the topography-based water accumulation tendency (0.3) and the surface water infiltration potential into the soil (0.2). The rationale behind the assignment of weights is that, firstly, the amount of excess water to be disposed is considered the primary factor influencing the susceptibility of the territory to remain inundated for a longer or shorter period (i.e., in principle, the more water volume to be drained, the longer it takes to be drained). Secondly, once floodwater has inundated the territory, it tends to move downstream following the topography and/or accumulate in depressed areas, causing water logging. Through controlling water movement, topographical aspects (mostly related to terrain slope) also influence water infiltration into the soil, with steeper slopes causing higher flow velocities that lead to less opportunity for water infiltration into the soil (e.g., Khan et al., 2016). Topography can be thus considered a more predominant factor with respect to water infiltration, being selected as the second factor in order of relevance. However, since water movement is likely to be rather slow in most of the case study area given its mostly flat nature, the relative weight assigned to this factor is only slightly higher than the one assigned to the water infiltration factor. Surface water infiltration potential according to soil characteristics also affects the rate and velocity it is drained into the soil, being the third factor (in order of relevance) influencing the susceptibility condition, with infiltration processes mainly occurring once the excess water movement and accumulation processes are in progress after the rising stage of inundation (Hou et al., 2021). This last factor,

Table 3.1 Intrinsic conditions considered for mapping the vulnerability factors affecting the susceptibility of the territory to remain inundated for a longer or shorter period during a flooding event and corresponding datasets used

Intrinsic conditions of the territory	Vulnerability factor	Spatial datasets used and source
Amount of excess water covering the territory during a flooding event	Estimated flooding extent and water depth, modeled for a specific flooding event	Modeled flooding extent and related water depth estimated for a 100-year-return-period flooding event from riverine and coastal inundations. Source: data modeled within the framework of the Flood Risk Management Plan of the Eastern Alps River Basin District (Autorità di bacino distrettuale delle Alpi Orientali, 2021) (available at: https://sigma.distrettoalpiorientali.it/portal/index.php/servizi/dati-terr/idraulica/)
Topography of the area that controls the movement and drainage of water	Tendency of an area to accumulate and retain versus to evacuate water	Topographic convergence index (TCI), defined as the logarithm of the ratio of flow accumulation and local slope. It is a dimensionless index that describes the tendency of an area to accumulate and retain water under steady-state conditions (higher values are associated with higher water accumulation potential). It balances contributing areas capturing the tendency to receive water versus slope angles capturing the tendency to evacuate water. It was calculated through the GRASS tool *r.terraflow* in QGIS software using a 10-m-resolution digital terrain model (DTM) as input data. The source of the DTM used is the regional geoportal of Friuli Venezia Giulia region (available at: https://irdat.regione.fvg.it/CTRN/ricerca-cartografia/). The TCI is calculated only for those non-water areas that are expected to be inundated to obtain a TCI variability that is fully representative of the flood-prone areas
Surface water infiltration potential into the soil	Surface water infiltration potential into the soil, which depends on: (i) Soil permeability characteristics, which allow for more or less rapid water infiltration into the soil (e.g., Rizzi et al., 2017) (ii) Amount of impervious surface impeding the water to reach and infiltrate into the soil	(i) Soil permeability classes based on saturated hydraulic conductivity, which represents the ease with which pores in a saturated soil transmit water (this data is not publicly available yet; it was provided by the Friuli Venezia Giulia Regional Agency for Rural Development) (ii) Imperviousness density data (reference year 2018), describing the percentage of impervious surfaces covering an area. Source: Copernicus, the European Union's Earth observation program (available at: https://land.copernicus.eu/pan-european/high-resolution-layers/imperviousness/status-maps)

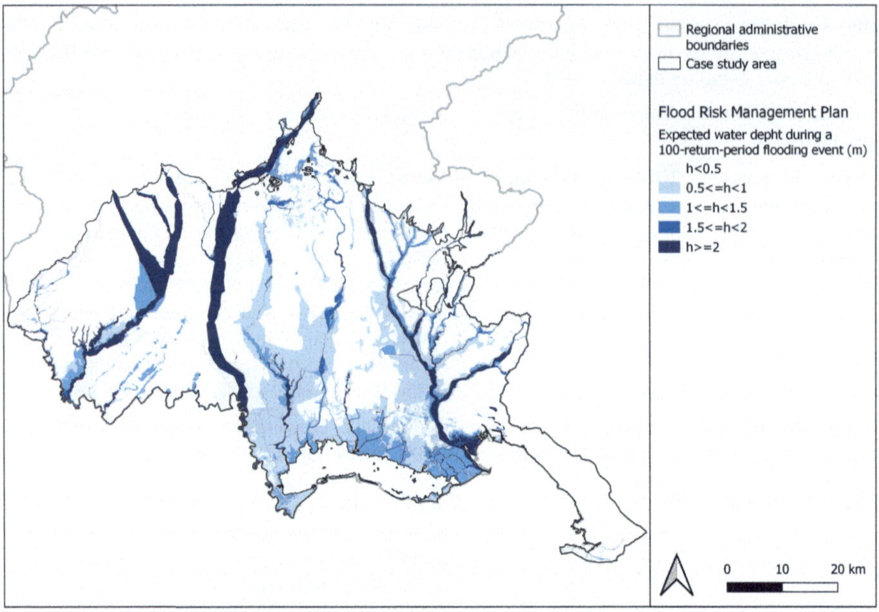

Fig. 3.3 Water depth estimated for a 100-year-return-period flooding event according to data modeled within the Flood Risk Management Plan of the Eastern Alps River Basin District (Autorità di bacino distrettuale delle Alpi Orientali, 2021)

in turn, depends on soil permeability characteristics and the amount of impervious surface covering the soil, which are attributed the same weight (0.5) during their spatial aggregation to assign them equal importance (neutral approach) since no information is available to further deepen their relative relevance in determining different susceptibility conditions (e.g., presence and sizing of drainage or water accumulation systems in highly impervious areas, or human interventions altering the infiltration capacity of soil, for example due to human-induced soil compaction in intensive agricultural areas, etc.). Figure 3.7 shows the criteria tree adopted in the MCDA.

The result of the progressive spatial combination through the weighted sum of the vulnerability factors according to the criteria tree hierarchy returns the overall vulnerability map representing the spatial variability of the potential vulnerability condition in the case study area, with scores from 1 (lowest vulnerability potential) to 5 (highest vulnerability potential). Table 3.2 lists the criteria (i.e., vulnerability factors) used in the MCDA according to their hierarchical level and the vulnerability scores assigned to their classes/values.

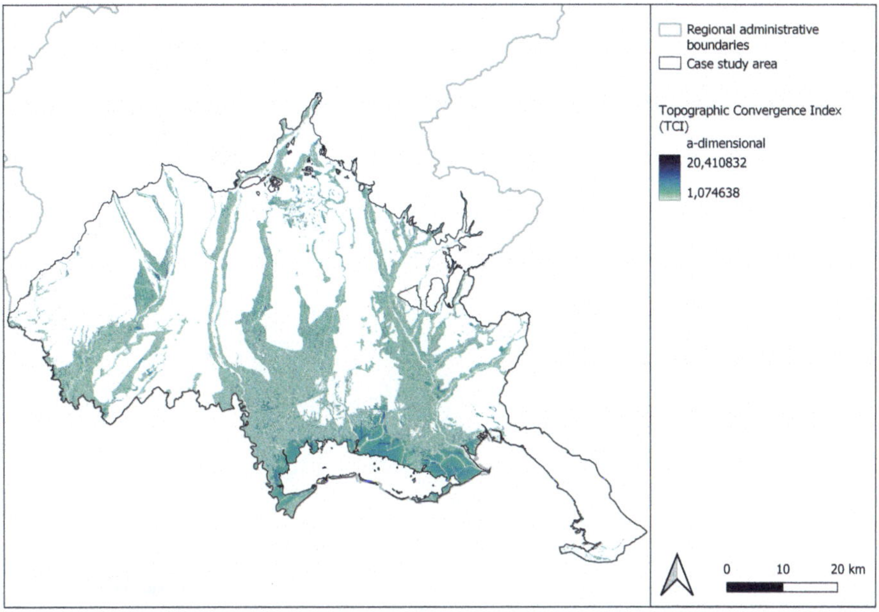

Fig. 3.4 Topographic convergence index (a-dimensional) calculated using the digital terrain model and representing water accumulation tendency (the higher the value the higher accumulation tendency) in flood-prone areas

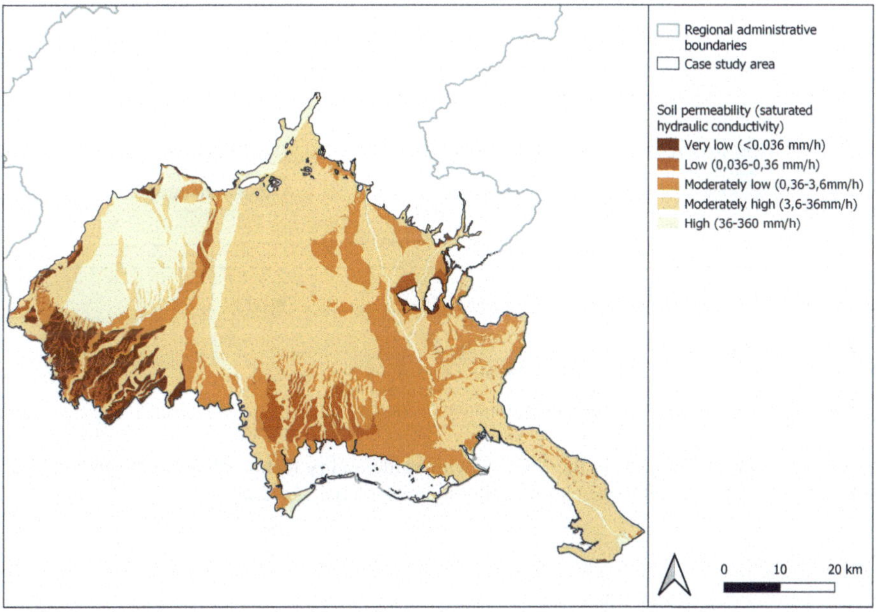

Fig. 3.5 Soil permeability characteristics based on saturated hydraulic conductivity

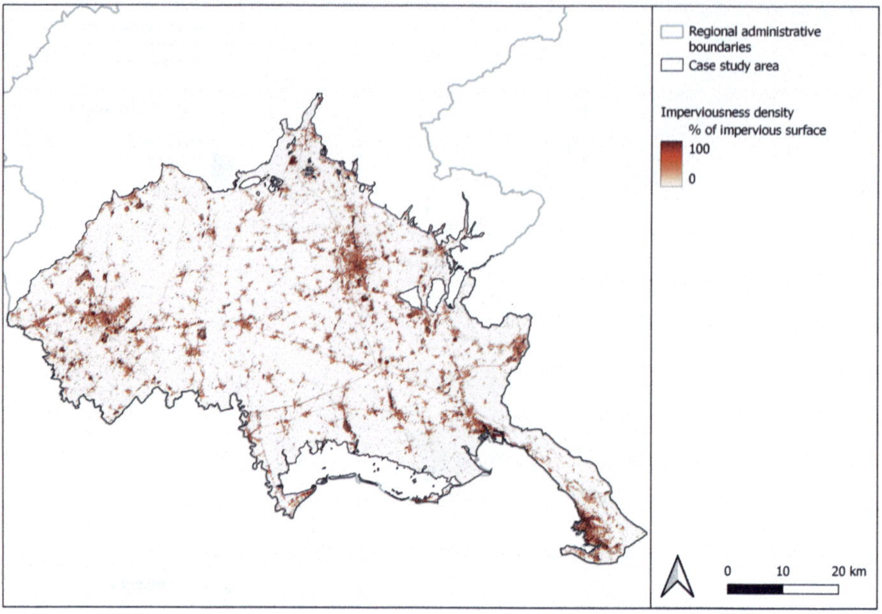

Fig. 3.6 Imperviousness density representing the percentage of impervious surface covering the soil

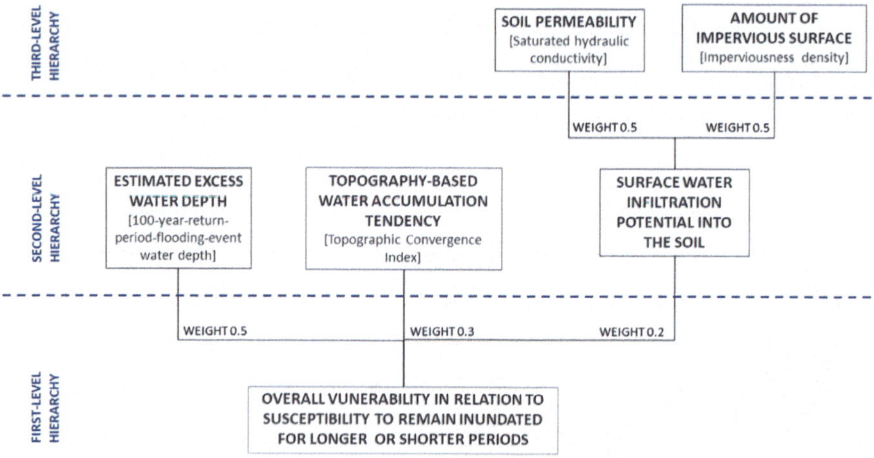

Fig. 3.7 Criteria tree adopted in the MCDA, with vulnerability factors organized in hierarchical levels and weights assigned to each of them during their spatial aggregation

Table 3.2 MCDA criteria (i.e., vulnerability factors) according to their hierarchical level and vulnerability scores assigned to their classes/values

First-level hierarchy	Second-level hierarchy	Third-level hierarchy	Classes/values of vulnerability factors used to assign vulnerability scores	Vulnerability scores
Overall vulnerability	Estimated excess water depth [100-year-return-period-flooding-event water depth] **Weight**: 0.5		< 0.5 m 0.5–1 m 1–1.5 m 1.5–2 m > 2 m	1 2 3 4 5
	Topography-based water accumulation tendency [Topographic convergence index] **Weight**: 0.3		< 1.91 (1st quintile) 1,91–2.97 2.97–4.56 4.56–6.61 > 6.61 (5th quintile)	1 2 3 4 5
	Surface water infiltration potential into the soil [A + B][1] **Weight**: 0.2	A. Soil permeability [Saturated hydraulic conductivity] **Weight**: 0.5	High (36–360 mm/h) Moderately high (3.6–36 mm/h) Moderately low (0.36–3.6 mm/h) Low (0.036–0.36 mm/h) Very low (< 0.036 mm/h)	1 2 3 4 5
		B. Amount of impervious surface [Imperviousness density] **Weight**: 0.5	0–20% 20–40% 40–60% 60–80% 80–100%	1 2 3 4 5

[1] The areas characterized by 100% imperviousness density are directly assigned with a vulnerability score of 5 concerning the factor 'surface water infiltration potential into the soil', without computing the weighted sum with the soil permeability criteria given the impossibility of water to reach and infiltrate into the soil.

3.3 Results

The spatial variability of the vulnerability condition of the territory in relation to the vulnerability factors considered (i.e., estimated flooding extent and water depth in a 100-years-return-period flooding event, topographical aspects controlling the movement and drainage of water according to the tendency of an area to accumulate and retain versus to evacuate water, and surface water infiltration potential into the soil according to soil permeability characteristics and amount of impervious surface) is represented in Figs. 3.8, 3.9 and 3.10.

The overall vulnerability map (Fig. 3.11), resulting from the spatial combination (i.e., weighted sum) of the three single vulnerability factors, reveals the vulnerability profile of the case study's flood-prone areas in relation to their susceptibility to remain inundated for longer or shorter periods during a flooding event with a return period of 100 years.

In general, since the intrinsic condition that was considered to be more relevant in the assessment is the amount of excess water expected to cover the territory during a flooding event, areas with higher vulnerability values are especially found in correspondence to the river network where water depth is expected to be higher (i.e., up to more than 2 m). In part of these areas, the presence of low surface water infiltration capacity by soil (due to soil characteristics) and/or of areas prone to accumulate and retain water due to local topography exacerbate the high vulnerability

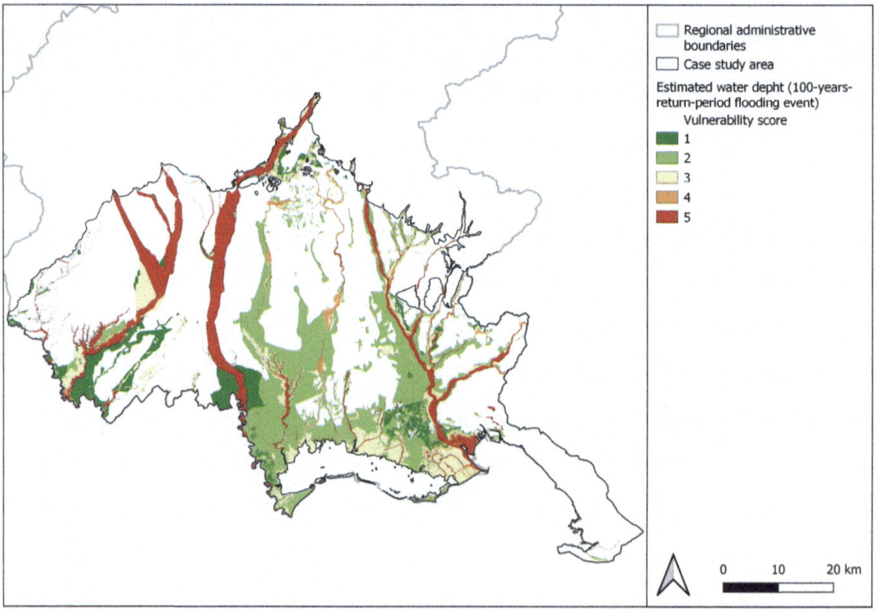

Fig. 3.8 Vulnerability scores according to the estimated water depth during a 100-years-return-period flooding event

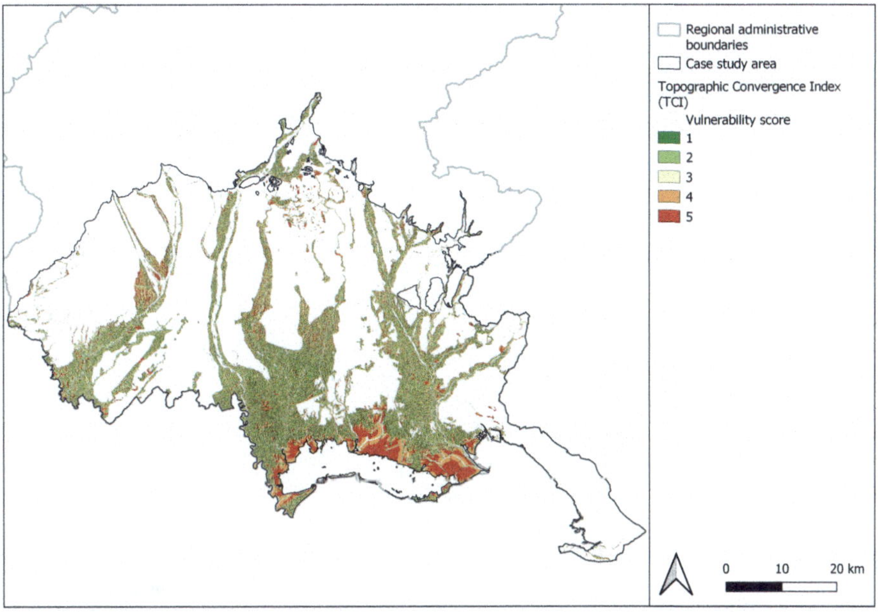

Fig. 3.9 Vulnerability scores according to the topographic convergence index (TCI) in flood-prone areas

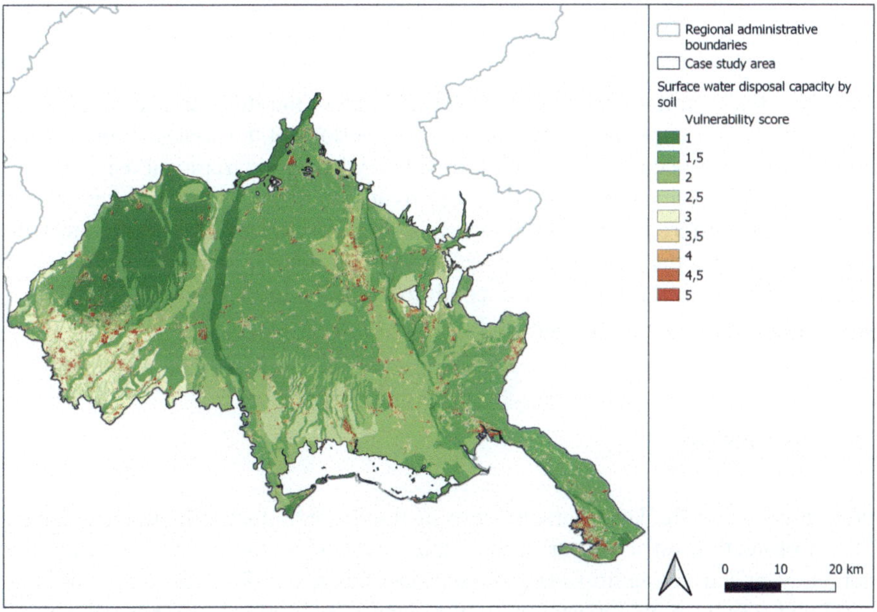

Fig. 3.10 Vulnerability scores according to the surface water infiltration potential into the soil

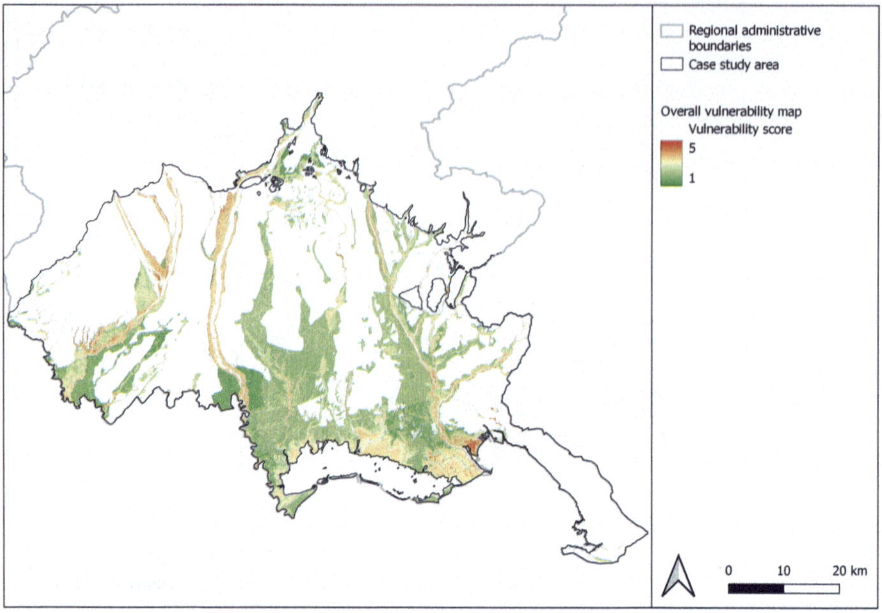

Fig. 3.11 Overall vulnerability map showing the vulnerability profile of the case study's flood-prone areas in relation to their susceptibility to remain inundated for longer or shorter periods during a 100-years-return-period flooding event (i.e., the higher the score, the higher the vulnerability condition)

condition. This is particularly true in the western part of the study area, characterized by soil with very low permeability, and in large portions of the area surrounding the coastal lagoon—especially on the eastern side—where topographical aspects (i.e., high values of TCI, meaning high tendency to accumulate and retain water) play a major role. At a more local scale, impervious surfaces covering the soil contribute to an increase of vulnerability compared to the surrounding pervious areas, when characterized by similar topographic conditions, soil characteristics, and expected water depth (see the zoom-in maps in Fig. 3.12 for an example).

3.4 Discussion

Duration is one of the key characteristics of flood events that influence the design of flood protection infrastructures, the deployment of rescue resources during the emergency, and affecting impacts and damages of exposed elements (e.g., buildings and related contents) and repartition of related costs in the aftermath (Serinaldi et al., 2018; Thieken et al., 2005).

The flooding event of May 2023 in the Emilia Romagna region (Italy) showed that some areas are likely to remain submerged by water for several days or weeks

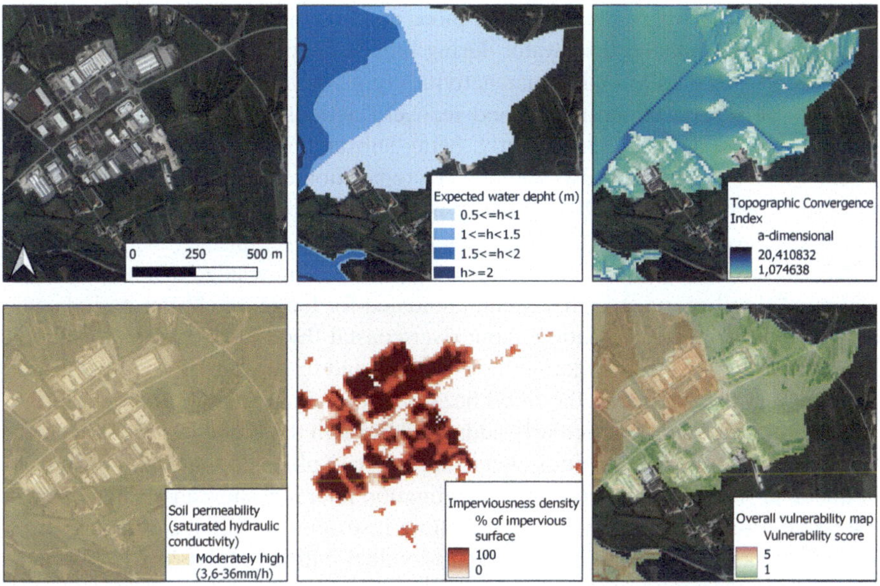

Fig. 3.12 Zoom-in maps showing a portion of a flood-prone area with impervious surfaces, the four single vulnerability factors considered, and the overall vulnerability map

after a major flooding event, according to the geomorphological characteristics of the flooded area. This is explained by the fact that flood duration is (also) a function of the topography of the area and surface dynamics, with areas most likely to remain submerged for longer periods that are not necessarily those located in the proximity of, e.g., a river, but those showing more difficulty in draining out the floodwater even if located farther from the river (Chyon et al., 2022). But this is not the only case. For example, in 2013 Alberta State (USA) experienced a significant flood event, triggered by heavy rainfall with large spatial coverage, with the initial flood-related damages occurring during the first week; however, some areas remained flooded even for several weeks, worsening those initial damages (Serinaldi et al., 2018).

These examples stress the importance of analyzing flood events and their differentiated duration within flood-prone areas to set up more effective mitigation and response strategies.

While water levels can be (more or less) easily provided by hydraulic modeling, the provision of flood durations demands more sophisticated models (Thieken et al., 2005), as well as very accurate knowledge of the context where the flood event occurs. The assessment based on MCDA presented in this paper does not pretend to substitute assessments based on hydraulic modeling but can provide indicative locations and vulnerability patterns that still can support certain decisions, both in emergency response and the risk mitigation planning phases.

For example, during the emergency phase (i.e., a flooding event), knowledge about what areas may indicatively remain inundated for a longer period can support decisions towards where there may be more need for hydraulic pump modules to pump away accumulated floodwater during inundations; or can support the identification of exposed areas and human activities (e.g., buildings, farmlands) that may require more efforts in terms of flood recovery actions because (more) prone to remain submerged (and, consequently, being substantially impacted) for a longer period, with consequences also on recovery repartition costs and insurance-related issues (e.g., Serinaldi et al., 2018). This type of assessment could thus be relevant to inform, e.g., civil protection and emergency plans.

Instead, during the risk mitigation planning phase, knowledge about the predisposition of flood-prone areas to remain inundated for longer or shorter periods may complement existing information about river/coastal flood hazards/risks to support the prioritization of risk mitigation interventions and the design of flood protection measures. On the one hand, the flood hazard/risk assessments and the risk mitigation planning aspect are particularly addressed through the Flood Risk Management Plans, with concrete implications also in subordinate plans that must conform to it. On the other hand, the planning of risk mitigation interventions can also be directly addressed in regional to local spatial planning instruments, which can benefit from the available knowledge about flood hazards, vulnerabilities, and risks to make more informed decisions (e.g., measures to avoid or minimize flood impacts).

In the case study of the Friuli Venezia Giulia region, most of the areas more susceptible to remain inundated for longer periods are found in cropland areas. The impacts that the agricultural sector may experience from a major flooding event may thus be substantially exacerbated due to the presence of farmland areas prone to remain submerged for a rather prolonged period. Flood duration is in fact among the main variables that define flood damage to agricultural lands, together with time of year of flood occurrence, water depth, flow velocity, and deposition of pollutants (Forster et al., 2008).

Furthermore, the approach proposed in this study is applicable at a rather large scale (e.g., from basin to regional scale), while more detailed assessments based on hydraulic models are usually provided at a more local scale (e.g., district to city scale (Zhang et al., 2023)), and their large-scale application would require much greater effort and resources (e.g., technical, time, economic). In this study, the assessment has been carried out considering both river and coastal floodings with a 100-year-return period simultaneously occurring in the whole case study area. While this is a very extreme case, although not 100% impossible, in some cases it could be reasonable to perform separately the assessment for the two typologies of flooding considered (river and coastal) and, for river floods, to disaggregate it at the scale of (each) river basin area. The approach proposed can be easily adjusted and replicated considering these different scales.

While the case study presented is focused on the assessment of flood-related susceptibility aspects that can support decisions to better prepare and manage the territory and the response strategies in the light of a flood event, it is worth mentioning that when it comes to flooding, 'zero risk' does not exist. Flood protection systems are in fact designed to withstand up to a certain volume of water, and even if we upgrade them to respond to higher and higher return period events, they cannot protect us from all floods, with climate change that will only make such extreme events more and more frequent (thus, the return period of flooding events is likely to be shorter in the future). Even in cases where the river runoff volume does not exceed local flow capacities, there may be the risk of flood defense systems' failure/collapse due to water pressure and erosion processes, among others. Events involving the collapse of flood protection systems during a flooding event are not so rare, but often they are ignored by decision-makers dealing with spatial planning decisions. For example, the presence of a levee is usually interpreted as a permanent safety barrier in the urban planning field, leading to building development that is based on an incorrect assumption (Autorità di bacino distrettuale delle Alpi Orientali, 2021).

To correctly frame what happened in the Emilia Romagna case (but this can be extended to all the Italian regions and beyond) in the context of socio-hydrology, namely the field of study focusing on society-river interactions, we can refer to the so-called 'levee effect': building flood-control structures reduces people perception of risk and, consequently, favors the development of human settlement in floodplain areas.

For this reason, the concept of residual risk needs to be always considered when planning territorial development and in decisions dealing with selecting locations for human activities, as well as for the design of effective flood protection measures (Fiori et al., 2023). Residual risk can be described as "the risk that remains in unmanaged form, even when effective disaster risk reduction measures are in place, and for which emergency response and recovery capacities must be maintained" (UNISDR, 2009). Also, the European Commission (2004) recommends always considering that the construction of flood protection systems does not constitute a fully safe condition and to include among the flood hazard scenarios the possibility of the protection system's failure/collapse. The flood hazard map used for the case study area assessment (see Fig. 3.2), which is based on the European Union's Flood Directive, accounts for this possibility.

In conclusion, it can be said that current and future flood impacts are the result of intertwined climatic and non-climatic dynamics (including spatial planning decisions) which make the population and economy more sensitive to flooding. For example, demographic factors determine the sensitivity of people to floods, with the elderly, young children, and people with health problems being the more vulnerable in principle, and with other several socio-economic factors influencing people's vulnerabilities, including population income, social status, housing type, type of households, number of social hot spots such as schools, hospitals, kindergartens, and pensioners' homes (Schauser et al., 2010). Economic and demographic developments are therefore very important drivers, and, while development and economic growth continue, the scale of potential losses will also increase (European Environment Agency, 2012).

3.5 Conclusion

This study proposes a rather simple and quick method to assess the vulnerability of flood-prone areas in relation to their susceptibility to remain inundated for shorter or longer periods during a flooding event that can be used as a decision support tool in emergency response and risk mitigation planning phases.

This temporal aspect of flooding, often neglected in flood hazard, susceptibility/vulnerability, risk, and/or impact/damage assessments (e.g., Merz et al., 2010; Serinaldi et al., 2018), constitutes a relevant factor that is of particular importance in the light of a flooding event. The longer a flooding lasts, the larger the material damage, especially due to service interruption. For example, for the period during which a portion of land remains inundated, everyday routines are distorted, firms are not able to produce, agricultural products cannot be cultivated, and reconstruction cannot begin, substantially influencing the indirect losses (Koks et al., 2015). The approach proposed, based on spatial MCDA, constitutes a preliminary assessment that could be further refined. For example, future steps may include the involvement of a panel of experts to review and eventually enhance the assessment method proposed, especially with regards to the identification of possible other criteria to integrate into the assessment and to the revision of the weights to apply (i.e., in terms of relative importance in determining the susceptibility condition). Assessments based on MCDA (sometimes using expert judgments and/or stakeholder preferences) have been widely used to develop decision support tools based on vulnerability and/or risk analysis (e.g., Pagano et al., 2023; Rizzi et al., 2017; Torresan et al., 2012). Examples of expert involvement in MCDA models related to flooding aspects are plenty (i.e., through the so-called analytical hierarchical process), but are especially found in studies aimed at spatially assessing flood hazard, susceptibility/vulnerability, and/or risk in areas with a lack of data coming from more complex hydraulic models (e.g., Nachappa et al., 2020; Zzaman et al., 2021). In the specific case presented in this study, the objective instead is to assess an aspect of flooding (i.e., the susceptibility to remain flooded for shorter or longer periods during a flooding event) that is not usually considered (e.g., the existing flood hazard and flood risk in the Friuli Venezia Giulia region are assessed as a function of modeled flood water depths and velocities), which can complement and/or be integrated with existing flood hazard, susceptibility/vulnerability, risk, and/or impact/damage assessments to further inform flood-related decisions (e.g., flood risk mitigation, response, and recovery strategies).

Acknowledgments This study was partly supported by the Friuli Venezia Giulia region (*Regione Autonoma Friuli Venezia Giulia*) under the framework of the Operative Agreement (2022–2023) "*Accordo operativo con il Servizio Pianificazione Paesaggistica, Territoriale e Strategica per attività di supporto scientifico e metodologico alla predisposizione di una Variante al Piano del Governo del Territorio (PGT) relativamente all'introduzione della tematica di adattamento al Cambiamento Climatico e Resilienza Territoriale negli strumenti urbanistici di area vasta*".

The author would like to thank the Regional Agency for Rural Development of the Friuli Venezia Giulia region (*ERSA FVG*) for sharing unpublished data that was used in this work.

References

Autorità di bacino distrettuale delle Alpi Orientali. (2021). Aggiornamento e revisione del Piano di Gestione del Rischio di Alluvioni 2021–2027.

Barredo, J. I. (2006). Major flood disasters in Europe: 1950–2005. *Natural Hazards, 42*(1), 125–148.

CIMA Research Foundation. (2023). The Italian floods of May 2023: A scientific analysis. Accessed November 6, 2023. https://www.cimafoundation.org/en/news/the-italian-floods-of-may-2023-a-scientific-analysis/

Da Lio, C., & Tosi, L. (2018). Land subsidence in the Friuli Venezia Giulia coastal plain, Italy: 1992–2010 results from SAR-based interferometry. *Science of the Total Environment, 633*, 752–764.

Dang, N. M., Babel, M. S., & Luong, H. T. (2011). Evaluation of food risk parameters in the Day River Flood Diversion Area, Red River Delta, Vietnam. *Natural Hazards, 56*(1), 169–194.

Dutta, D., Srikantha, H., & Katumi, M. (2003). A mathematical model for flood loss estimation. *Journal of Hydrology, 277*, 24–49.

Chyon, M. S. A., Biswas, S., Mondal, M. S., Roy, B., & Rahman, A. (2022). Integrated assessment of flood risk in Arial Khan floodplain of Bangladesh under changing climate and socioeconomic conditions. *Journal of Flood Risk Management, 16*(2), e12876. https://doi.org/10.1111/jfr3.12876

ERCC—Emergency Response Coordination Centre. (2023). ECHO Daily Map of 26 May 2023. Accessed November 6, 2023. https://erccportal.jrc.ec.europa.eu/ECHO-Products/Maps#/maps/4501

European Commission. (2004). *Communication from the commission to the council, the European Parliament, the European economic and social committee and the committee of the regions.* Flood risk management—Flood prevention, protection and mitigation. Brussels, COM(2004)472 final.

European Environment Agency. (2012). *Urban adaptation to climate change in Europe. Challenges and opportunities for cities together with supportive national and European policies.* EEA Report No 2/2012. European Environment Agency, Copenhagen.

Fiori, A., Mancini, C. P., Annis, A., Lollai, S., Volpi, E., Nardi, F., & Grimaldi, S. (2023). The role of residual risk on flood damage assessment: A continuous hydrologic-hydraulic modelling approach for the historical city of Rome, Italy. *Journal of Hydrology: Regional Studies, 49*, 101506. https://doi.org/10.1016/j.ejrh.2023.101506

Forster, S., Kuhlmann, B., Lindenschmidt, K.-E., & Bronstert, A. (2008). Assessing flood risk for a rural detention area. *Natural Hazards and Earth Systems Sciences, 8*, 311–322.

Genovese, E., et al. (2007). *An assessment of weather-related risk in Europe: Maps of flood and drought risks.* EC Joint Research Center.

Haraguchi, M., & Lall, U. (2015). Flood risks and impacts: A case study of Thailands floods in 2011 and research questions for supply chain decision making. *International Journal of Disaster Risk Reduction, 14*, 256–272.

Hou, J., Zhang, Z., Zhang, D., Shi, B., Chen, G., & Zhang, H. (2021). Study on the influence of infiltration on flood propagation with different peak shape coefficients and duration. *Water Policy, 23*(4), 1059–1074. https://doi.org/10.2166/wp.2021.193

Khan, M. N., Gong, Y., Hu, T., Lal, R., Zheng, J., Justine, M. F., Azhar, M., Che, M., & Zhang, H. (2016). Effect of slope, rainfall intensity and Mulch on erosion and infiltration under simulated rain on purple soil of South-Western Sichuan Province, China. *Water, 8*, 528. https://doi.org/10.3390/w8110528

Koks, E. E., Bočkarjova, M., Moel, H., & Aerts, J. C. J. H. (2015). Integrated direct and indirect flood risk modeling: Development and sensitivity analysis. *Risk Analysis, 35*(5), 882–900.

ISPRA—Istituto Superiore per la Protezione e la Ricerca Ambientale. (n.d.). *Landslide and flood hazard/risk Italian geoportal with data from "Rapporto ISPRA 2021 su Dissesto idrogeologico in Italia: pericolosità e indicatori di rischio"*. Accessed November 6, 2023. https://idrogeo.isprambiente.it/app/pir/r/6?@=45.896438559619156,13.398723307339827,5

Lazzarin, T., Viero, D. P., Molinari, D., Ballio, F., & Defina, A. (2022). A new framework for flood damage assessment considering the within-event time evolution of hazard, exposure, and vulnerability. *Journal of Hydrology, 615*, 128687.

Lugeri, N., Kundzewicz, Z. W., Genovese, E., Hochrainer, S., & Radziejewski, M. (2010). River flood risk and adaptation in Europe—Assessment of the present status. *Mitigation and Adaptation Strategies for Global Change, 15*(7), 621–639.

Merz, B., Kreibich, H., Schwarze, R., & Thieken, A. (2010). Assessment of economic flood damage. *Natural Hazards and Earth Systems Sciences, 10*, 1697–1724.

Nachappa, T. G., Piralilou, S. T., Gholamnia, K., Ghorbanzadeh, O., Rahmati, O., & Blaschke, T. (2020). Flood susceptibility mapping with machine learning, multicriteria decision analysis and ensemble using Dempster Shafer Theory. *Journal of Hydrology, 590*, 125275. https://doi.org/10.1016/j.jhydrol.2020.125275

Pagano, M., Fernetti, M., Busetti, M., Ghribi, M., & Camerlenghi, A. (2023). Multicriteria GIS-based analysis for the evaluation of the vulnerability of the marine environment in the Gulf of Trieste (north-eastern Adriatic Sea) for sustainable blue economy and maritime spatial planning. *People and Nature, 00*, 1–20.

Rizzi, J., Torresan, S., Zabeo, A., Critto, A., Tosoni, A., Tomasin, A., & Marcomini, A. (2017). Assessing storm surge risk under future sea-level rise scenarios: A case study in the North Adriatic coast. *Journal of Coastal Conservation, 21*, 453–471.

Serinaldi, F., Loecker, F., Kilsby, C. G., et al. (2018). Flood propagation and duration in large river basins: A data-driven analysis for reinsurance purposes. *Natural Hazards, 94*, 71–92. https://doi.org/10.1007/s11069-018-3374-0

Schauser, I., Otto, S., Schneiderbauer, S., Harvey, A., Hodgson, N., Robrecht, H., Morchain, D., Schrander, J.-J., Khovanskaia, M., Celikyilmaz-Aydemir, G., Prutsch, A., & McCallum, S. (2010). *Urban regions: vulnerabilities, vulnerability assessments by indicators and adaptation options for climate change impacts—scoping study*. ETC/ACC Technical Paper, European Topic Centre on Air and Climate Change, Bilthoven.

Thieken, A. H., Müller, M., Kreibich, H., & Merz, B. (2005). Flood damage and influencing factors: New insights from the August 2002 flood in Germany. *Water Resources Research, 41*, W12430. https://doi.org/10.1029/2005WR004177

Tingsanchali, T., & Promping, T. (2022). Comprehensive Assessment of flood hazard, vulnerability, and flood risk at the household level in a municipality area: A case study of Nan Province, Thailand. *Water, 14*, 161. https://doi.org/10.3390/w14020161

Titley, H. A., Cloke, H. L., Harrigan, S., Pappenberger, F., Prudhomme, C., Robbins, J. C., et al. (2021). Key factors influencing the severity of fluvial flood hazard from tropical cyclones. *Journal of Hydrometeorology*.

Torresan, S., Critto, A., Rizzi, J., & Marcomini, A. (2012). Assessment of coastal vulnerability to climate change hazards at the regional scale: The case study of the North Adriatic Sea. *Natural Hazards and Earth Systems Sciences, 12*, 2347–2368.

UNISDR. (2009). *Terminology on disaster risk reduction*. Geneva.

Usman Kaoje, I., Abdul Rahman, M. Z., Idris, N. H., Razak, K. A., Wan Mohd Rani, W. N. M., Tam, T. H., & Mohd Salleh, M. R. (2021). Physical flood vulnerability assessment using geospatial indicator-based approach and participatory analytical hierarchy process: A case study in Kota Bharu, Malaysia. *Water, 13*, 1786.

Zhang, X., Kang, A., Ye, M., Song, Q., Lei, X., & Wang, H. (2023). Influence of terrain factors on urban pluvial flooding characteristics: A case study of a small watershed in Guangzhou, China. *Water, 15*, 2261. https://doi.org/10.3390/w15122261

Zzaman, R. U., Nowreen, S., Billah, M., Islam, A. S. (2021). Flood hazard mapping of Sangu River basin in Bangladesh using multi-criteria analysis of hydro-geomorphological factors. *Journal of Flood Risk Management*, 1–19.

Chapter 4
Climate Change Adaptation and Regional Planning: The Liguria Region Case

Filippo Fraschini, Chiara Franciosi, and Marina Morando

Abstract Local adaptation strategies and plans are essential tools to tackle the ongoing impact generated by climate change and increase the adaptive capacity of local communities to possible future climate-related challenges. Fondazione CIMA has teamed up with other research institutions, with wide expertise on natural risks and climate change policies, to support the Liguria Region in the development of its adaptation strategy. In this paper, we present the methodology applied, focusing on three key milestones: (i) the identification of climate change tendencies and expected scenarios; (ii) the application of the "impact chains" methodology ("Conceptual models describing climate impacts as cause-effects relationships within socio-ecological system"—Zebisch et al., 2021") with the aim to engage local stakeholders in the identification of context specific climate-related risks and adaptation objectives; (iii) the definition of matrixes connecting risks, adaptation objectives and measures for every regional sector affected by climate change. Each step of our approach has been characterised by the integration between quantitative data coming from the climate and impact modelling activities and qualitative knowledge coming from the engagement of local stakeholders, with the aim to reduce the inevitable uncertainty connected to climate forecasts and increase the tailoring of adaptation goals and solutions to the local needs and development priorities. In this paper, we outline the key elements of our approach, pinpointing strengths and weaknesses and the most relevant lessons learnt.

Keywords Climate change adaptation · Regional planning · Climate risk assessment · Uncertainty

---

F. Fraschini (✉) · C. Franciosi · M. Morando
CIMA Foundation, Savona, Italy
e-mail: filippo.fraschini@cimafoundation.org

F. Fraschini
CNR-IRCrES, Consiglio Nazionale delle Ricerche, Istituto di Ricerca sulla Crescita Economica Sostenibile, Torino, Italy

M. Granceri Bradaschia et al. (eds.), *Climate Change Adaptation, Flood Risk, and Beyond*, Planning for Climate Proof Cities,
https://doi.org/10.1007/978-3-031-65463-3_4

4.1 Introduction

The worsening of climate change impacts and the slow pace of mitigation efforts at the global level are increasing the urgency of adaptation policies, focused at preparing our economies and social systems, increasing resilience and promoting sustainable development (Intergovernmental Panel on Climate Change—IPCC, 2018, 2019, 2022a, 2022b). The latest report by IPCC (IPCC, 2023) warned that due to climate change the average global surface temperature has increased by 1.09 °C, rising faster since the 1970s. The severity of the impacts generated by climate change is worsening (e.g. the average rate of sea level rise is accelerating, starting from 1.3 mm yearly between 1901 and 1971, to 1.9 mm yearly between 1971 and 2006, and further increasing to 3.7 mm yearly between 2006 and 2018). IPCC (2022a, 2022b) highlights that the extent and magnitude of climate change impacts are larger than estimated in the previous assessments, leading to some irreversible effects or other consequences approaching irreversibility. Evidence of observed changes in extremes such as heatwaves, heavy precipitation, droughts and tropical cyclones, and, in particular, their attribution to human influence, has further strengthened since the previous Assessment Report by IPCC (IPCC, 2022a, 2022b). Moreover, these effects will continue in the coming years, regardless of the mitigation policies developed by states. The EU Commission maintains in its New EU Strategy on Adaptation to Climate Change (2021): "*Halting all greenhouse gas emissions would still not prevent the climate impacts that are already occurring. These will continue for decades, even if global and European efforts to cut greenhouse gas emissions prove effective.*" Furthermore, these climate hazards interact between one other, as well as with other non-climate related hazards, generating complex risks, transferring through interconnected systems and across regions (IPCC, 2022a, 2022b).

In order to face such complex risks and the ever-growing cross-cutting impacts generated or intensified by climate change, adaptation policies have been recognised as a key vehicle for sustainable development (United Nations—UN, 2015; United Nations Framework Convention on Climate Change—UNFCCC, 2015; United Nations Office for Disaster Risk Reduction—UNDRR, 2015). The 2030 Agenda for Sustainable Development (United Nations, 2015) identified a specific task for adaptation: "*Strengthen resilience and adaptive capacity to climate related hazards and natural disasters in all countries*". Furthermore, according to the Paris Agreement (article 7), adaptation is recognised as a "*global challenge faced by all with local, subnational, national, regional and international dimensions*", while the European Green Deal (2019) highlights the urgency of adaptation policies: "*Strengthening the efforts on climate-proofing, resilience building, prevention and preparedness is crucial*".

In the last Assessment Report—AR6 (IPCC, 2022a, 2022b) adaptation is defined as: "*The process of adjustment to actual or expected climate and its effects, in order to moderate harm or exploit beneficial opportunities*" (p. 2216). Adaptation is thus aimed at both present or expected climate-related impacts, requiring a balance between policies oriented towards the ongoing impacts of climate change and policies dedicated to possible expected effects, where anticipatory planning effort is necessary, preparing the target community for the most severe and complex future impacts. Pro-active policies are also essential for preparing for more transformative measures, avoiding lock-ins generated by high sunk cost investments defined without taking on board the possible effects of climate change. Therefore, a climate change adaptation strategy becomes an important instrument to identify a comprehensive set of far-sighted policies with the aim to organise the knowledge on climate change impacts, at the local level and to define anticipatory solutions, besides the reactive policies designed for urgent adaptation needs. Adaptation strategies can also be an opportunity to develop cross-cutting policies between the sectors of a public administration, co-ordinating the development priorities and adaptation needs of the various sectors.

However, despite the presence of a multitude of adaptation policy guidance or tools, the framework of the adaptation strategies and plans remains uneven. In Italy, there are diverse approaches to adaptation policymaking at a local level: (i) regions developing an adaptation strategy, or directly an adaptation plan; (ii) regions focusing on both adaptation and mitigation; (iii) regions designing adaptation strategies for a single sector (e.g. Regione Siciliana—Guidelines for adaptation policies in the agriculture sector) or (iv) integrating adaptation objectives and measures inside a broader Sustainable Development Strategy. Moreover, the structure of these documents and the balance between quantitative and qualitative approaches are also uneven (e.g. the analysis of climate scenarios and the assessment of climate-related risks are approached differently in various contexts).

In this paper, we discuss the approach we have followed to support the Liguria Region administration in designing its own climate change adaptation strategy. The paper presents a real application of the methodologies developed at EU and National level, with the aim to discuss the main challenges encountered in our activities and the solutions we applied for a tailoring of the adaptation policies to the regional context. We focus on the key steps and milestones of our case study, rather than examining results, such as outcomes from the climate predictions, or presenting in detail objectives and measures. We discuss key strengths and weaknesses of the approach followed, in order to pinpoint some crucial insights to support decision-makers or practitioners in the process to design their adaptation policies. Moreover, we focus specifically on the participatory activities we have carried out for the identification of location-specific climate-related risks and adaptation objectives emerging from local climate adaptation needs. In our approach, we have applied a light-touch version of the impact chains methodology with the aim to assess the climate related risks, in the attempt to engage the public administration officials and to find a balance between quantitative modelled knowledge and qualitative assessments, enhancing both technical and empirical expertise.

4.2 The Liguria Region Adaptation Strategy

The Liguria Region is squeezed in between the Ligurian Sea, the Alps and the Apennines. The region has 1.5 million inhabitants mainly concentrated alongside the coastline, characterised by an extremely high soil consumption, especially in the area of the Genoa municipality, where lies 50% of the regional population. According to ISPRA (2022), Liguria is in eighth place among the Italian regions with the highest percentage of land consumption (i.e. 7.2%, meaning 39,260 ha), with particularly high values alongside the coastline and the hinterland of the main urban centres where Ligurian industrial activity was concentrated in the past. Land consumption within 300 m from the coast reaches 47% in Liguria, more than double the national average value which stands at 22% (ISPRA, 2022). Liguria has a baseline of high flood and coastal flood risk, and severe climate change impacts are expected in the coming years. The average temperature is increasing, and the distribution of cumulative precipitations is changing over the years, with decreasing precipitations during spring and increasing amounts of rainfall in terms of both cumulative values and intensity during autumn (ARPAL, 2013). The Liguria Region is also highly vulnerable to climate-related hazards due to the characteristics of its population. Liguria has the highest average age in Italy—i.e. 49.3 years, with the national average of 46—and it is among the oldest regions in Europe, with worsening tendencies over the coming years.

In 2018, the Region started the activities for the development of the Regional Sustainable Development Strategy (RSDS), through the involvement of local institutions and stakeholders, with the aim to identify its specific sustainable development objectives. The Regional Climate Change Adaptation Strategy has been identified by the regional administration as a priority field for integrated action towards the implementation of the RSDS.

The technical support for the definition of the Liguria Region Adaptation Strategy has been financed by the "SNSVS2 Call"—Call for the promotion of research projects to support the implementation of the National Sustainable Development Strategy (Ministero dell'Ambiente e della Tutela del Territorio e del Mare, 2017). A partnership among CIMA Foundation, the Department of Architecture and Design of the University of Genoa and the Research Centre for Ponente Ligure of the University of Genoa has been reached to establish a Scientific Working Group (SWG) to support the regional administration. This research group has been created with the aim to aggregate a variety of competencies and skills, to face the complex and wider framework of adaptation policymaking. The point of reference inside the regional administration was the Environment and Civil Protection Department—Environmental Impact Assessment and Sustainable Development Sector. The technical support for the definition of the Regional Adaptation Strategy was structured following the methodological frameworks proposed at National and European level. At the National level, guidelines were designed by CReIAMO PA (CReIAMO PA, 2020), a project developed by the Italian Ministry for the Environment, aimed at empowering the public administrations to tackle environmental and climate change issues. The LIFE Master Adapt project also developed another key methodological document (Ballarin Denti et al., 2019). This project was aimed at developing methods and tools to activate coherent and effective mainstreaming processes. The project presented climate change adaptation regional strategies as key framework tools to guide and coordinate the adaptation mainstreaming process. Guidelines from Master Adapt focused on risk and vulnerability assessment and the steps to identify a set of adaptation goals and objectives. Furthermore, the EU Commission produced its own Guidelines on Developing Adaptation Strategies (EU Commission, 2013), as an attached document to the first EU Adaptation Strategy to Climate Change (Fig. 4.1). These guidelines provide a common understanding of key features relevant to the process for the definition of adaptation policies, building on the experience available in the EU. These guidelines are the pillars of the Adaptation Support Tool by the European Climate Adaptation Platform Climate-ADAPT.

In supporting the regional administration in developing its own Climate Change Adaptation Strategy, we focused on the first three main steps of the EU approach: (i) Preparing the Ground for Adaptation; (ii) Assessing Risk and Vulnerabilities to Climate Change; (iii) Identifying Adaptation Options.

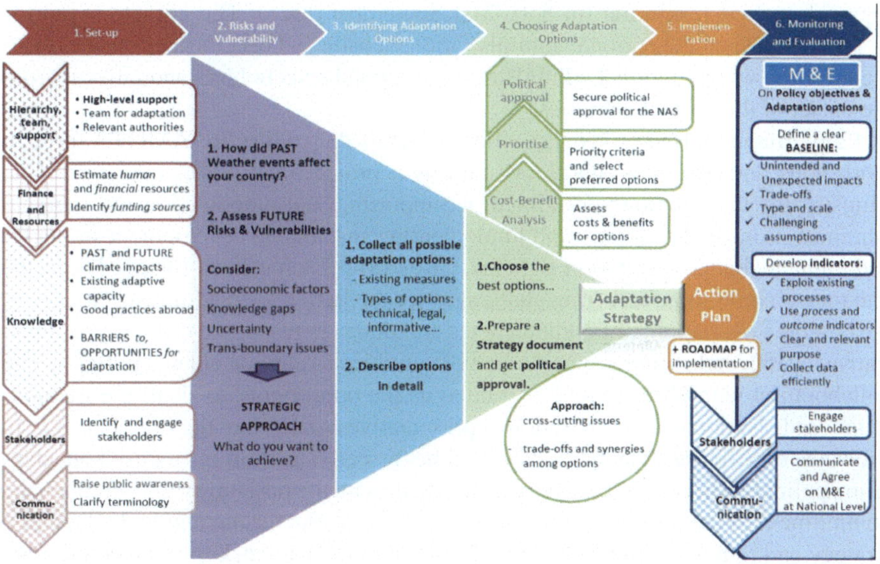

Fig. 4.1 The EU Commission six-steps methodology for the development of adaptation strategies. *Source* EU Commission (2013)

4.2.1 Preparing the Ground for Adaptation

As described in the EU Commission methodology and in the CReIAMO PA guidelines, a set of preliminary activities is crucial in order to pave the way for an effective process to design an adaptation strategy. We focused mainly on three key tasks: (i) obtaining high level support; (ii) setting up a working group dedicated to the definition of the adaptation strategy; (iii) creating a common epistemic framework.

The first activity was focused to obtain high level support by the public administration to the whole adaptation process. High-level support can be triggered by recommendations from a higher level of governance, or triggered by legal obligations, even coming from a single sector (EU Commission, 2013). In our approach we engaged both the high-level regional decision-makers and officials and the national level, from the very beginning of our activities. In a preliminary workshop, we presented the first results of our climate change modelling activity related to climate change scenarios and expected impacts. We engaged the directors of the regional departments in this workshop, who are most affected by climate change and the head of the Environment and Civil Protection sector in the Regional Council. As an output of this meeting, the climate change scenarios modelled by CIMA Foundation were approved by a regional council resolution (n. 850, 21st September 2021). The climate scenarios were published on the regional administration website and became a reference document for the whole regional administration, regional stakeholders and citizens from the outset of the project activities.

Furthermore, during project activities, we teamed up with CReIAMO PA through workshops and meetings. CReIAMO PA is a project (2017–2023) of the Italian Ministry for the Environment aimed at supporting and empowering the regional administrations in developing and implementing sustainable development strategies and climate change adaptation policies. CReIAMO PA has been financed with the aim to overcome some recurring challenges for the Italian local administrations (e.g. the complex overlapping of competencies and duties between the different administrative levels), dedicating a specific line of activities to climate change policies. The collaboration with CReIAMO PA granted us the opportunity to share our methodology with National administration representatives and other regions, but also to exchange experiences and methods with other research institutions focused on adaptation policies. This twofold activity provided us with the possibility to strengthen the commitment of the regional administration on the works for the regional adaptation strategy and to have high level technical and political feedback to the methodology proposed.

The establishment of a Regional Working Group for the Adaptation Strategy has been another essential task during the preliminary activities. The officials of the Environment and Civil Protection Department adopted the working group (WG) previously committed to the definition of the Regional Sustainable Development Strategy (RSDS) and created a Regional Working Group for the Climate Change Adaptation Strategy (established with Decree of the Secretary General n. 3975/2021 and updated with n. 4462/2021). This working group (WG) was composed by both officials of the regional administrations, and experts from the regional technical agencies (e.g. ARPAL—the regional Agency for the Protection of the Environment and IRE—the regional Agency for the Energy). The continuity between the members of the WG for the RSDS and RCCAS has proved effective in pursuing synergies between the two documents. Furthermore, the link between the two strategies can be crucial for the implementation of adaptation objectives and measures. The definition of a RSDS is indeed mandatory for regional administrations in Italy, as foreseen by article 34 of legislative decree n.152 of 2006, and regions must define their contribution to the achievement of the national strategy objectives, identifying priorities and the actions to be undertaken. On the contrary, climate change adaptation strategies are currently voluntary and the link with the RSDS is therefore an opportunity to find tools and instruments for the effective implementation of measures.

Furthermore, coherently with the EU Commission methodology, a key preliminary activity has been dedicated to creating a common epistemic framework between the Regional Working Group and the Scientific Working Group. Even though the Intergovernmental Panel on Climate Change harmonised the scientific literature on adaptation in its Assessment Reports, several authors still present the abundance of conceptualisations and definitions about adaptation (Hall, 2017; Fankhauser, 2017; Schipper, 2007; Sherman et al., 2016; UNFCCC LDC Expert Group, 2012). There is no universal agreement between decision-makers on adaptation (Berrang-Ford et al., 2019; UNEP, 2018), and the understanding of the climate risk concepts is uneven. Therefore, as one of the first preliminary activities, we discussed key words and concepts related to climate change and adaptation policymaking and the University of Genoa created a shared glossary, which has been published on the website of the Liguria Region.

4.2.2 Assessing Risks and Vulnerabilities to Climate Change

The assessment of risks and vulnerabilities to climate change has been the second key step for the definition of the Liguria region adaptation strategy. The importance of climate change risk assessment has been recognised through the existing global political and scientific framework (UNFCCC, Warsaw International Mechanism for Loss and Damage, Sendai Framework for DRR). In our approach, we organised this activity over two main phases: (i) the analysis of climate scenarios and impacts on natural resources; (ii) the assessment of climate related risks.

The impacts generated by climate change are inevitably specific to a local context, emerging from climate dynamics at a local level and from both the natural and socio-economic characteristics of a territory. These climate-induced effects interact with the locally exposed elements/communities and their characteristics in terms of vulnerability (composed by sensitivity, coping capacity and adaptive capacity), generating risks for the regional sectors and for the local community. IPCC (2022a, 2022b) defines risk as "*The potential for adverse consequences for human or ecological systems, recognising the diversity of values and objectives associated with such systems*". This definition highlights the relevance of distinctive local values and objectives in determining specific local risks. Each community or public administration has its own development priorities, with particular consideration to a specific sector, economic activities or natural resources. Therefore, the engagement of the local community becomes essential in order to have a more accurate assessment of climate related risks, integrating "top-down" assessments based on predetermined attributes and indicators. Moreover, forecasts on climate change scenarios and expected impacts are inevitably uncertain, especially when dealing with regional or local territories and with long-term climate change effects. There are different layers of uncertainty regarding the evaluation of expected climate change impacts, generating the so-called "uncertainty cascade" (Wilby & Dessai, 2010). Uncertainty on expected impacts of climate change and the effectiveness of adaptation policies can hamper the identification of proper adaptation solutions. Thus, local community and local public institutions are an important source of knowledge on: (i) the ongoing effects generated by climate change on the local environment and society, (ii) the locations and assets exposed to climate hazards; (iii) the traditional/autonomous coping strategies developed by the local community. Therefore, with the aim to improve the assessment of climate-related risk for the regional adaptation strategy, we considered two types of information in our approach: (i) quantitative data, such as modelled data on climate variables or statistical data on socio-economic indicators; (ii) qualitative data, such as local knowledge emerging from interviews or workshops with regional officials or local stakeholders. The approach is consistent with the indications coming from the scientific literature and relevant institutions. The Paris Agreement Article 7.5 maintains: "*Parties acknowledge that adaptation action […] should be based on and guided by the best available science and, as appropriate, traditional knowledge, knowledge of indigenous peoples and local knowledge systems […]*".

4.2.2.1 Definition of Climate Scenarios and Impact Analysis on Natural Resources

Adaptation policies should be focused on both current and expected impacts generated or exacerbated by climate change. Therefore, the analysis of both ongoing tendencies and long-term climate scenarios and related impacts is a first essential activity for the definition of adaptation objectives and measures. In order to evaluate such ongoing climate tendencies and effects generated by climate change on the regional territory, we have considered the analyses made by the local Agency for the Protection of the Environment (ARPAL). ARPAL has developed a network of thermometers and rain gauges covering the whole regional territory, monitoring temperature and precipitation values since 1961. ARPAL is monitoring and assessing both average and extreme values indicators (e.g. average variation in the mean annual precipitations—Fig. 4.2), and it organised data on maps, collected in the "Atlante Climatico della Liguria" (ARPAL, 2013). This layer of information presented the opportunity to identify the main tendencies generated by climate change and to pinpoint seasonal and territorial differences (Fig. 4.2).

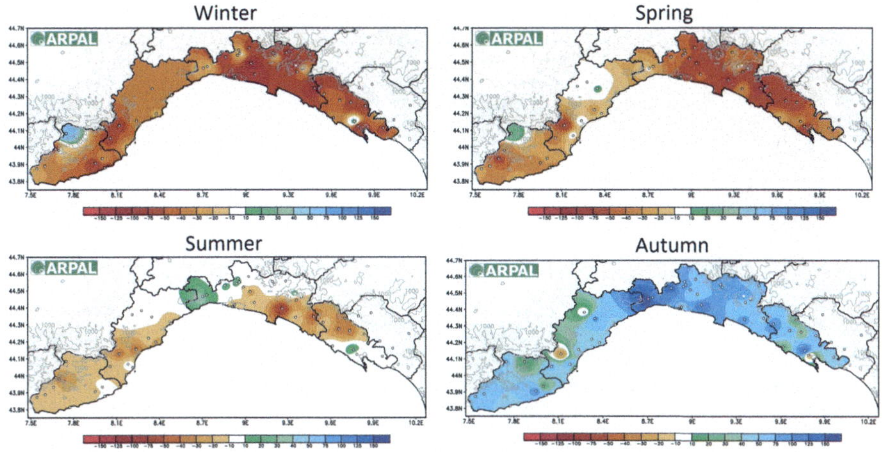

Fig. 4.2 Variation in Liguria seasonal cumulative precipitations, 1960/1990—1980/2010. *Source* ARPAL (2013)

In order to assess the future climate scenarios, CIMA Foundation has instead developed two climatic runs over two periods of interest, one historical (1979–2008) and a future one (2038–2068) on the Alpine area. The purpose of the study has been the evaluation of the impact of climate change by comparing climate statistics modelled on the historical period with those during the future period. The atmospheric model adopted was the Weather Research and Forecasting Model (WRF) version 3.9.1.1, with initial conditions provided by the EC-Earth model at 25 km^2. Climate forecasts has a 4 km^2 resolution. The future climate scenario has been estimated using the Representative Concentration Pathway 8.5 (RCP 8.5): in this scenario, a + 4.3 °C increase in global temperature is estimated between pre-industrial levels and the end of the century (2081–2100). This analysis has been structured over seven indicators, characterising both extreme and average parameters on temperature and precipitation expected tendencies. These indexes are: mean annual temperature (Fig. 4.3), frost days, winter and summer cumulative precipitations, extreme precipitations, consecutive dry days, 95th percentile ("P95") of daily precipitation over all wet days. These climate predictions were an important source of knowledge for the development of the regional adaptation strategy, giving high resolution predictions on key dimensions for the regional climate. However, this analysis relies on one single climate model, and it has been made considering just a single climate scenario, the most pessimistic one, partially ignoring the high level of uncertainty inevitably characterising climate predictions. In an attempt to face the uncertainty of these predictions, we have developed two kinds of activities: (i) we compared these evaluations with the climate predictions from the Italian National Climate Change Adaptation Plan (Ministero dell'Ambiente e della Sicurezza Energetica, 2023) and we made a survey of national and European projects focused on climate and impact predictions for Liguria; (ii) we engaged the regional administration, sharing a questionnaire to the

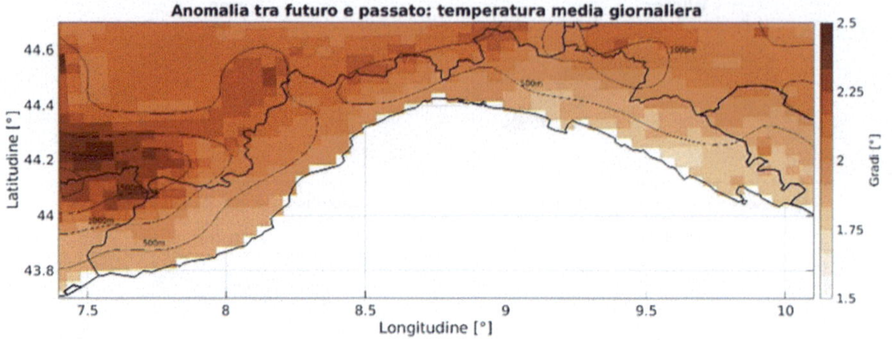

Fig. 4.3 Expected variation in the mean annual temperature 1981/2010—2041/2070. *Source* CIMA Foundation

regional working group for the adaptation strategy with the aim to investigate: (i) the local perception on climate change tendencies, (ii) the memory of past extreme events, (iii) predictions and narratives on the possible effects generated by climate change in the future on their specific sector and (iv) specific needs of the public administration in terms of knowledge on climate change and expected impacts.

The second step for the identification of climate risks has been the analysis of the ongoing and expected impacts of climate change on the local environment and natural resources. The activities were organised over three main tasks: (i) questionnaires and dedicated workshop with the RWG, (ii) the review of the state-of-the-art knowledge from ongoing or past national and European projects on climate impacts in Liguria; (iii) a modelling assessment on water resources conducted by CIMA Foundation. This analysis estimated indicators on the expected variations in terms of water availability, starting from a bias correction of climate scenario data and an additional downscaling activity, from 4 km^2 to 0.5/1 km^2. The hydrological model CONTINUUM has then been used to assess four key indicators: mean annual (and seasonal) river discharge variation (Fig. 4.4), snow availability variation, soil moisture variation and maximum annual river discharge. Water resource is a key element for the Liguria Region as both a source of conflict between water demands from different uses and risks emerging from floods. Thus, this analysis has been essential to characterise climate change related risks for many sectors.

4.2.2.2 Climate Risk Assessment and the Identification of Adaptation Objectives

The climate change risk and vulnerability assessment (CRVA) is an essential activity to design an adaptation strategy and to identify measures to tackle the specific local impacts of climate change, coherently with the needs of local communities. LIFE Master Adapt guidelines maintain that: "*An essential condition for planning aimed at governing the critical issues linked to climate change is the knowledge of the*

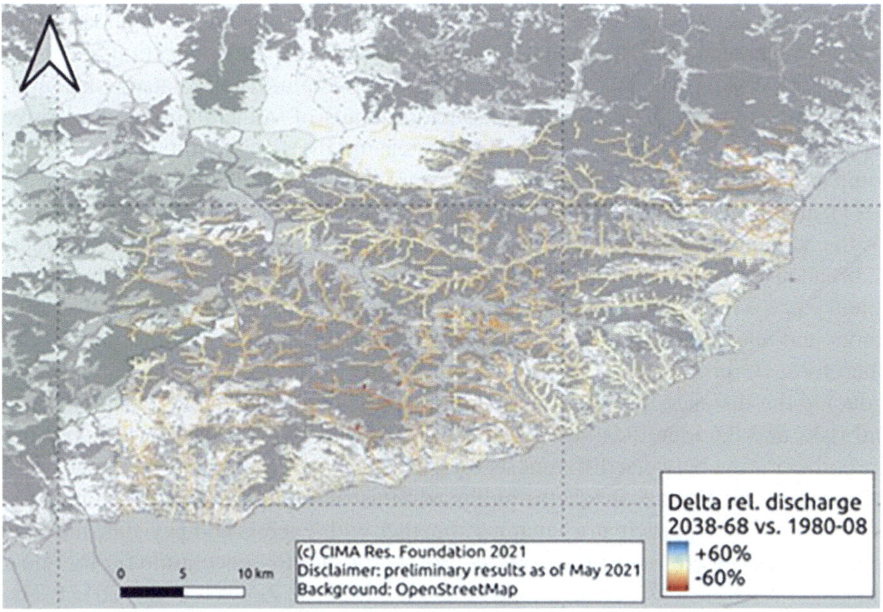

Fig. 4.4 Relative delta discharge in river flows in Liguria 2038/2068—1980/2008. *Source* CIMA Foundation

environmental, social, and economic elements that determine the vulnerability of the territory and understanding their interaction with the changing climate". Several approaches have been proposed to risk and vulnerability assessments throughout the years, some providing a general framework, some focused on certain sectors (e.g., agriculture, infrastructure, etc.) or environments (e.g., urban areas, coastal areas, alpine environments). In this work, we followed the framework set by the last IPCC report (IPCC, 2022a, 2022b) and by the more consolidated disaster risk reduction community framework (UN, 2016), where risk emerges by the intersection between hazard/exposure/vulnerability and is defined as "*the potential for adverse consequences for human or ecological systems, recognising the diversity of values and objectives associated with such systems*".

CIMA Foundation supported the regional administration in the risk assessment activity using the impact chains methodology (IC) (Fig. 4.5). This approach is an emerging mixed quantitative–qualitative methodology for the assessment of climate risks (Arabadzhyan et al., 2021; Fritzsche et al., 2014; Schneiderbauer et al., 2013; Zebisch et al., 2017 and 2022), an analytical tool to understand better, systemise and prioritise the factors that drive risk and vulnerability in a given system (cause-and-effect relationship) (Zebisch et al., 2017). Zebisch et al. (2021) define IC as: "*cause-effect models that describe the relationship between climate change induced hazards, exposed elements and their vulnerability, and resulting impacts*". IC crucially counts on the engagement of local stakeholders ("*a person or an organization that has a legitimate interest in a project or entity, or would be affected by a particular action or policy*"—IPCC, 2007, p. 881), communities, private entities, and public departments officials responsible for the policies, in order to identify the connections between climate drivers (e.g. the increase of the average temperature) and related risks, highlighting the context and sector-specific determinants of vulnerability and adaptive capacity. This approach is recognised as increasing the acceptability of the CRVA, compared to the top-down approach with generic indicator schemes (Zebisch et al., 2021), and it sets the scene for a participatory identification of potential entry points for the adaptation measures, also allowing the integration of different data sources. IC broadens the understanding of the factors that drive climate change risk in a given country or system, giving the opportunity for tailoring to the local context of indicators and adaptation measures (Zebisch et al., 2017; Ballarin Denti et al., 2019). Therefore, IC provides the opportunity to improve the decision-making process, reducing the distance between the scientific assessments on climate change effects and risks and the identification of context specific adaptation measures, increasing the collaboration between different sectors, highlighting development priorities, and enhancing competencies inside the public administration. It is indeed recommended to develop IC in a participatory manner together with experts and key stakeholders, trying to elaborate a realistic picture of the sector and to concentrate on the most important relationships between factors (Zebisch, 2021; Zebisch et al., 2017). The cause-and-effect nature of an IC can stimulate discussions among the stakeholders and thereby promote active stakeholder engagements. It could also promote collaboration and strengthen the existing networks among public and private stakeholders (Estoque et al., 2022). The enhancement of collaborations inside the public administration and the integration of different knowledge sources (e.g. scientific data, experiential knowledge, …) on both climate change local impacts and on the effectiveness of adaptation solutions can improve the implementation of the selected policies. IC methodology has been initially developed by Schneiderbauer et al. (2013) to assess climate vulnerability in the Alps, then, it has been applied in the context of international co-operation (Fritzsche et al., 2014), to the National Vulnerability Assessment of Germany (Buth et al., 2017) and it has now been systematised in the International Organisation for Standardization 14092 Standard (ISO, 2020), where IC is presented as one of the initial steps in conducting climate change risk assessment.

We applied IC to the climate risk assessment activity of the Liguria Region. According to the categories of IC application proposed by Menk et al (2022), our IC activity mainly falls in the group "*Qualitative models of cause and effect: Understanding system links through shared knowledge*", where the risks assessment is predominantly based on the representation of cause and effect relations between climate drivers and site specific risks, integrating expert and stakeholder knowledge, data from existing literature and modelled data (even though in a minor role). However, albeit we did not estimate a quantitative risk index and neither did we generally did highlight priority geographical areas for adaptation measures, our IC application was grounded on modelled quantitative data and information on climate tendencies and expected scenarios coming from the literature and dedicated research projects.

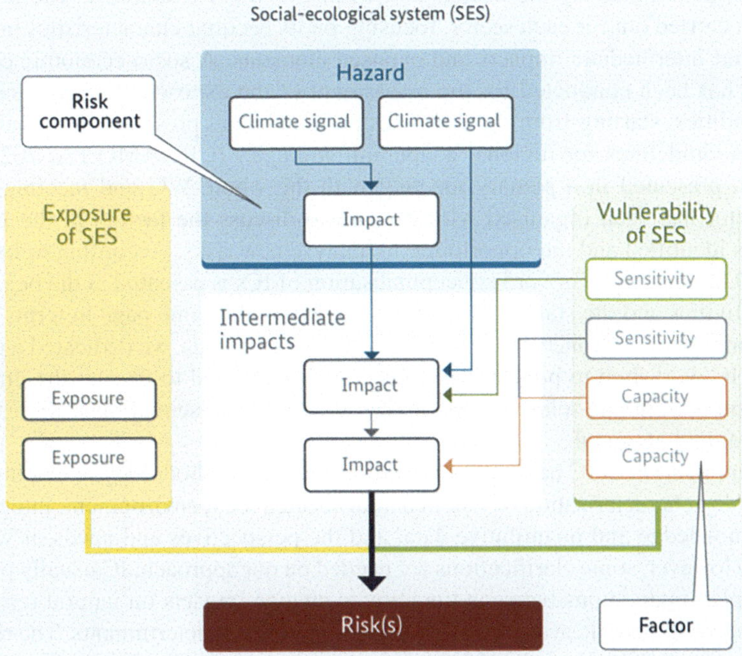

Fig. 4.5 Structure of an Impact Chain according to the IPCC approach. *Source* Zebisch et al. (2017)

We structured our approach over three main phases: (i) a preliminary draft proposal of IC for every sector, (ii) a plenary meeting and bilateral workshop discussion, (iii) a final presentation of the analysis to the WG. Thus, we initially made IC for each sector of the regional administration. We considered twelve sectors: Agriculture, Forests, Fishing and Marine Ecosystems, Water Resources, Human Health, Tourism, Urban Areas, Civil Protection and Natural Extreme Events, Energy, Transport, Coastal Areas, Ecosystems. The activity has been organised over five main steps (Figs. 4.6 and 4.7): (i) Identification of the most important climatic drivers for each sector, indicating ongoing climate change trends and their expected evolution on the regional territory (e.g. possible future scenarios on the evolution of the pattern of precipitations during the year, both cumulative and extreme values); (ii) Identification of climate change intermediate impacts, analysing the possible effects on key local natural resources (e.g. decreasing of water resources available during summer); (iii) Assessment of the main elements exposed (e.g. greenhouse crops and related revenues); (iv) Characterisation of the elements exposed according to their attributes in terms of vulnerability, focusing on both sensitivity and adaptive capacity elements (e.g. diversification in the farm source of revenues, presence of highly efficient irrigation systems); (v) Identification of the main risks generated or intensified by climate change (e.g. reduction in agricultural production and related revenues, determined by the decrease of available water resources). The analysis has been carried out for each sector, focusing on its peculiar characteristics in terms of relevant intermediate impacts and exposed elements. A socio-economic context analysis has been conducted for the assessment of the exposed elements and their vulnerabilities, starting from the attributes and indicators presented by CReIAMO PA in its Guidelines for regional adaptation strategies (CReIAMO PA, 2020). IC has been presented in a plenary meeting with the whole WG and meetings with each sector has been organised with the aim to discuss the factors of the IC, the key risks identified and the opportunity to analyse new risks. According to Estoque et al. (2022), a straightforward conceptualisation of ICs is essential at the beginning of IC activities and the stakeholders should all be on the same page in terms of the understanding of the conceptual framework to be used. Thus, we dedicated a part of the first IC workshop to present the framework clearly and to discuss the different dimensions of climate-related risks. IC has then been presented once more to the WG for a final approval.

The application of IC has proven the efficacy of the methodology in dealing with the complex characterisation of risk in a multifaceted local environment, integrating expert knowledge and quantitative data, and the perspectives and needs of various sectors. However, some clarifications are needed on our approach. IC usually present the complex interactions between the climate change impacts on natural resources and the socio-economic systems in an intricate network of determinants. The recommendation from Zebisch et al. (2021) is to move from relatively linear and sectoral IC to "*impact webs, which include feedback relations and cross-connections*". Climate risks emerge from numerous interacting factors indeed, where for every incremental

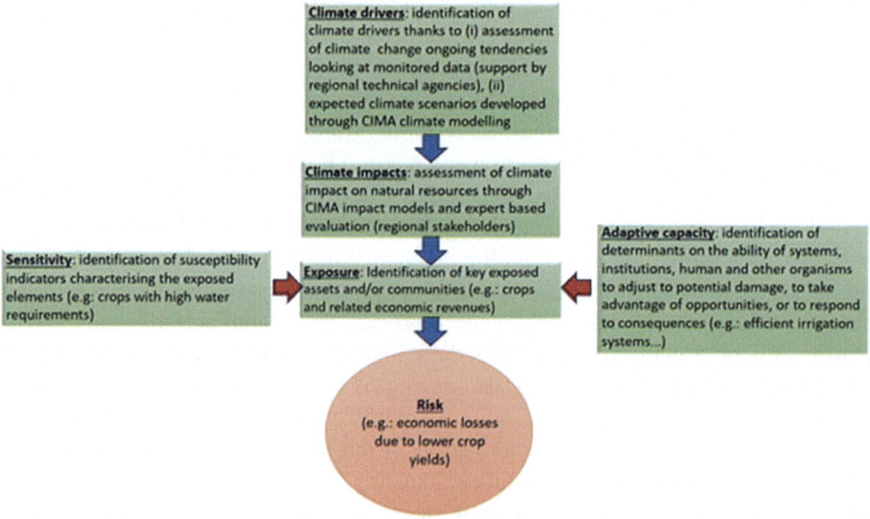

Fig. 4.6 The approach for the identification of climate risks in the Liguria Region. *Source* CIMA Foundation

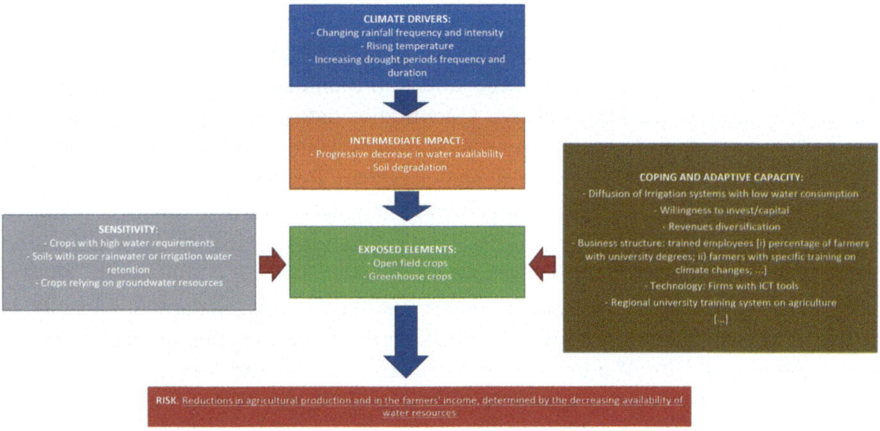

Fig. 4.7 The identification of a risk generated or exacerbated by climate change in Liguria Region, example from the agriculture sector. *Source* CIMA Foundation

change in greenhouse gasses and temperature, there are diverse responses in climatological, ecological, hydrological and other biophysical systems (Adger et al., 2018). Even though we considered the importance to propose a more integrated approach between the various impacts, covering different sectors, in this first application of the method, we organised IC in a more linear representation, considering each sector separately. This approach did not allow the representation of the full complexity of the climate impacts; however, it equally emerged as an effective approach for the identification of cause-effect relationships specific to the local context and to clarify the dimensions and the terminology of climate change with the local stakeholders. Officials of the regional administrations fund IC approach proved effective in analysing risks generated by climate change. They added elements to exposure and vulnerability dimensions, and they suggested new risks, in relation to their knowledge on the sector and the ongoing challenges they are facing. However, in an eventual update to this work, we should likely focus on integrating different sectors in workshops on a few common topics, deeply analysing the most critical climate related issues (e.g. water resource availability, extreme events, …) and highlighting synergies between regional departments and cross-sectoral policies. Furthermore, we should also consider possible feedback between the various components of the IC dimensions (hazard, exposure, vulnerability), instead of identifying single direction connections between climate-related hazards, exposed elements, their characteristics in terms of sensitivity and adaptive capacity and the resulting risks. Lastly, a refined version of the IC methodology may consider other two dimensions: (i) the eventual positive impacts generated by climate change; (ii) possible future evolutions in exposure and vulnerability determinants (in this IC application we merely considered the expected evolution of the local population in terms of the number of citizens and average age).

Even though we characterised each indicator with specific regional values (comparing the performance of the region with other regions at national and European level) we did not aggregate these values in a unique risk index, ranking risks and identifying intervention priorities. Quantitative data was collected especially on climate drivers and intermediate impacts on water resources and the socio-economic dimensions, according to the indexes suggested by CReIAMO PA.

Conversely, in our approach we specifically focused on the assessment of adaptive capacity. Capacity is defined by UNDRR (United Nations, 2016) as "*the combination of all the strengths, attributes and resources available within a community, society or organization that can be used to achieve agreed goals*". Adaptive capacity is instead a concept related to climate change, introducing dynamic elements associated to the availability of a community or institution to adapt to changing climate conditions. Adaptive capacity is defined by the IPCC as: "*The ability of systems, institutions, humans and other organisms to adjust to potential damage, to take advantage of opportunities, or to respond to consequences*"; or defined as: "*set of determinants that reflect the political, economic, technological and infrastructural capacity and willingness to adapt to the impacts of climate change*" (Juhola et al., 2013). A system's adaptive capacity is mostly determined by a local set of resources and conditions that constrain or facilitate the ability of the system to adapt successfully to changes in climate (Adger et al., 2005). Adaptive capacity is a key dimension in order to start defining: (i) location specific adaptation goals; (ii) the local barriers to adaptation; (iii) indicators to monitor progress of adaptation strategies over time.

Scientific research regarding adaptive capacity has rapidly grown since 2001, when the Intergovernmental Panel on Climate Change recognised adaptive capacity as a key component of vulnerability. According to Siders (2019) adaptive capacity is a highly interdisciplinary concept, covering different scales, sectors, and geographical locations. Different methodologies and metrics have been used to evaluate this dimension. The research field is rather fragmented and there are few cross-references between the articles dealing with this topic, giving the dimension of a research field that is still not very robust. Frequently, new research papers do not refer to previous articles and case studies analysed in previous works. It is also an area addressed by diverse scientific disciplines and the scale of the analysis varies continuously, with a higher concentration of case studies focused on communities, municipalities, and regions, while the national or inter-regional level is less frequent. Some studies analyse the effects of a single hazard (e.g. fires) on a plurality of sectors or natural resources, while others focus on a single sector or resource and consider a plurality of climate change effects to which social or natural systems respond.

In order to assess the adaptive capacity of the regional sectors, we combined two approaches: (i) Evaluation of specific indicators for each sector (e.g. Agricultural sector: productive diversification of companies, ICT diffusion, percentage of company managers with degrees and/or training on CCs); (ii) participatory analysis of the adaptation measures implemented by the regional administration, with the aim to focus on the real capacity of the region to design and implement adaptation policies (indications on possible adaptation measures found in PNACC). Stakeholder engagement has been here effective in collecting views on adaptive capacity and barriers to action. This is recognised as a valuable approach that can be taken in the absence of comprehensive adaptation data (Adger et al., 2018). The analysis of the existing policies developed by the Liguria Region to face climate change effects granted us indications on the institutional ability/willingness to promote adaptation, focusing on planned or implemented measures. Furthermore, this assessment has been important to identify programs, plans and other normative instruments used by the public administration to implement policies (e.g. Rural Development Programme, Regional Plan for Water Resources Protection, EU Interreg Programmes, …). However, following this approach, two kinds of challenges emerged: (i) information on the effects/effectiveness of the measure is not always available; (ii) the assessment of existing adaptation efforts is not an indication of the willingness to promote adaptation in the future (the presence of adaptation policies might be connected to the willingness of a specific political party or to ad-hoc economic resources from national or international administrations). In order to increase the accuracy of our approach, we integrated the participatory evaluation of adaptive capacity with an assessment of indexes and indicators, starting with the assessment made by the Italian National Climate Change Adaptation Plan (version June 2018), focused on provinces. In these works, adaptive capacity indicators are aggregated into five dimensions: knowledge and awareness, technology, infrastructures, institutions, economic resources. However, these two indexes assess the general adaptive capacity of the local territories, without a specific focus on sectors. Therefore, we integrated these results with adaptive capacity indicators specific for each sector. These indicators have been found in scientific literature or in other regional adaptation strategies (e.g. the Sardinia Region Climate Change Adaptation Strategy).

4.2.3 Risk/objectives/measures Matrixes for Each Sector

In the last step of our approach, CIMA supported the WG in the identification of adaptation objectives and measures and in developing risk/objectives/measures matrixes for each sector. The identification of adaptation objectives has been guided by two key elements: (i) the risks analysed in the impact chains activity; (ii) the development priorities and adaptation needs emerged by the workshops with the regional sectors. Starting from these objectives the WG has identified adaptation measures for each sector.

In a preliminary activity, we analysed the state of the art of decision criteria and principles to guide the identification of climate change adaptation measures (IPCC, 2022a, 2022b; Climate Adapt—Urban Adaptation support tool—prioritising adaptation options; Castellari et al., 2014). A series of recurring adaptation principles has been recognised in order to guide the selection of adaptation measures: (i) constantly update, promote and disseminate the scientific knowledge on climate change; (ii) promote the mainstreaming of adaptation objectives and measures inside the public administration decision-making processes, programs and plans; (iii) engage local stakeholders and communities to foster participatory processes for tailoring of adaptation policies; (iv) identify cross-cutting measures to integrate needs and competencies of different sectors; (v) promote adaptation solutions in synergy with mitigation policies; (vi) prioritise flexible measures that can be adjusted according to the evolution of climate change patterns and the new scientific findings on climate scenarios and expected impacts; (vii) design robust policies to promote adaptation strategies effective in the long term regardless of the way risks might evolve with global warming. The adaptation measures have then been organised in matrixes for each sector, connecting risks/objectives and measures. According to the Italian National Plan for Adaptation to Climate Change (Ministero dell'Ambiente e della Tutela del Territorio e del Mare, 2017), adaptation measures have been divided into five categories: governance, knowledge and information, infrastructures, nature-based solutions, participatory processes (Table 4.1).

According to the recommendations from national and international institutions (Castellari et al., 2014; UNDP-UNEP, 2011; EU Commission, 2013), we followed a mainstreaming approach for the development and planning of adaptation policies. Climate change adaptation mainstreaming is defined by UNDP-UNEP (2011) as "*the iterative process of integrating considerations of climate change adaptation into policy-making, budgeting, implementation and monitoring processes at national, sector and subnational levels*". Mainstreaming should be aimed at improving the implementation process of adaptation measures, also ensuring that future development plans and programs are actively designed to reduce the vulnerability to climate change (Commission on Climate Change and Development, 2008). Moreover, mainstreaming can provide a more efficient use of resources. In order to support the mainstreaming of adaptation measures in the regional plans/programs and in the decision-making processes, the Department of Architecture and Design of the University of Genoa led an analysis on the normative framework of the Liguria Region. They examined the most relevant regional plans and programs potentially affected by climate change or able to promote adaptation solutions. The analysis was focused on the following dimensions: (i) description of the main goal and objectives of the plan; (ii) an evaluation on the presence of considerations on the climate change impacts; (iii) an analysis regarding eventual adaptation objectives and measures proposed in the document; (iv) eventual recommendations from the European or National level regarding adaptation-related issues. Furthermore, this review activity was also aimed at identifying roles and responsibility for implementing the normative instruments and the adaptation options, as well as the timeframe for the revision of regional programs and plans, with the aim to identify possible entry points for adaptation policies and integrate climate change knowledge (EU Commission, 2013). In the matrix, a column has been reserved to the identification of possible policy-making instrument and plans for the implementation of the adaptation measures proposed. The Liguria Region has then identified a series of indicators for monitoring the implementation of the adaptation measures and the accomplishment of the strategic objectives, linking these indicators with those in the Regional Sustainable Development Strategy.

Table 4.1 Excerpt of the risk/objectives/measures matrix for the agriculture sector

Agriculture			
Impacts	Strategic objectives	Measures	Plans/programs for the implementation
1. Reduction in agricultural production and the related income determined by the decrease of available water resources	– Increase the resilience of agricultural companies to the reduction of available water resources – Reduce water consumption by local agricultural companies Improve education and training activities with the aim to improve the management of water resources in the agricultural sector	Governance: – Promote the income diversification of local agricultural companies, integrating agricultural activity with the provision of alternative services (e.g. agritourism, educational farms, …) – Promote insurances on agricultural companies yields – Provide incentives to encourage generational turnover within the agricultural companies Information: – Increase the awareness and the knowledge of farmers on the availability of water resources, focusing on the possible impact of climate change – Improve the collaboration between the local scientific community (CREA, Regional Institute of Floriculture and CERSAA) and the Ligurian agricultural production sector Infrastructures and systems adaptation and improvements: – Disseminate water saving practices, such as the use of water gauges to measure water consumption – Disseminate more efficient irrigation practices, such as micro-irrigation systems – Test innovative techniques for reusing water resources for irrigation – Promote water purification systems Solutions based on ecosystem services: – Promote pilot projects to evaluate the effectiveness of Nature Based Solutions (NBS) to reduce the impact of the agricultural sector on the quality of regional water resources	– Rural Development Programme – Regional Plan for Water Resources Protection – Regional plan for animal health – Regional Territorial Plan – EU Interreg. Programmes – LR 60/1986 (olive growing) – National Strategy for Internal Areas – National Recovery and Resilience Plan

Source Liguria Region Climate Change Adaptation Strategy

4.3 Conclusions

According to the scientific literature, the implementation of effective, transformative and pro-active adaptation solutions is proceeding at a slow pace, despite the proliferation of strategies and plans dedicated to adaptation policymaking. In order to increase the effectiveness of adaptation strategies, to improve the ownership of adaptation measures identified and to pave the way for the implementation process, we supported the Liguria Region in developing its own adaptation strategy through a structured participatory approach. Activities aimed at designing climate scenarios and impact assessments were developed mixing high resolution modelled data with local knowledge coming from the engagement of high-level regional officials, technical agencies and key local stakeholders. A light-touch version of the Impact Chains methodology was applied for the assessment of climate-related risks. Even though we framed regional risks through a rather linear representation, this participatory approach emerged as an effective tool to characterise the regional specific attributes leading to climate related risks and to create a common understanding of the climate change and adaptation epistemic framework inside the regional administration. IC emerged also as an effective approach to bridge the gap between complex scientific information and data on present and expected climate change impacts and the local knowledge on climate change effects and adaptation policies effectiveness, in order to: (i) improve the tailoring of the identified solutions to the local context; (ii) pave the way to the implementation process. Future developments of this approach may be further dedicated to: (i) an improved characterisation of connections and feedback between climate change impacts; (ii) an integration between the IC of different sectors to pinpoint critical cross-cutting impacts and integrated adaptation solutions; (iii) a broader engagement of local stakeholders, also considering communities, local NGOs, firms and research institutions. The focus on adaptive capacity has been an opportunity to encourage the mainstreaming of adaptation policies inside the regional policymaking processes, identifying indicators for monitoring the adaptation strategy.

Acknowledgements The Authors thank Katia Perini, Michela Gallo, Silvia De Angeli, Angela Celeste Taramasso, Adriano Magliocco, Maria Canepa from the University of Genoa for teaming up in the activities of the project Tools, governance and training for the climate change adaptation strategy towards the sustainable development financed by the "SNSVS2 Call"—Call for the promotion of research projects to support the implementation of the National Sustainable Development Strategy (2019). The Authors thank Antonio Parodi, Simone Gabellani, Anna Napoli and Francesco Avanzi from CIMA Foundation for their research and quantitative findings on the climate change scenarios and related impacts on the water resources in the Liguria region.

References

Adger, W. N., Arnell, N. W., & Tompkins, E. L. (2005). Successful adaptation to climate change across scales. *Global Environmental Change, 15*(2), 77–86.

Adger, W. N., Brown, I., & Surminski, S. (2018). Advances in risk assessment for climate change adaptation policy. *Philosophical Transactions of the Royal Society A: Mathematical, Physical and Engineering Sciences, 376*(2121), 20180106.

Arabadzhyan, A., Figini, P., García, C., González, M. M., Lam-González, Y. E., & León, C. J. (2021). Climate change, coastal tourism, and impact chains—A literature review. *Current Issues in Tourism, 24*(16), 2233–2268.

ARPAL—Centro Funzionale Meteoidrologico di Protezione Civile. (2013). *Atlante Climatico della Liguria*. Grafica KC.

Ballarin Denti, A., Barbieri, L., Baruzzi, V., Fraschini, F., Freixo Santos, T., Giordano, F., Lapi, M., Litt, G., Lucia, V., Luise, D., Magni, F., Marras, S., Oliveri, S., Pregnolato, M., Suppa, A., Zambrini, M., & Zuin, M. (2019). *Linee guida per le strategie regionali di adattamento ai cambiamenti climatici*.

Berrang-Ford, L., Biesbroek, R., Ford, J. D., Lesnikowski, A., Tanabe, A., Wang, F. M., Chen, C., Hsu, A., Hellmann, J. J., Pringle, P., Grecequet, M., Amado, J. C., Huq, S., Lwasa, S., & Heymann, S. J. (2019). Tracking global climate change adaptation among governments. *Nature Climate Change, 9*(6), 440–449.

Buth, M., Kahlenborn, W., Greiving, S., Fleischhauer, M., Zebisch, M., Schneiderbauer, S., & Schauser, I. (2017). *Guidelines for climate impact and vulnerability assessments: Recommendations of the interministerial working group on adaptation to climate change of the German Federal Government*. Umweltbundesamt, Section I 1.6-KomPass-Climate Impacts and Adaptation in Germany.

Castellari, S., Venturini, S., Giordano, F., Ballarin Denti, A., Bigano, A., Bindi, M., Bosello, F., Carrera, L., Chiriacò, M.V., Danovaro, R., Desiato, F., Filpa, A., Fusani, S., Gatto, M., Gaudioso, D., Giovanardi, O., Giupponi, C., Gualdi, S., Guzzetti, F., Lapi, M., Luise, A., Marino, G., Mysiak, J., Montanari, A., Pasella, D., Pierantonelli, L., Ricchiuti, A., Rudari, R., Sabbioni, C., Sciortino, M., Sinisi, L., Valentini, R., Viaroli, P., Vurro M., & Zavatarelli, M. (2014). *Elementi per una Strategia Nazionale di Adattamento ai Cambiamenti Climatici*. Ministero dell' Ambiente e della Tutela del Territorio e del Mare, Roma.

Commission on Climate Change and Development. (2008). *Overview of adaptation mainstreaming initiatives*.

CReIAMO PA. (2020). *Metodologie per la definizione di strategie e piani regionali di adattamento ai cambiamenti climatici*.

Estoque, R. C., Ooba, M., Togawa, T., Yoshioka, A., Gomi, K., Nakamura, S., Tsuji, T., Hijioka, Y., Watanabe, M., & Kitahashi, M. (2022). Climate impact chains for envisaging climate risks, vulnerabilities, and adaptation issues. *Regional Environmental Change, 22*(4), 133.

EU Commission. (2013). *Commission staff working document—Guidelines on developing adaptation strategies*. Brussels.

EU Commission. (2019). *The European green deal*. Brussels.

EU Commission. (2021). *Forging a climate-resilient Europe—The new EU Strategy on Adaptation to Climate Change*. Brussels.

Fankhauser, S. (2017). Adaptation to climate change. *Annual Review of Resource Economics, 9*, 209–230.

Fritzsche, K., Schneiderbauer, S., Bubeck, P., Kienberger, S., Buth, M., Zebisch, M., Kahlenborn, W. (2014). *The vulnerability sourcebook: Concept and guidelines for standardised vulnerability assessments*. Deutsche Gesellschaft fur Internationale Zusammenarbeit Rep., 180 pp.

Hall, N. (2017). What is adaptation to climate change? Epistemic ambiguity in the climate finance system. *International Environmental Agreements, 17*(1), 37–53.

IPCC. (2007). *Climate change: Impacts, adaptation and vulnerability* (M. L. Parry et al., Eds.) (976 pp). Cambridge University Press.

IPCC. (2018). Summary for policymakers. In *Global warming of 1.5 °C. An IPCC special report on the impacts of global warming of 1.5 °C above pre-industrial levels and related global greenhouse gas emission pathways, in the context of strengthening the global response to the threat of climate change, sustainable development, and efforts to eradicate poverty* [Masson-Delmotte, V., Zhai, P., Pörtner, H.-O., Roberts, D., Skea, J., Shukla, P. R., Pirani, A., Moufouma-Okia, W., Péan, C., Pidcock, R., Connors, S., Matthews, J. B. R., Chen, Y., Zhou, X., Gomis, M. I., Lonnoy, E., Maycock, T., Tignor, M., Waterfield, T. (Eds.)].

IPCC. (2019). *Summary for policymakers.* In: Climate change and land: An IPCC special report on climate change, desertification, land degradation, sustainable land management, food security, and greenhouse gas fluxes in terrestrial ecosystems [Shukla, P. R., Skea, J., Calvo Buendia, E., Masson-Delmotte, V., Pörtner, H.-O., Roberts, D. C., Zhai, P., Slade, R., Connors, S., van Diemen, R., Ferrat, M., Haughey, E., Luz, S., Neogi, S., Pathak, M., Petzold, J., Portugal Pereira, J., Vyas, P., Huntley, E., Kissick, K., Belkacemi, M., Malley, J. (Eds.)].

IPCC. (2022a). Climate change 2022: Impacts, adaptation and vulnerability. In *Contribution of working group II to the sixth assessment report of the intergovernmental panel on climate change* [Pörtner, H.-O., Roberts, D. C., Tignor, M., Poloczanska, E. S., Mintenbeck, K., Alegría, A., Craig, M., Langsdorf, S., Löschke, S., Möller, V., Okem, A., Rama, B. (Eds.)] (p. 3056). Cambridge University Press, Cambridge University Press.

IPCC. (2022b). Climate change 2022: Impacts, adaptation, and vulnerability. In *Contribution of working group II to the sixth assessment report of the intergovernmental panel on climate change* [Pörtner, H.-O., Roberts, D. C., Tignor, M., Poloczanska, E. S., Mintenbeck, K., Alegría, A., Craig, M., Langsdorf, S., Löschke, S., Möller, V., Okem, A., Rama, B. (Eds.)]. Cambridge University Press (In Press).

IPCC. (2023). Summary for policymakers. In *Climate change 2023: Synthesis report. Contribution of working groups I, II and III to the sixth assessment report of the intergovernmental panel on climate change.*

ISO. (2020). ISO/TS 14092:2020(en) *Adaptation to climate change—Requirements and guidance on adaptation planning for local governments and communities*

ISPRA. (2022). *Consumo di suolo, dinamiche territoriali e servizi ecosistemici (Edizione 2022).*

Juhola, S., Peltonen, L., & Niemi, P. (2013). Assessing adaptive capacity to climate change in European regions. *European Climate Vulnerabilities and Adaptation: A Spatial Planning Perspective*, 113–130.

Menk, L., Terzi, S., Zebisch, M., Rome, E., Lückerath, D., Milde, K., & Kienberger, S. (2022). Climate change impact chains: A review of applications, challenges, and opportunities for climate risk and vulnerability assessments. *Weather, Climate, and Society, 14*(2), 619–636.

Ministero dell'Ambiente e della Sicurezza Energetica. (2023). *Piano Nazionale di adattamento ai cambiamenti climatici (Italian National climate change adaptation plan)*

Ministero dell'Ambiente e della Tutela del Territorio e del Mare. (2017). *Strategia Nazionale per lo Sviluppo Sostenibile (National Sustainable Development Strategy)*

Schipper, E. L. F. (2007). *Climate change adaptation and development: Exploring the linkages.* Tyndall Centre for Climate Change Research Working Paper.

Schneiderbauer, S., Pedoth, L., Zhang, D., & Zebisch, M. (2013). Assessing adaptive capacity within regional climate change vulnerability studies—An Alpine example. *Natural Hazards, 67*, 1059–1073.

Sherman, M., Berrang-Ford, L., Lwasa, S., Ford, J., Namanya, D. B., Llanos-Cuentas, A., & Ihacc Research Team. (2016). Drawing the line between adaptation and development: A systematic literature review of planned adaptation in developing countries. *Wiley Interdisciplinary Reviews: Climate Change, 7*(5), 707–726.

Siders, A. R. (2019). Adaptive capacity to climate change: A synthesis of concepts, methods, and findings in a fragmented field. *Wiley Interdisciplinary Reviews: Climate Change, 10*(3), e573.

United Nations. (2015). *A/RES/70/1 transforming our world: The 2030 agenda for sustainable development.*

United Nations. (2016). *Report of the open-ended intergovernmental expert working group on indicators and terminology relating to disaster risk reduction, A/71/644*

United Nations Environment Programme. (2018). *The adaptation gap report 2018*. United Nations Environment Programme (UNEP).

United Nations Environment Programme & United Nations Development Programme. (2011). *Mainstreaming climate change adaptation into development planning: A guide for practitioners*.

United Nations Framework Convention on Climate Change. (2015). *Paris agreement*.

United Nations Framework Convention on Climate Change LDC Expert Group. (2012). *National adaptation plans: Technical guidelines for the national adaptation plan process*.

United Nations Office for Disaster Risk Reduction. (2015). *Sendai framework for disaster risk reduction 2015–2030*.

Wilby, R. L., & Dessai, S. (2010). Robust adaptation to climate change. *Weather, 65*(7), 180–185.

Zebisch, M., Schneiderbauer, S., Fritzsche, K., Bubeck, P., Kienberger, S., Kahlenborn, W., Schwan, S., & Below, T. (2021). The vulnerability sourcebook and climate impact chains—A standardised framework for climate vulnerability and risk assessment. *International Journal of Climate Change Strategies and Management, 13*(1), 35–59.

Zebisch, M., Schneiderbauer, S., Renner, K., Below, T., Brossmann, M., Ederer, W., & Schwan, S. (2017). *Risk supplement to the vulnerability sourcebook: Guidance on how to apply the vulnerability sourcebook's approach with the new IPCC AR5 concept of climate risk* (68 pp). Deutsche Gesellschaft fur Internationale Zusammenarbeit Rep.

Zebisch, M., Terzi, S., Pittore, M., Renner, K., & Schneiderbauer, S. (2022). *Climate impact chains—A conceptual modelling approach for climate risk assessment in the context of adaptation planning* (pp. 217–224). Springer International Publishing.

Chapter 5
Spotting Deficit Areas: Mapping Ecosystem Service Demand and Supply to Define Local Preparedness to Floods

Alessandra Longo, Linda Zardo, and Davide Longato

Abstract Adaptation planning can increase climate resilience by reducing the level of hazard exposure of people and assets and increasing the adaptive capacity of the city to respond to shocks. Among the measures that planning practice can take to build resilience to climate change, the promotion and management of ecosystem services (ESs) have been recognised as a promising, effective, win–win, multipurpose, and cost-effective solution. The integration of ESs into spatial plans is considered an indicator of a city or region's ability to implement adaptation actions toward sustainability and resilience. Unfortunately, the level of integration of ES-related concepts into spatial planning is still below expectations. A major limitation preventing the widespread integration of ESs into planning practices for climate adaptation relates to the gap between science and practice, that is, between the production of scientific knowledge about ESs and its actual use to inform and support decision-making. In particular, ESs demand mapping is critical to support decision-makers in understanding where and which ESs are (most) needed in relation to the needs of target beneficiaries, as it can be used to identify where and which ESs are most needed. The objective of this paper is to provide a step-by-step methodology to identify areas where there is a deficit of ESs in response to flooding. The work develops a GIS method to identify and map ES demand areas (or areas requiring services) that do not benefit from ESs supply to support ecosystem-based climate adaptation policies. Several types of results are provided: (i) an approach that can be replicated in other regional contexts, (ii) insights from the application to a case study, and (iii) considerations on the usefulness of specific concepts to serve research and practice in the field of ES-based adaptation planning. The novelty of this study is that it provides a categorisation of deficit areas based on the physical characteristics of the land, which provides a key entry point for planners to begin incorporating ES-based adaptation into practice.

A. Longo (✉) · L. Zardo · D. Longato
Department of Architecture and Arts, Iuav University of Venice, Venice, Italy
e-mail: alongo@iuav.it

M. Granceri Bradaschia et al. (eds.), *Climate Change Adaptation, Flood Risk, and Beyond*, Planning for Climate Proof Cities,
https://doi.org/10.1007/978-3-031-65463-3_5

Keywords Climate change adaptation · Ecosystem services · Friuli Venezia Giulia · Italy · Flood protection

5.1 Introduction

Floods are one of the challenges that cities have always had to face. Particularly, in Europe floods represent the costliest natural disaster in terms of human and economic losses (Dottori et al., 2020). However, because of climate change, they are becoming more frequent and severe (Marta Suárez et al., 2018). On top of this, the sprawling urbanization (implying soil impermeabilization and land take) has worsened their effects (World Bank, 2021). Cities and territories need urgently to re-think and re-build their resilience. Making cities and territories more resilient means strengthening urban and territorial systems against turbulence and building flexible and elastic functions and structures to make them less vulnerable during crises while maintaining the capacity for learning and transformation (Kabisch, 2019).

The way of tackling the challenges and building resilience highly depends on the institutional capacities (Tanner et al., 2017; World Economic Forum & McKinsey & Company, 2023). Thus, the resilience of cities to floods and, more in general, to climate change refers to a broad set of resources and includes the degree to which local authorities have integrated (or are in the process of doing so) climate change considerations into their long-term planning practices and development governance, policies and processes (Kabisch, 2019; Rözer et al., 2022; Srivastava & Sahay, 2023). In particular, *adaptation* planning (words in italics in the text refer to an entry in the glossary) can boost climate resilience by reducing the level of exposure of people and assets to hazards, as well as increasing the adaptive capacity of the city to respond to shocks (Cortinovis & Geneletti, 2018).

Among the measures that the planning practice can adopt to build climate change resilience, the promotion and management of ecosystem services (ESs) has been recognized to be a promising, effective, win–win, multipurpose and affordable solution (United Nations Environment Programme, 2021, 2022). ESs are fundamental to the well-being of people, and their benefits contribute among others to adapting to climate change effects (Lavorel et al., 2020). Among the main ESs to respond to floods, flood protection, runoff regulation and erosion control can be mentioned (Haines-Young & Potschin, 2013). Such ESs have the capacity to maintain and control relevant ecosystem processes (MEA, 2003) and directly reduce the severity of extreme events (Munang et al., 2013; UNEP & WCMC, 2020).

The benefits that we derive from the natural environment in a particular area depend on the availability, spatial distribution, and condition of the ecosystems (Vallecillo Rodriguez et al., 2022). Therefore, making strategic decisions about how to allocate land use during planning processes is crucial for the sustainable management of natural resources (Cortinovis & Geneletti, 2018). When spatial plans include ecosystem services, it is an indication that cities and regions are taking adaptation measures towards sustainability and resilience (Frantzeskaki et al., 2016). In this sense, considering the ES concept in planning practice, through its integration into plans and policy design, can lead to more sustainable and informed outcomes (Ruckelshaus et al., 2015; Van der Biest et al., 2020). The latter is among the reasons why policymakers are increasingly calling for the integration of the ES concept at all levels of decision-making (Geneletti et al., 2020) awaiting further insights into the importance of biodiversity and ESs for society (Pörtner et al., 2021) and, thus, more support for strategies and measures for sustainable landscape development (Sun et al., 2022; von Haaren et al., 2019).

Despite a growing interest by both scientists and experts in using the ES concept for real-world applications to support sustainable landscape development and climate change adaptation (Albert et al., 2016; Marques et al., 2022), the level of integration of ES-related concepts in spatial planning is still limited (Di Marino et al., 2019; Longato et al., 2021). Several literature reviews have examined the obstacles that hinder the diffused integration of ESs in planning practices for climate adaptation. A major limitation preventing it highlighted by the scientific literature concerns a gap between the creation of scientific knowledge on ESs and its application in decision-making processes (Bertuol-Garcia et al., 2018; Bitoun et al., 2022; Olander et al., 2017; Qiu et al., 2022; Wei & Zhan, 2023). Carmen et al. (2018) suggest that knowledge becomes useful when it addresses the problems driving its demand. Therefore, the need for useful knowledge in planning practices and policies varies from case to case (Brunet et al., 2018). To facilitate communication, engagement, and participation among stakeholders, user-friendly decision-support tools can be adopted, leading to more informed and inclusive decision-making processes (Grêt-Regamey et al., 2017; Zulian et al., 2018). Since knowledge regarding Ess is complex (Evans, 2019), it is crucial for the approach to be flexible enough to adapt to the evolving requirements of end-users and potential emerging obstacles (Nesshöver et al., 2016). Collaboration across disciplines in the planning context can help achieve effective adaptive land management (Brunet et al., 2018; Lerouge et al., 2017).

Another barrier that contributes to the gap between science and practice is the fact that ESs knowledge is often produced solely for academic purposes (Bitoun et al., 2022). Furthermore, the focus is mainly on the analysis of supply at the expense of demand (Dworczyk & Burkhard, 2021). To assist decision-makers in understanding where and which ESs are most needed in relation to the targeted beneficiaries' requirements, mapping ESs demand is essential (Verhagen et al., 2017). The lack of the use of ESs demand spatial assessments in planning practice may undermine the effectiveness of planning decisions related to the allocation and prioritization of, for instance, climate adaptation interventions (Longato et al., 2023). Being spatially explicit in considering both ESs supply and demand in addition to supplies can link essential information on ESs balance, which can assist in defining a level of ESs preparedness for a specific climate change hazard.

To address this gap, the present study aims to identify areas where there is a deficit of ESs in response to floods to support ecosystem-based climate adaptation policies. A stepwise methodological process (outlined in Sect. 5.3) is designed to quantify and map areas with a demand for flood protection ES that do not benefit from its supply, i.e. where there is demand but no supply. Specifically, this study addresses three primary research questions:

- How can the demand (for ES) associated with the possible occurrence of climate change impacts be spatially defined?
- How can the correspondence between ES supply and demand be assessed?
- How can ESs knowledge support adaptation planning at the local level?

The methodology is applied to a coastal area of an Italian region as an illustrative case study (Sect. 5.2).

Glossary

Adaptation: The IPCC AR6 WGII Glossary defines it as follows: "In human systems, the process of adjustment to actual or expected climate and its effects, in order to moderate harm or exploit beneficial opportunities. In natural systems, the process of adjustment to actual climate and its effects; human intervention may facilitate adjustment to expected climate and its effects." (IPCC, 2022). In this study, the term "adaptive" therefore refers to a type of planning that promotes climate change adaptation.

Expected impacts of climate change: the term "impact" is defined by IPCC as the "consequences of realised risks on natural and human systems, where risks result from the interactions of climate-related hazards (including extreme weather/ climate events), exposure, and vulnerability" (IPCC, 2022). In general, the added term "expected" refers to those impacts that are predicted to occur at a given location in the future according to climate-related studies. In this work, it specifically refers to the list of impacts identified by the Regional Agency for Environmental Protection of Friuli Venezia Giulia (ARPA FVG) Report (ARPA FVG, 2018) for the regional territory. These impacts are organised into impacted sectors.

Impacted sector: one of the 18 categories of natural systems and socio-economic sectors for which ARPA FVG (2018) recognised specific climate change-related impacts. This work considers only 16 of them as it focuses on terrestrial ecosystems. These sectors are: Quantity and quality of water resources, Desertification, land degradation and droughts, Hydrogeological instability, Terrestrial ecosystems, Inland and transitional water ecosystem, Health, Forests, Agriculture and food production, Aquaculture, Energy, Coastal areas, Tourism, Urban settlements, Cultural heritage, Transport and infrastructure, Dangerous industries and infrastructure. The other two that are excluded are Marine ecosystems and Marine fisheries.

Landscape area: one of the 12 administrative areas identified by the Regional Landscape Plan of FVG according to the following delimitation criteria: (a) hydro-geomorphological; (b) environmental-ecological; (c) identity-historical-cultural; (d) administrative-managerial; (e) permanence of historical territorialisation; (f) coherence with aggregated settlement-territorial systems. See Sect. 5.2.

5.2 Case Study

The *landscape area* (words underlined in the text refer to an entry in the glossary) number 12, named "Lagoon and coastline" (Fig. 5.1), occupies the southern part of the Friuli Venezia Giulia (FVG) Autonomous Region, Italy, and covers approximately 40000 ha. FVG, which is located in the north-east of Italy, shares borders with Austria to the north, Slovenia to the east and the Adriatic Sea to the south. It extends for about 8000 km^2 and has a population of 1.2 million inhabitants (RAFVG, 2022). The moderately continental climate is influenced by the geomorphological variability of the region, which is characterised, from north to south, by mountains, hills, plains (high and low), lagoon and karst.

Back to the case study, the area is strongly diversified due to its high geomorphological heterogeneity: fluvial, coastal, lagoon, marine and karst environments are found in it. These include the lagoon of Grado and Marano, with the mouth of the Stella and the Isonzo rivers, the resurgences of the Timavo river and the delta of the Tagliamento river that flows into the Adriatic Sea. About 46% (18221 ha) of the territory is covered by various protected areas, including regional nature reserves and vast areas included in the Natura 2000 network, both as Special Protection Areas (SPAs) and Special Areas of Conservation (SACs).

Over the years, large portions of the backshore plain have been affected by widespread land reclamation processes that have transformed natural areas (including reed beds, marshes, salt meadows and wet woods) into cultivated ones. These land use transformations have led to the alteration of the ecological balances and ecosystem processes of the remaining natural environments. Moreover, in more recent times, reclamation processes have also affected the lands bordering the coast as they have been driven by mass beach tourism.

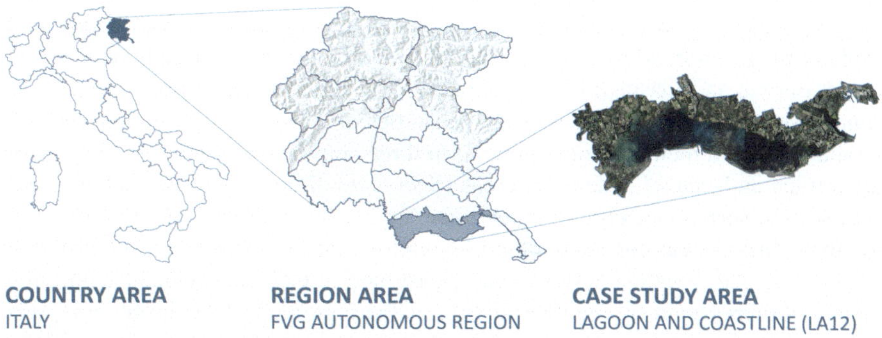

Fig. 5.1 Map and location of the case study area

At either end, between the lagoon and the sea, are the two main regional seaside resorts: Lignano Sabbiadoro and Grado. These can be associated with the polarised and densely populated settlement system, with an intensive land use specifically for seaside tourism. On the eastern border is also the port of Monfalcone, a crucial hub for connections to the Trieste metropolitan system, the hinterland and the foothills of the Karst.

The lagoon landscape is one of the most delicate and fragile environments in the entire regional context and is subject to multiple natural and anthropic pressures. These include marine inundation, subsidence, erosion, saline wedge rise in the lagoon and along waterways, water pollution from agricultural, urban and industrial discharges, and proposed expansion of tourist centres to increase local supply.

Although the coastal strip is the least rainy area in the region, it is exposed to a high hydraulic risk determined by floods as a result of river flooding phenomena and by the possible difficulties in the disposal of rainwater by the complex water network, both in urban and agricultural areas. According to the studies produced by the Regional Agency for Environmental Protection of FVG (ARPA FVG) in 2018 and focusing on current and future climate trends and their impacts, the area is experiencing an increase in marine flooding mainly caused by the joint action of sea level rise and subsidence, resulting in erosion and salt wedge ingression phenomena. This increase could lead to serious socio-economic losses given the considerable importance of coastal municipalities to the regional economy. The FVG coast is also strongly affected by storm surge events, also known as high waters and typical in autumn and winter, whose impact is considerable in the form of flooding and erosion.

5.3 Materials and Methods

The methodology of this study is based on the spatial comparison of the demand and supply of the ES of flood protection. The demand arises from the need to address the impacts of river and coastal flooding on the regional territory. The supply side follows a land cover approach that assesses the capacities of ecosystems to provide the ES. The materials and methods are structured in three consecutive steps (Fig. 5.2). Step 1 quantifies and maps ES demand areas (Sect. 3.1). Step 2 identifies deficit areas through the overlapping of ES demand and supply areas (Sect. 3.2). Finally, step 3 defines cluster areas for adaptation planning action based on the results of step 2 (Sect. 3.3). All mapping activities are performed in GIS environment within the case study represented by the coastal area of the region. As a main outcome, this work presents a step-by-step methodology that employs an ecosystem perspective to recognise common traits of flood risk situations affecting different impacted sectors. Accordingly, this methodology can assist policymakers in proposing cross-sectoral policies for adaptation planning.

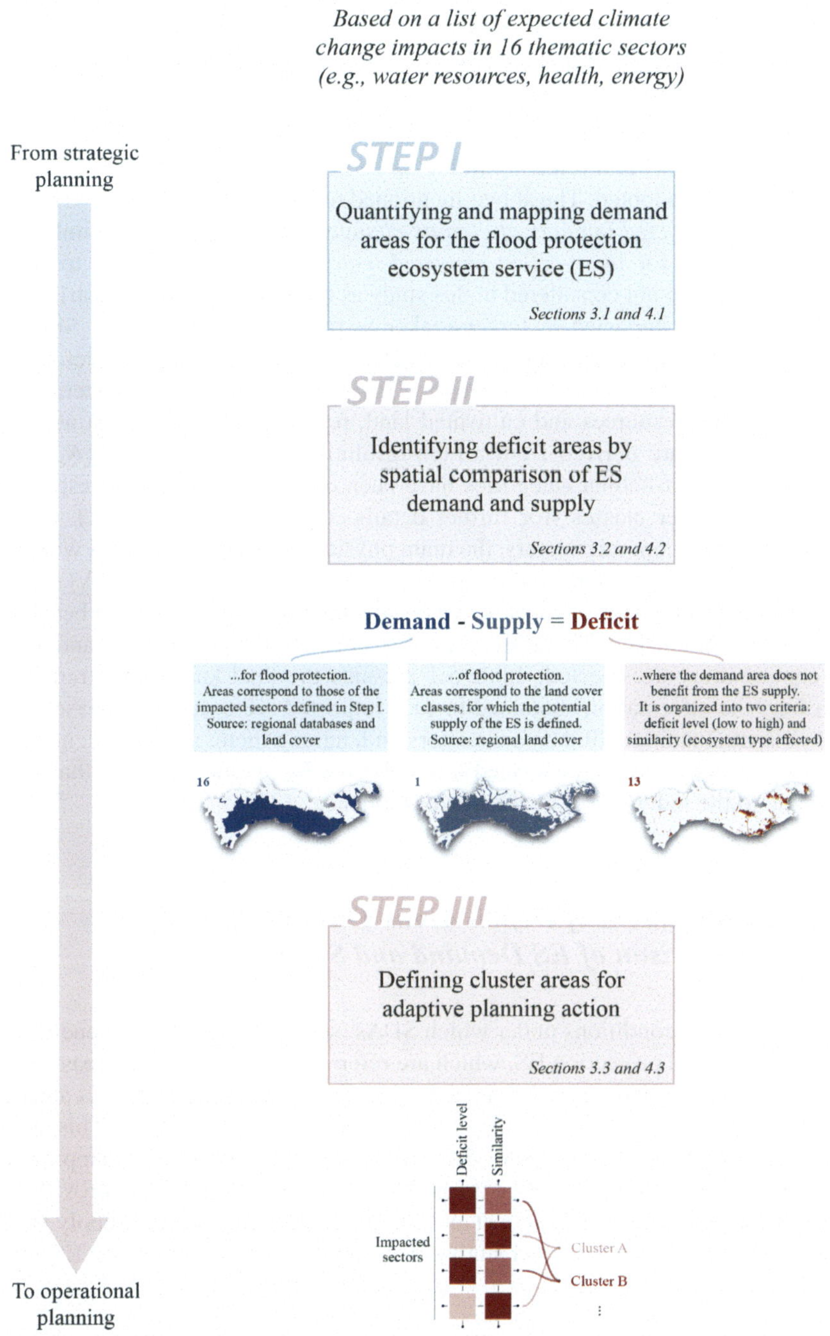

Fig. 5.2 Structure of the three methodological steps

5.3.1 *Quantification and Mapping of Demand Areas for Flood Protection ES*

In this paper, the concept of Service Demanding Area (SDA) proposed by Dworczyk and Burkhard (2021) and referring to the "area in which people's needs or desires can be located" is adopted. The above-mentioned report by ARPA FVG (2018) identifies a list of *expected impacts of climate change*, whose causes are mainly heat- and water-related, for 18 so-called *impacted sectors*. Of these, two refer to marine ecosystems and are not considered in this study as it focuses only on terrestrial ones. The remaining 16 impacted sectors are taken as proxies for ES demand. Since the impacted sectors are thematic and not spatially explicit, the first step is to define their spatial dimension. For all sectors relating to natural and semi-natural systems, such as forests, water resources and cultivated land, regional land cover data are used. Land cover data are derived from a high-resolution habitat map (ISPRA, 2021), where the different habitat categories have been converted into the corresponding Corine Land Cover classes (for further details, see Longo et al., 2024). For the remaining socio-economic sectors, the main physical elements associated with each impacted sector are investigated through a content analysis of the ARPA FVG report. To give an example, in the case of Tourism, specific reference is made to beach and mountain winter tourism. Therefore, the data from coastal municipalities and mountain communities are selected. Table 5.1 presents the final knowledge framework describing the information contained in each impacted sector and the datasets used for their spatialization. All data are vectors and, among them, geometric entities of points and polylines are transformed into polygons by creating a buffer that varies according to the individual's likely area of influence.

5.3.2 *Identification of Deficit Areas Through Spatial Comparison of ES Demand and Supply*

Step 2 detects the conditions under which SDAs (see Table 5.1) do not benefit from the supply of flood protection ES, which are referred to here as deficit areas. In this study, it is assumed that where SDAs and ES supply areas overlap, there is a benefit (i.e., demand is satisfied); where they do not overlap, there is a deficit. This spatial comparison, performed with QGIS 3.26 Buenos Aires, enables the mapping and subsequent quantification of deficit areas. The capacities of the land to provide ESs were previously assessed in Longo et al. (2024) (see the ESs potential supply matrix displayed in Fig. 5.3 of the cited study). The supply of flood protection ES in the

Table 5.1 Knowledge framework for ES demand spatialisation by impacted sector

Impacted sector	Description of the information	Source of the data
Quantity and quality of water resources	Freshwaters (rivers, lakes and glaciers)	Regional land cover (Longo et al., 2024)
Desertification, land degradation and droughts	All terrestrial ecosystems	Regional land cover (Longo et al., 2024)
Hydrogeological instability	Hydraulic and landslide risk areas	Alpi Orientali's database (https://distrettoalpiorientali.it/) Catalogo IRDAT FVG (https://irdat.regione.fvg.it/)
Terrestrial ecosystems	All terrestrial ecosystems	Regional land cover (Longo et al., 2024)
Inland and transitional water ecosystems	Inland (wetlands, rivers and lakes) and transitional (intertidal flats, salt marshes, estuaries and the lagoon) water ecosystems	Regional land cover (Longo et al., 2024)
Health	Urban areas	Regional land cover (Longo et al., 2024)
Forests	Forest ecosystems	Regional land cover (Longo et al., 2024)
Agriculture and food production	Agricultural areas	Regional land cover (Longo et al., 2024)
Aquaculture	Allocated areas for aquaculture	Catalogo IRDAT FVG (https://irdat.regione.fvg.it/)
Energy	Urban areas, transport, solar farms, hydropower plants, power lines	Regional land cover (Longo et al., 2024) Catalogo IRDAT FVG (https://irdat.regione.fvg.it/) PPR FVG (https://webgiscarnia.regione.fvg.it/it/map/ppr_fvg/#)
Coastal areas	Municipalities that border the lagoon and maritime shoreline	Catalogo IRDAT FVG (https://irdat.regione.fvg.it/)
Tourism	Coastal municipalities and mountain communities	Catalogo IRDAT FVG (https://irdat.regione.fvg.it/)
Urban settlements	Urban areas	Regional land cover (Longo et al., 2024)
Cultural heritage	Artistic and natural assets of supra-regional relevance: UNESCO sites, Natura 2000 Network, protected areas (parks and reserves), archaeological areas, buildings of historical, artistic and architectural interest	Catalogo IRDAT FVG (https://irdat.regione.fvg.it/) PPR FVG (https://webgiscarnia.regione.fvg.it/it/map/ppr_fvg/#)

(continued)

Table 5.1 (continued)

Impacted sector	Description of the information	Source of the data
Transport and infrastructure	Road infrastructure (primary and secondary roads and related areas), railway infrastructure, airports, ports	Catalogo IRDAT FVG (https://irdat.regione.fvg.it/) Corine Land Cover 2018 (https://land.copernicus.eu/en)
Dangerous industries and infrastructure	Municipalities where dangerous infrastructures are present	Catalogo IRDAT FVG (https://irdat.regione.fvg.it/)

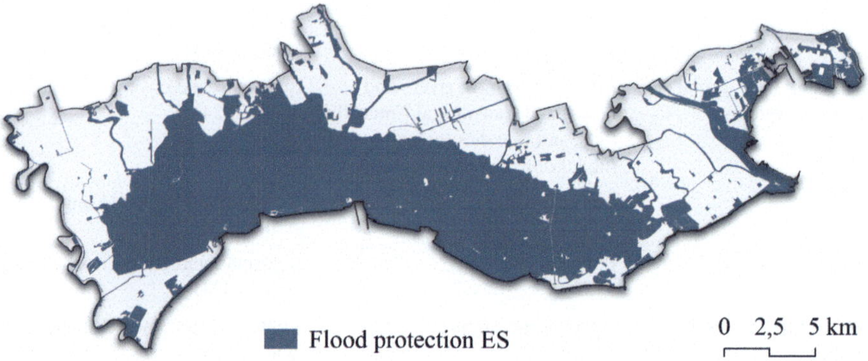

Fig. 5.3 Supply map of the flood protection ecosystem service (previously assessed in Longo et al., 2024)

coastal landscape area, filtered on the medium–high consensus level,[1] is presented in Fig. 5.3.

5.3.3 *Definition of Cluster Areas for Adaptation Planning*

This section aims to identify similar risk situations among the different impacted sectors to guide the design of adaptation policies. A 2-step assessment is proposed to organise the deficit areas into clusters. To this end, two criteria are selected and compared: deficit level and similarity. For the former, thresholds are defined concerning the percentage of land that does not benefit from the flood protection ES. The percentage is divided into four equal ranges, whose values indicate the following deficit levels: < 25%—low, 25–50%—medium–low, 50–75% medium–high and > 75% high. Second, the similarity criterion investigates the ecosystem types (e.g.,

[1] The assessment of ESs supply by different land cover types in Longo et al. (2024) was based on the work by Bordt and Saner (2019). They combined nine input studies and produced a matrix that links priority ecosystem types with the provision of specific ESs. The level of consensus is defined as the number of input studies that agree on the importance of a specific ecosystem/ES linkage.

grassland, rivers and lakes, wetlands) causing the deficit and their presence in each sector. To do so, the deficit areas of the impacted sectors are overlapped with regional land cover data (with the exception of those obtained from these data that already contain the information) and the ecosystem types not providing the ES are outlined. This results in a matrix that groups the impacted sectors with the same combination of these two criteria into clusters.

5.4 Results

5.4.1 Quantification and Mapping of Demand Areas for Flood Protection ES

The two main results of applying the methods of Sect. 3.1 to the illustrative case study are presented. Figure 5.4 shows the mapping of the SDAs for each impacted sector and Table 5.2 presents their quantification in both absolute terms (hectares) and percentages. Overall, half of the impacted sectors (8 out of 16) have a SDA covering less than or equal to 10% of the total surface of the landscape area; 6 impacted sectors have a SDA covering between 41 and 56% of the area; and the remaining 2 impacted sectors—i.e., Coastal areas and Tourism—100%. More in details, the SDA of Quantity and quality of water resources, Inland and transitional water ecosystems and Aquaculture are limited to water ecosystems. In the case of Health, urban areas are adopted as a proxy. This type of land cover is also used for Urban settlements and part of the Energy sector, to which power lines, hydropower plants and solar farms are added. Local transport crosses urban areas and connects them through a network of primary and secondary roads, railways, and ports. The Forests impacted sector is derived by filtering the regional land cover data on forest and woodland ecosystems (4%), which include broad-leaved and coniferous forest. The impacted sectors of Desertification, land degradation and droughts, Hydrogeological instability and Terrestrial ecosystems almost coincide as their SDA concerns the hinterland, which comprises the following ecosystems: cropland, grassland, forest and woodland, heathland and shrub and sparsely vegetated land. Cultivated land is used to describe the agricultural sector. The SDA of Cultural heritage, which includes UNESCO sites, protected areas and archaeological zones, accounts for about half of the landscape area. For Coastal areas, Tourism and Dangerous industries and infrastructure, municipalities are used as proxies. In the first two cases, all municipalities intersecting the maritime and lagoon shoreline are selected. Since this is a coastal area, municipalities cover its entire extent, which explains why their SDA covers 100% of the territory. In the third case, the municipalities concerned only partially touch the area because they are located along the border.

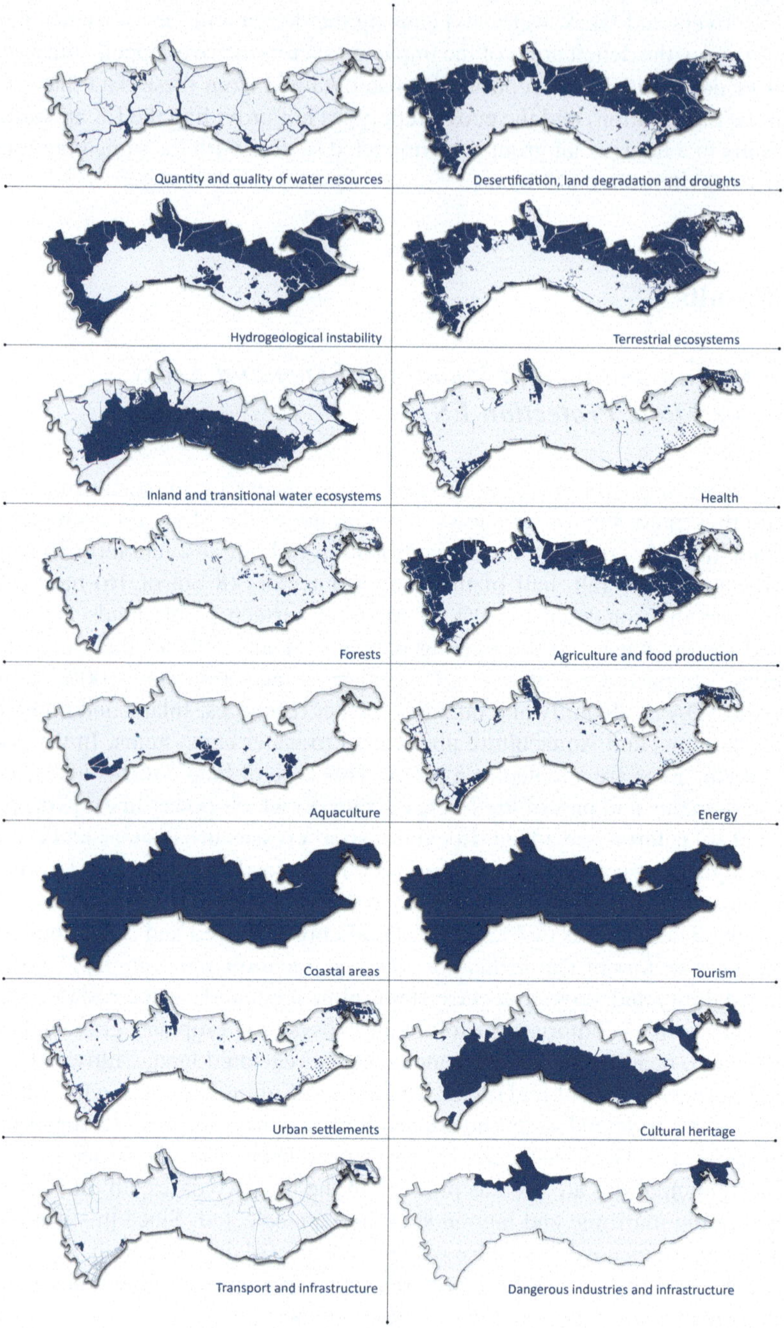

Fig. 5.4 Map of the SDAs of the impacted sectors

Table 5.2 Quantification of the SDAs of the impacted sectors in absolute values (hectares) and percentages

Impacted sector	Service demanding areas (SDA)	
	Area (ha)	% of SDA out of the total case study area
Quantity and quality of water resources	2116.9	5
Desertification, land degradation and droughts	18961	48
Hydrogeological instability	22043.6	56
Terrestrial ecosystems	18961	48
Inland and transitional water ecosystems	17636.8	45
Health	2969.2	8
Forests	1541.3	4
Agriculture and food production	16192.1	41
Aquaculture	2283.9	6
Energy	2989.4	8
Coastal areas	39547.8	100
Tourism	39547.8	100
Urban settlements	2969.2	8
Cultural heritage	19470.2	49
Transport and infrastructure	991.2	3
Dangerous industries and infrastructure	3920.7	10

5.4.2 *Identification of Deficit Areas Through Spatial Comparison of ES Demand and Supply*

The SDAs obtained in Sect. 4.1 are compared for each impacted sector with the supply of the flood protection ES (Fig. 5.3). The total percentage of land within the case study area involved in the supply of the ES is about 52%. It refers to grasslands, forests, moors and heathlands, bare rocks, beaches, dunes, sand plains, inland marshes, water courses and bodies, and marine inlets and transitional waters (i.e., the lagoon, estuaries, intertidal flats, and salt marshes).

The deficit areas emerging from the overlap between ES demand and supply are spatialised in Fig. 5.5 and consequently quantified in Table 5.3. It is evident that compared to Fig. 5.4 (Sect. 4.1), the number of impacted sectors mapped is smaller (13). This is because, for the sectors of Quantity and quality of water resources, Inland and transitional water ecosystems and Forests, ES demand and supply coincide, generating no deficit areas. These are followed by the areas of the Aquaculture and Cultural heritage sectors, whose deficit covers less than 10% of the total SDA.

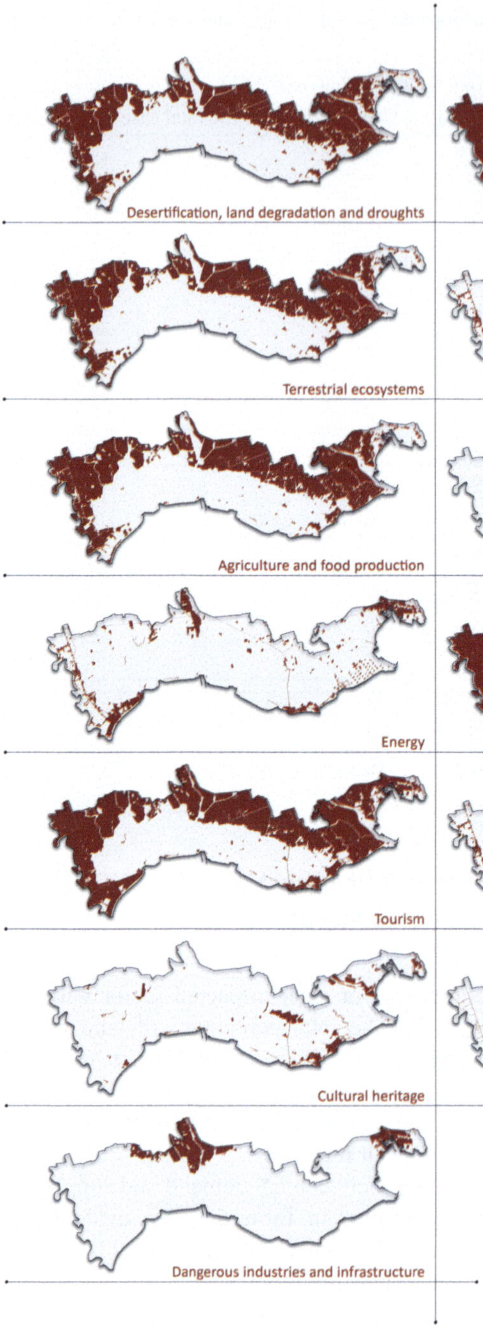

Fig. 5.5 Deficit areas spatialisation by impacted sector

Table 5.3 Quantification of deficit areas by impacted sectors in absolute values (hectares) and percentages

Impacted sector	Deficit areas (SDA—flood control ES supply)	
	Area (ha)	% of deficit area out of the SDA
Quantity and quality of water resources	0	0
Desertification, land degradation and droughts	16192.1	85
Hydrogeological instability	18104.3	82
Terrestrial ecosystems	16192.1	85
Inland and transitional water ecosystems	0	0
Health	2969.2	100
Forests	0	0
Agriculture and food production	16192.1	100
Aquaculture	50	2
Energy	2986.5	100
Coastal areas	19147.5	48
Tourism	19147.5	48
Urban settlements	2969.2	100
Cultural heritage	1496.5	8
Transport and infrastructure	771.3	78
Dangerous industries and infrastructure	2909.7	74

On the other hand, for 9 of the impacted sectors, the percentage of land not benefiting from the ES is more than 70%. Among them, 4 of the impacted sectors have a maximum deficit, i.e. 100% (Table 5.3). The total percentage of land with this maximum deficit within the case study area is 48.4% (Fig. 5.6). The value results from overlapping deficit areas and not from the sum of the individual ones, thus avoiding double counting. Generally, the impacted sectors that do not benefit from the ES are those where the SDAs correspond to urban and agricultural ecosystems.

5.4.3 Definition of Cluster Areas for Adaptation Planning

Results from Sect. 3.3 are summarised in Fig. 5.7. Since the impacted sectors of Quantity and quality of water resources, Inland and transitional water ecosystems and Forests do not present deficit areas, they are excluded from the list. Based on the results obtained from the combination of the two criteria, the classes low to medium–low and medium–high to high are merged, as shown in the legend. In fact, the application of the first criterion returns only one impacted sector in the medium–high deficit level, namely Dangerous industries and infrastructure, resulting in the impossibility of falling into a cluster. Moreover, 61.5% (8) of the impacted sectors

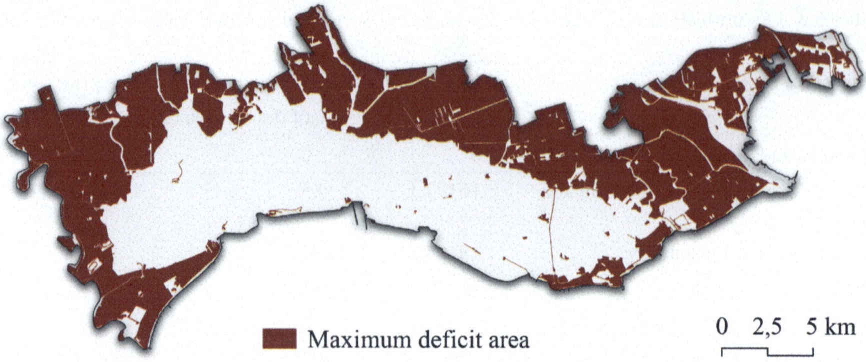

Fig. 5.6 Map of the area of maximum deficit for those impacted sectors where demand and supply do not coincide at all

present the highest level of deficit, about 15% (2) the low and the same percentage the medium–low.

In this section (see also Sect. 3.3), the similarity criterion is based on ecosystem types. The analysis returns only two ecosystem types, i.e. urban and cropland, as these are the only ecosystems that do not provide the service according to the ESs supply matrix in Longo et al. (2024). Hence, from this analysis, these ecosystem types emerge as critical and deficient. The second criterion therefore shows three situations: the presence of both ecosystem types (urban and cropland), only the urban ecosystem or only the cropland ecosystem (see legend in Fig. 5.7).

Four clusters are identified, within which the impacted sectors are quite evenly distributed: all clusters group three sectors, except for cluster D which counts four. The clusters are described as follows. Cluster A gathers three impacted sectors with the highest deficit level and only covered by agricultural land. The sectors in cluster B show a medium–high to high level of deficit and both types of ecosystems. In particular, Hydrogeological instability and Dangerous industries and infrastructure are mainly covered by cropland, vice versa the Transport and infrastructure sector by urban areas. Cluster C contains three of the four sectors with the maximum level of deficit (100%) and is characterised by urban ecosystems. From the analysis of land cover types, the Energy sector shows 1% occupied by agricultural land. However, looking at the spatial data, this is an error in the overlap of the areas; therefore, it is considered as urban exclusively. Finally, Cluster D groups four impacted sectors with a medium–low to low deficit level and in which both ecosystem types are present.

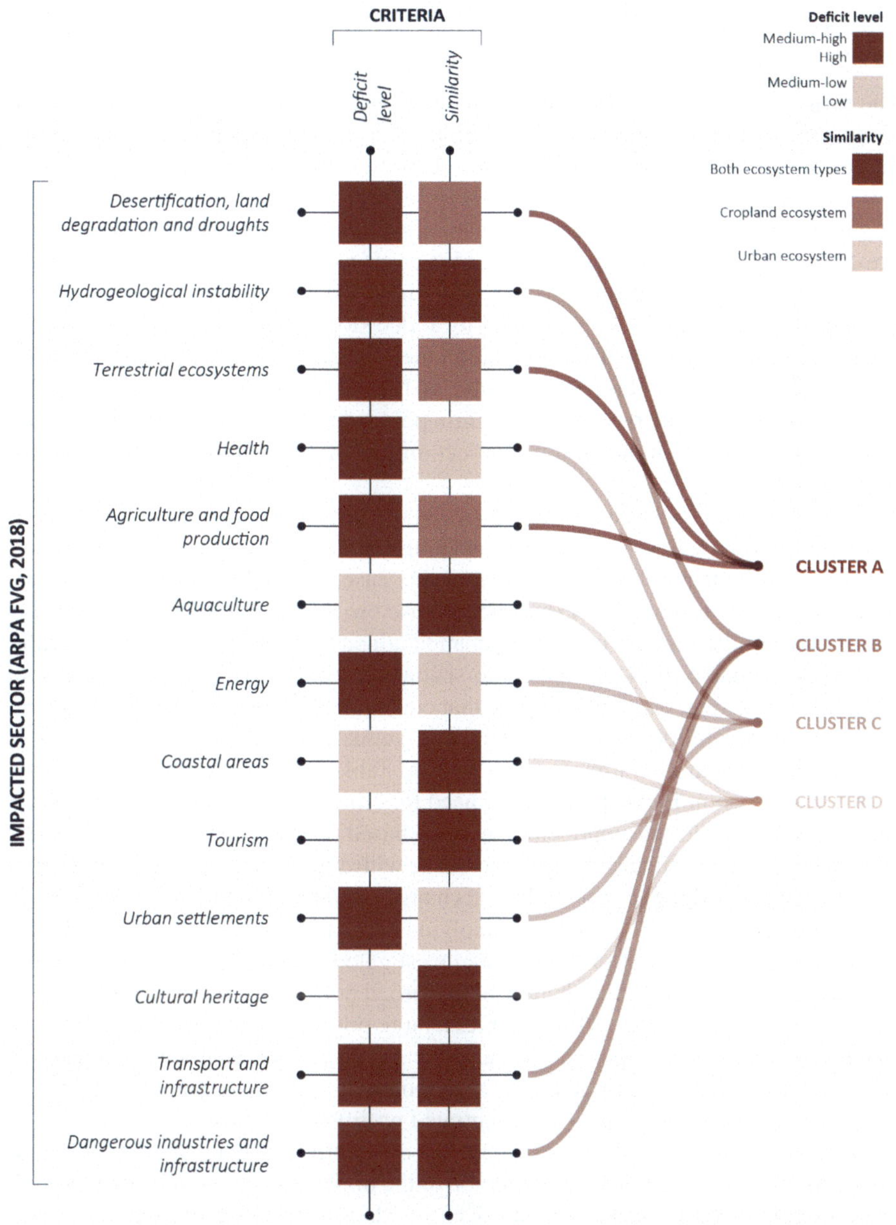

Fig. 5.7 Matrix that combines for each impacted sector the two criteria of level of deficit and similarity, and groups them into clusters

5.5 Discussion

The primary objective of this study is to support decision-makers and planners in identifying deficit areas that need to be addressed for effective climate change adaptation. Specifically, a focus on flood adaptation is proposed. This research offers several types of results: (i) it provides a step-by-step approach that can be replicated in any regional context in Italy and adapted to other EU countries as well; (ii) it provides insights from the application of the case study in terms of trends and highlights that may be of interest to expert reasoning; (iii) the study provides results for the specific case study in the service of local decision-makers. These results can be used to design appropriate flood adaptation measures that are tailored to the specific needs of the region; (iv) finally, the study sheds light on the usefulness of particular concepts to guide future research and inform policy decisions and practice in the field of ES-based adaptation planning. This section discusses the above contributions in more detail.

Section 3.1 and its results (Sect. 4.1) focus on quantifying and mapping SDAs. It is worth spending a few words on the meaning of SDAs. SDAs are portions of land that represent a demand for ES, which, in this case, is the flood protection. While it may seem like a rough first experiment to use the impacted sectors as proxies for ES demand in the context of climate change adaptation, we believe it is critically important to enable decision-makers to spatialise demand. This exercise offers an alternative to normal demand mapping that concentrates, where assessed, on people and their vulnerability, neglecting economic goods and the spatial distribution of different systems and sectors. For this reason, Table 5.1 provides decision-makers with a starting point for spatialising the need for climate change adaptation to which they can subsequently respond with measures based on ecosystem services. In addition, Table 5.1 links climate change knowledge with already familiar data to planners, thus creating a bridge for proper climate change adaptation in planning practice.

Regarding the specific results of applying the case study to identify and map SDAs, it appears that half of the sectors investigated occupy a small portion of the landscape area while the other half occupy a broad portion of its territory. Of course, the fact that one impacted sector occupies a large portion of territory does not imply that this sector is more important than another: this aspect has profound implications in terms of the measures to be taken, assuming that sectors are equally relevant to the economy and society but are concentrated or diffuse in space.

Section 3.2 and its results (Sect. 4.2) focus on identifying and mapping the deficit areas. At this stage, the need for adaptation of the impacted sectors indicated through the concept of SDA (Sects. 3.1 and 4.1), overlaps with areas that provide flood protection ES. Some limitations need to be mentioned. First, this study assumes that the supply areas coincide with the beneficiary areas and does not consider the spatial behaviour of ESs supply or ESs flows. Nevertheless, regulating ecosystem services depend on interactions with adjacent ecosystems, and this is also the case for flood protection services whose supply and demand areas must be closely connected (Fisher et al., 2009; Nedkov & Burkhard, 2012). Furthermore, this synthetic approach presents an easily replicable methodology, making it critical for bridging theory and practice in ES-based adaptation plans. One observation that stands out from the study's results is that 13 sectors have deficit areas because specific areas correspond to specific land covers. This correlation leads to an 'automatic' deficit for sectors usually situated in land covers that do not provide flood protection ES. Conversely, the same correlation between function (sector) and physical characteristics of an area (land cover type) leads to 'automatic' satisfaction for those sectors usually located in land covers that provide flood protection ES. In summary, it emerges that sectors that tend to coincide with urban areas or cultivated land tend to be deficient, as these land cover types are vulnerable in terms of providing this specific ES.

Since the ultimate goal of this study is to set the foundation for ES-based adaptation plans and to inform the decision-makers involved, the most interesting insights come from Sect. 3.3 and its results. Sections 3.3 and 4.3 focus on categorising deficit areas by relating the two criteria of deficit level and similarity (ecosystem type). By "deficit level" we mean the percentage of areas in the sector that are in a deficit of ES supply despite needing it for climate change adaptation (to flooding only). Thus, a "high level of deficit" is not a question of intensity, nor the sector's relevance to the region: it is only a question of the amount of territory (i.e. the area) that needs ES supply but does not benefit from it. By linking deficit levels with ecosystem types, it is possible to identify categories, referred to here as clusters, that allow planners to design different solutions. For example, Cluster A collects the impacted sectors with the highest deficit level and is covered only by agricultural land. Planning actions to improve adaptation for Cluster A will be designed differently from those for Cluster C, which contains areas with the highest deficit level (100%) and involves urban ecosystems. The purpose of Sect. 4.3 is not to formulate precise solutions but to go beyond identifying a problem area or a "safe area" and to ask, based on the recognition of different types of ecosystems that cause a deficit, how this can help us get out of a "one zone". In addition, the application of case studies provides insights into the location of these categories and clusters, thus providing a preliminary geography for interventions.

5.6 Conclusion

Planning to reduce the impacts of climate change calls for a sound synergy between planning tools and cross-sectoral cooperation. Integrating climate adaptation issues into planning processes is necessary. Using the spatially explicit assessment tools elaborated in this study can support several planning decisions to improve resilience to climate change. For example, they can assist in identifying priority areas for implementing adaptation planning policies and interventions across the deficit areas, e.g., through the enhancement of existing or creation of new ecosystems able to provide the required ESs; through the development of specific risk reduction measures; or through the promotion of conservation measures to secure the ESs provided by existing ecosystems and the related benefits (for reducing flood risk in this case) (e.g., Longato et al., 2022).

Although the general ES concept is still scarcely applied in planning instruments, it is argued that its inclusion can support decision-making and bring added value to spatial planning—especially in relation to land use management—in terms of sustainability, ecosystem integrity, and better exploitation of natural resources (Cortinovis & Geneletti, 2018). Moreover, it is believed to provide flexibility to planning by making it sensitive to the non-linear relationships that exist between ecosystem changes and performances (Botequilha-Leitão & Díaz-Varela, 2020).

The proposed methodology for identifying and mapping deficit areas to support ES-based adaptation sheds light on the importance of including both sides of supply and demand in ESs planning. More importantly, it proposes a user-friendly method to facilitate their inclusion in existing planning tools. Overall, the profound innovation of this study is to provide a categorisation of deficit areas based not on the level of climate risk nor the level of importance or impact but on the physical characteristics of the territory.

While applying the methodology to other case studies could provide comparable results and valuable insights for future research, the greater interest lies in the level of usability and appreciation of this tool by decision-makers. To what extent does this tool return a recognisable image of the area for its experts? And can the level of information of deficit areas be supplemented with dimensions related to the relevance of deficits? The next steps of this research will focus on further developments of this work in terms of policy design.

References

Albert, C., Galler, C., Hermes, J., Neuendorf, F., von Haaren, C., & Lovett, A. (2016). Applying ecosystem services indicators in landscape planning and management: The ES-in-planning framework. *Ecological Indicators, 61*, 100–113. https://doi.org/10.1016/j.ecolind.2015.03.029

ARPA FVG. (2018). *Studio conoscitivo dei cambiamenti climatici e di alcuni loro impatti in Friuli Venezia Giulia*. https://www.arpa.fvg.it/temi/temi/meteo-e-clima/pubblicazioni/studio-conoscitivo-dei-cambiamenti-climatici-e-di-alcuni-loro-impatti-in-friuli-venezia-giulia/

Bertuol-Garcia, D., Morsello, C., El-Hani, C., & Pardini, R. (2018). A conceptual framework for understanding the perspectives on the causes of the science—Practice gap in ecology and conservation. *Biological Reviews, 93*(2). https://doi.org/10.1111/brv.12385

Bitoun, R. E., Trégarot, E., & Devillers, R. (2022). Bridging theory and practice in ecosystem services mapping: A systematic review. *Environment Systems and Decisions, 42*(1). https://doi.org/10.1007/s10669-021-09839-7

Bordt, M., & Saner, M. A. (2019). Which ecosystems provide which services? A meta-analysis of nine selected ecosystem services assessments. *One Ecosystem, 4*. https://doi.org/10.3897/oneeco.4.e31420

Botequilha-Leitão, A., & Díaz-Varela, E. R. (2020). Performance based planning of complex urban social-ecological systems: The quest for sustainability through the promotion of resilience. *Sustainable Cities and Society, 56*. https://doi.org/10.1016/j.scs.2020.102089

Brunet, L., Tuomisaari, J., Lavorel, S., Crouzat, E., Bierry, A., Peltola, T., & Arpin, I. (2018). Actionable knowledge for land use planning: Making ecosystem services operational. *Land Use Policy, 72*. https://doi.org/10.1016/j.landusepol.2017.12.036

Carmen, E., Watt, A., Carvalho, L., Dick, J., Fazey, I., Garcia-Blanco, G., Grizzetti, B., Hauck, J., Izakovicova, Z., Kopperoinen, L., Liquete, C., Odee, D., Steingröver, E., & Young, J. (2018). Knowledge needs for the operationalisation of the concept of ecosystem services. *Ecosystem Services, 29*. https://doi.org/10.1016/j.ecoser.2017.10.012

Cortinovis, C., & Geneletti, D. (2018). Ecosystem services in urban plans: What is there, and what is still needed for better decisions. *Land Use Policy, 70*, 298–312. https://doi.org/10.1016/j.landusepol.2017.10.017

Di Marino, M., Tiitu, M., Lapintie, K., Viinikka, A., & Kopperoinen, L. (2019). Integrating green infrastructure and ecosystem services in land use planning. Results from two Finnish case studies. *Land Use Policy, 82*, 643–656. https://doi.org/10.1016/J.LANDUSEPOL.2019.01.007

Dottori, F., Mentaschi, L., Bianchi, A., Alfieri, L., & Feyen, L. (2020). *Adapting to rising river flood risk in the EU under climate change. KJ-NA-29955-EN-N (online)*. https://doi.org/10.2760/14505 (online).

Dworczyk, C., & Burkhard, B. (2021). Conceptualising the demand for ecosystem services—An adapted spatial-structural approach. *One Ecosystem, 6*. https://doi.org/10.3897/oneeco.6.e65966

Evans, N. M. (2019). Ecosystem services: On idealization and understanding complexity. *Ecological Economics, 156*. https://doi.org/10.1016/j.ecolecon.2018.10.014

Fisher, B., Turner, R. K., & Morling, P. (2009). Defining and classifying ecosystem services for decision making. *Ecological Economics, 68*(3). https://doi.org/10.1016/j.ecolecon.2008.09.014

Frantzeskaki, N., Kabisch, N., & McPhearson, T. (2016). Advancing urban environmental governance: Understanding theories, practices and processes shaping urban sustainability and resilience. *Environmental Science & Policy, 62*, 1–6. https://doi.org/10.1016/j.envsci.2016.05.008

Geneletti, D., Esmail, B. A., Cortinovis, C., Arany, I., Balzan, M., van Beukering, P., Bicking, S., Borges, P. A. V., Borisova, B., Broekx, S., Burkhard, B., Gil, A., Inghe, O., Kopperoinen, L., Kruse, M., Liekens, I., Lowicki, D., Mizgajski, A., Mulder, S., Veidemane, K., et al. (2020). Ecosystem services mapping and assessment for policy-and decision-making: Lessons learned from a comparative analysis of European case studies. *One Ecosystem, 5*. https://doi.org/10.3897/oneeco.5.e53111

Grêt-Regamey, A., Altwegg, J., Sirén, E. A., van Strien, M. J., & Weibel, B. (2017). Integrating ecosystem services into spatial planning—A spatial decision support tool. *Landscape and Urban Planning, 165*. https://doi.org/10.1016/j.landurbplan.2016.05.003

Haines-Young, R., & Potschin, M. (2013). Common international classification of ecosystem services (CICES): Consultation on version 4, August-December 2012. *EEA Framework Contract No EEA/IEA/09/003, December 2012*.

IPCC. (2022). Climate change 2022—Impacts, adaptation and vulnerability. In *Contribution of working group II to the sixth assessment report of the intergovernmental panel on climate change* [Pörtner, H.-O., Roberts, D. C., Tignor, M., Poloczanska, E. S., Mintenbeck, K., Alegría, A., Craig, M., Langsdorf, S., Löschke, S., Möller, V., Okem, A., Rama, B. (Eds.)]. Cambridge University Press. https://doi.org/10.1017/9781009325844

ISPRA—Istituto Superiore per la Protezione e la Ricerca Ambientale: Carta della Natura. (2021). Retrieved July 5, 2023, from https://www.isprambiente.gov.it/it/servizi/sistema-carta-della-natura

Kabisch, S. (2019). Urban transformations to pursue sustainability through resource efficiency, quality of life and resilience: A conceptual approach. *Geographia Technica, 14*. https://doi.org/10.21163/GT_2019.141.23

Lavorel, S., Locatelli, B., Colloff, M. J., & Bruley, E. (2020). Co-producing ecosystem services for adapting to climate change. *Philosophical Transactions of the Royal Society B: Biological Sciences, 375*(1794). https://doi.org/10.1098/rstb.2019.0119

Lerouge, F., Gulinck, H., & Vranken, L. (2017). Valuing ecosystem services to explore scenarios for adaptive spatial planning. *Ecological Indicators, 81*. https://doi.org/10.1016/j.ecolind.2017.05.018

Longato, D., Cortinovis, C., Albert, C., & Geneletti, D. (2021). Practical applications of ecosystem services in spatial planning: Lessons learned from a systematic literature review. In *Environmental science and policy* (Vol. 119, pp. 72–84). Elsevier Ltd. https://doi.org/10.1016/j.envsci.2021.02.001

Longato, D., Cortinovis, C., Balzan, M., & Geneletti, D. (2022). Identifying spatial opportunities for nature-based solutions planning in cities: A case study in the area of Valletta, Malta. *Lecture Notes in Civil Engineering, 242 LNCE*. https://doi.org/10.1007/978-3-030-96985-1_12

Longato, D., Cortinovis, C., Balzan, M., & Geneletti, D. (2023). A method to prioritize and allocate nature-based solutions in urban areas based on ecosystem service demand. *Landscape and Urban Planning, 235*. https://doi.org/10.1016/j.landurbplan.2023.104743

Longo, A., Zardo, L., Maragno, D., Musco, F., & Burkhard, B. (2024). Let's do it for real: Making the ecosystem service concept operational in regional planning for climate change adaptation. *Sustainability, 16*(2), 483. https://doi.org/10.3390/su16020483

Marques, A. L., Alvim, A. T. B., & Schröder, J. (2022). Ecosystem services and urban planning: A review of the contribution of the concept to adaptation in urban areas. In *Sustainability (Switzerland)* (Vol. 14, Issue 4). https://doi.org/10.3390/su14042391

Marta Suárez Marta, E., Fernández de Manuel, B., Méndez-Fernández, L., Oneindia, M., & Gómez-Baggethun, E. (2018). Nature-based solutions and resilience as complementary strategies for urban governance and planning: A review of assessment methodologies. *Reframing Urban Resilience Implementation*.

MEA. (2003). Millennium ecosystem assessment: Ecosystems and human well-being—A framework for assessment. In *Millennium ecosystem assessment*.

Munang, R., Thiaw, I., Alverson, K., Liu, J., & Han, Z. (2013). The role of ecosystem services in climate change adaptation and disaster risk reduction. In *Current opinion in environmental sustainability* (Vol. 5, Issue 1). https://doi.org/10.1016/j.cosust.2013.02.002

Nedkov, S., & Burkhard, B. (2012). Flood regulating ecosystem services—Mapping supply and demand, in the Etropole municipality, Bulgaria. *Ecological Indicators, 21*. https://doi.org/10.1016/j.ecolind.2011.06.022

Nesshöver, C., Vandewalle, M., Wittmer, H., Balian, E. V., Carmen, E., Geijzendorffer, I. R., Görg, C., Jongman, R., Livoreil, B., Santamaria, L., Schindler, S., Settele, J., Sousa Pinto, I., Török, K., van Dijk, J., Watt, A. D., Young, J. C., & Zulka, K. P. (2016). The Network of knowledge approach: Improving the science and society dialogue on biodiversity and ecosystem services in Europe. *Biodiversity and Conservation, 25*(7). https://doi.org/10.1007/s10531-016-1127-5

Olander, L., Polasky, S., Kagan, J. S., Johnston, R. J., Wainger, L., Saah, D., Maguire, L., Boyd, J., & Yoskowitz, D. (2017). So you want your research to be relevant? Building the bridge between ecosystem services research and practice. In *Ecosystem services* (Vol. 26). https://doi.org/10.1016/j.ecoser.2017.06.003

Pörtner, H.-O., Scholes, R. J., Agard, J., Archer, E., Arneth, A., Bai, X., Barnes, D., Burrows, M., Chan, L., Cheung, W. L., Diamond, S., Donatti, C., Duarte, C., Eisenhauer, N., Foden, W., Gasalla, M. A., Handa, C., Hickler, T., Hoegh-Guldberg, O., Ngo, H., et al. (2021). *Scientific outcome of the IPBES-IPCC co-sponsored workshop on biodiversity and climate change*. https://doi.org/10.5281/ZENODO.5101125

Qiu, L., Dong, Y., & Liu, H. (2022). Integrating ecosystem services into planning practice: Situation, challenges and inspirations. In *Land* (Vol. 11, Issue 4). https://doi.org/10.3390/land11040545

Regione Autonoma Friuli Venezia Giulia. Statistica: Demografia. (2022). Retrieved July 5, 2023, from https://www.regione.fvg.it/rafvg/cms/RAFVG/GEN/statistica/FOGLIA11/

Rözer, V., Mehryar, S., & Surminski, S. (2022). From managing risk to increasing resilience: A review on the development of urban flood resilience, its assessment and the implications for decision making. *Environmental Research Letters, 17*(12). https://doi.org/10.1088/1748-9326/aca8bc

Ruckelshaus, M., McKenzie, E., Tallis, H., Guerry, A., Daily, G., Kareiva, P., Polasky, S., Ricketts, T., Bhagabati, N., Wood, S. A., & Bernhardt, J. (2015). Notes from the field: Lessons learned from using ecosystem service approaches to inform real-world decisions. *Ecological Economics, 115*. https://doi.org/10.1016/j.ecolecon.2013.07.009

Srivastava, A. K., & Sahay, A. (2023). From safety against floods to safety at floods*: Theory of urban resilience to flood adaptation and synergy with mitigation. *Anthropocene Science*. https://doi.org/10.1007/s44177-023-00064-2

Sun, X., Yang, P., Tao, Y., & Bian, H. (2022). Improving ecosystem services supply provides insights for sustainable landscape planning: A case study in Beijing, China. *Science of the Total Environment, 802*. https://doi.org/10.1016/j.scitotenv.2021.149849

Tanner, T., Bahadur, A., & Moench, M. (2017). Challenges for resilience policy and practice. In *ODI working paper* (Issue August).

United Nations Environment Programme. (2021). *Guidelines for integrating ecosystem-based adaptation into national adaptation plans: Supplement to the UNFCCC NAP technical guidelines*. https://wedocs.unep.org/20.500.11822/36703

United Nations Environment Programme. (2022). *Harnessing nature to build climate resilience: Scaling up the use of ecosystem-based adaptation*. https://www.unep.org/resources/report/harnessing-nature-build-climate-resilience-scaling-use-ecosystem-based-adaptation

United Nations Environment Programme, & World Conservation Monitoring Centre. (2020). *Making EbA an effective part of balanced adaptation strategies: Introducing the UNEP EbA briefing notes—briefing note 1*. https://wedocs.unep.org/xmlui/handle/20.500.11822/28174

Vallecillo Rodriguez, S., Maes, J., Teller, A., Babi Almenar, J., Barredo Cano, J., Trombetti, M., Abdul Malak, D., Paracchini, M., Carré, A., Addamo, A., Czucz, B., Zulian, G., Marando, F., Erhard, M., Liquete Garcia, M., Romao, C., Polce, C., Pardo Valle, A., Jones, A., Gumbert, A., et al. (2022). *EU-wide methodology to map and assess ecosystem condition* (Issue KJ-NA-31-226-EN-N (online)). Publications Office of the European Union. https://doi.org/10.2760/13048 (online).

Van der Biest, K., Meire, P., Schellekens, T., D'hondt, B., Bonte, D., Vanagt, T., & Ysebaert, T. (2020). Aligning biodiversity conservation and ecosystem services in spatial planning: Focus on ecosystem processes. *Science of the Total Environment, 712*. https://doi.org/10.1016/j.scitotenv.2019.136350

Verhagen, W., Kukkala, A. S., Moilanen, A., van Teeffelen, A. J. A., & Verburg, P. H. (2017). Use of demand for and spatial flow of ecosystem services to identify priority areas. *Conservation Biology, 31*(4). https://doi.org/10.1111/cobi.12872

von Haaren, C., Lovett, A. A., & Albert, C. (2019). *Landscape planning with ecosystem services: Theories and methods for application in Europe* (Landscape Series, Vol. 24). Springer.

Wei, F., & Zhan, X. (2023). A review of ES knowledge use in spatial planning. *Environmental Science and Policy, 139*. https://doi.org/10.1016/j.envsci.2022.11.003

World Bank. (2021). *A catalogue of nature-based solutions for urban resilience*. World Bank Group.

World Economic Forum, & McKinsey & Company. (2023). *Seizing the momentum to build resilience for a future of sustainable inclusive growth*. World Economic Forum, McKinsey & Company.

Zulian, G., Stange, E., Woods, H., Carvalho, L., Dick, J., Andrews, C., Baró, F., Vizcaino, P., Barton, D. N., Nowel, M., Rusch, G. M., Autunes, P., Fernandes, J., Ferraz, D., Ferreira dos Santos, R., Aszalós, R., Arany, I., Czúcz, B., Priess, J. A., Viinikka, A., et al. (2018). Practical application of spatial ecosystem service models to aid decision support. *Ecosystem Services, 29*. https://doi.org/10.1016/j.ecoser.2017.11.005

Part II
From Policy to Action

Chapter 6
Adapting to Adaptation: Methodological Support for Multiscalar Adaptation of the Venice Lagoon

Filippo Magni and Katia Federico

Abstract The impacts of climate change and the increasing frequency of resulting extreme events in recent years have given rise to significant environmental, societal, and economic repercussions in a delicate and highly susceptible region such as Venice. Precisely from this perspective, which considers areas particularly vulnerable to the effects of the climate, the "Venezia2021" scientific research program is included, coordinated by CORILA—Consortium for coordinating research relating to the Venice lagoon system. The research in question involves a complex process of identification and integration of innovative tools, data processing and analysis, and evaluation of impacts, to contribute to maintaining a correct balance of the lagoon ecosystem in a perspective of greater climate resilience. The general objective of the research was to build a strategic, accurate, and shared vision of the challenges that await the protection of the city and its lagoon (a world heritage site), in consideration of climate change scenarios. Specifically, this contribution analyzes the research experience of thematic axis n.5 which led to the drafting of the Climate Change Adaptation Plan for the Venice lagoon. The operational path that led to the construction of the plan was guided by an in-depth territorial study and development of an integrated system of analysis, evaluation, planning, management, and monitoring of the Venetian area capable of supporting the city and the activities that operate through coordinated adaptation actions aimed at increasing sustainability and resilience. This plan represents a model of how communities can work together to address the challenges of climate change. This integrated approach, based on solid scientific data and active stakeholder involvement, could be and become the basis for future adaptation plans in other vulnerable areas. This is because the plan development process demonstrated the effectiveness of a multidisciplinary approach, integrating scientific data, socio-economic analysis, and stakeholder involvement. Initially, detailed scientific analyses were conducted on the vulnerability of the Venice Lagoon to climate change, focusing on heat waves and flooding. These analyses provided a solid basis for understanding climate risks and the specific challenges faced by territories. It

F. Magni (✉) · K. Federico
Department of Architecture and Arts, University IUAV of Venice, Venice, Italy
e-mail: fmagni@iuav.it

M. Granceri Bradaschia et al. (eds.), *Climate Change Adaptation, Flood Risk, and Beyond*, Planning for Climate Proof Cities,
https://doi.org/10.1007/978-3-031-65463-3_6

represents an example of how interdisciplinarity, participation of local communities, and collaboration with stakeholders can contribute to creating sustainable solutions to address the impacts of climate change in other similar regions. The lesson learned from this experience is that preparation and planning are key to increasing climate resilience and protecting vulnerable territories.

Keywords Spatial planning · Climate change adaptation · Mainstreaming · GIS · Venice

Premise

At this crucial moment for choices on the fate of Venice and its lagoon, scientific research is called upon to provide the essential elements to allow political decision-makers to make choices based on knowledge. A forward-thinking administration depends on a continuous flow of information to understand, anticipate environmental and socioeconomic challenges, and mitigate future changes.

Venezia2021 is a scientific research program coordinated by CORILA,[1] which provides an integrated complex of observational tools and processing of the data collected, at the state of the art of knowledge, to help keep the correct balance of the lagoon ecosystem under control. In this way, it will be possible to protect the natural ecological functioning of the lagoon and at the same time safeguard the benefits for society (ecosystem services) provided by the lagoon (SNPA, 2019). Furthermore, in the current testing phase of the MOSE System, to defend against high water, Venezia2021 will also be able to develop new scenarios, to support defense strategies both for urban monumental buildings and for the so-called "minor buildings". The Venezia2021 research program was started in 2018 and ended in 2022, and is financed by the Interregional Supervision for Veneto, Trentino-Alto Adige, Friuli-Venezia Giulia of the Ministry of Infrastructure and Transport, as part of the works to safeguard Venice and the lagoon and also coordinates in an operational sense with them.

Venezia2021 involves more than 200 people, researchers, and collaborators belonging to CORILA member bodies, namely: Ca' Foscari University of Venice, IUAV University of Venice, University of Padua, National Research Council, and National Institute of Oceanography and Experimental Geophysics. An external and independent evaluation mechanism (ex-ante, in itinere, and ex-post) is established in the CORILA Scientific Technical Committee, composed of scientists of recognized

[1] CORILA promotes and coordinates research on the Venice lagoon, including at the international level; facilitates interaction with the scientific community; and studies the physical system, environmental, architectural and landscape, economic and social aspects of the lagoon and settlements, strongly interconnected components of a complex system. CORILA is an association between Ca' Foscari University of Venice, IUAV University of Venice, University of Padua, National Research Council and National Institute of Oceanography and Experimental Geophysics. It is supervised by the Ministry of Education University and Research.

international value, not involved in the execution of the research, and in a position of the absence of conflict of interest.

Venezia2021 is structured into 5 homogeneous themes, each of which is divided into Lines, for a total of 15 Research Lines (see Fig. 6.1). In every single topic, the aim was to obtain an effective integration of skills and knowledge, with contributions from the different research groups belonging to different bodies. The research activities are many and different from each other. They are carried out by the research groups of the CORILA member bodies, in some cases with the collaboration of other research groups and other interested PPAAs. The activities range from the collection of environmental data and samples, laboratory chemical analyses, implementation and integration of mathematical models, creation of short-medium and long-term scenarios, formulation of guidelines for the management of the future lagoon regulated by the MOSE, as well as methods for the protection of architectural heritage. The coordination offered by the CORILA operational structure supports and supports researchers in their relationships with each other and with the Administrations and provides tools for the management of data and information.

The "Venice 2021" adaptation plan is proposed as a useful and versatile climate change adaptation tool for the Venice Lagoon and more precisely for the 11 Municipalities that make up its drainage basin. This contribution derives from analyses conducted within thematic line 5.2, which focuses on the phenomenon of heat waves and flooding. In the initial phase, these analyses were integrated with in-depth socio-economic research specific to each municipality and meetings with stakeholders.

Unfortunately, the pandemic situation, which occurred during the project development period, severely limited interactions with stakeholders which were limited to the development of the first phase of the project. In this phase, however, it was possible to find important information that allowed the development of stakeholder and governance mapping. Furthermore, the meetings with the Municipalities immediately highlighted how the issue of flooding was central to the policies and actions for securing the territory, while, on the contrary, the phenomenon of heat waves was not perceived in the slightest. Due to this initial evidence, the study, after an initial phase, focused almost exclusively on the topic of flooding. The study highlighted how the lagoon territory is, at the same time, uniform and heterogeneous. Uniform from the more purely territorial and environmental aspects, but extremely heterogeneous and varied from a socio-economic point of view. This led the plan to have to analyze and propose different adaptation design alternatives, scalable and manageable by the Municipalities in very different areas and with very different critical issues.

In the second phase, the project worked on the preparation of the actual adaptation plan, that is, for each interested municipality, strategies, objectives, actions, and measures capable of overcoming territorial limits and vulnerabilities were identified. These measures have been defined and described in a basic abacus that supports the zoning of the areas and provides concrete possibilities to political decision-makers.

The Adaptation Plan is therefore composed of three main phases:

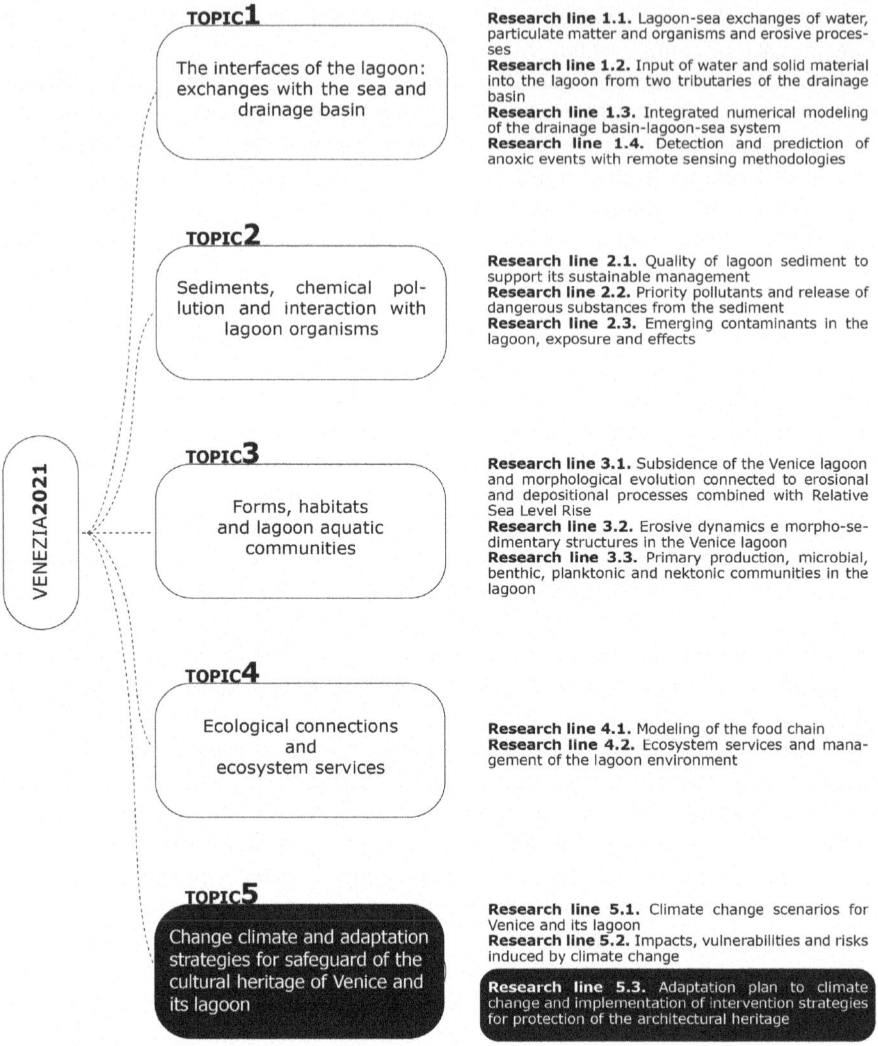

Fig. 6.1 Topics and lines of research of the Venezia2021 project. The figure shows the research topics (5) and the details of the subsequent research line related to each topic

(i) analysis of vulnerabilities and relevant territorial and management characteristics;
(ii) identification and zoning of the lagoon territory based on vulnerabilities and territorial characteristics;
(iii) proposal of an abacus of strategies, policies, actions, and measures to guarantee decision-makers alternatives. In this way, the plan remains a flexible instrument, not rigid, but capable of adapting to the different realities and sensitivities of the territory, without, however, losing effectiveness or limiting local initiatives.

The Venice Lagoon Adaptation Plan must be seen as a tool that does not replace ordinary planning but supports and implements it through specific strategies, objectives, measures, and actions. Furthermore, the plan is accompanied by two other supporting documents: the Emergency Plan, which responds to the specific needs of emergencies that climate impacts create on the territory, and the Decision Support System, which through a series of indicators, supports decision-makers in choice of strategies, objectives, actions and measures and their monitoring.

6.1 Safeguarding the Cultural Heritage of Venice and Its Lagoon from the Impacts of Climate Change: An Innovative Path Between Research and Experimentation

The Venetian territory is fragile and highly vulnerable. The impacts of climate change in recent years have widely manifested their effects on different aspects (see Fig. 6.2), ranging from hydrogeological risk to the overheating of urban areas (UHI) to coastal and lagoon defense problems, up to the deterioration of historic buildings (Musco, 2016). All this has significant environmental, social, and, in an environment like Venice, above all economic consequences which, if not adequately managed, can undermine the development of economic activities and the well-being of citizens. Theme 5 fits into this framework with the general objective of developing an integrated system of analysis, evaluation, planning, management, and monitoring of the Venetian area and its historical, artistic, and cultural heritage, capable of supporting the city and its activities that operate in it, through coordinated adaptation actions that aim to increase sustainability and resilience. The topic is divided into 3 specific lines of study[2]: Line 5.1. Climate change scenarios for Venice and its lagoon; Line 5.2. Impacts, vulnerabilities, and risks induced by climate change; Line 5.3. Adaptation plan to climate change and implementation of intervention strategies for the protection of architectural heritage (this contribution is the result of this last line of research). These three lines have specific objectives that are closely related and consequential. Model forecasts relating to climate scenarios and the rise of the mean sea level were developed and used to evaluate the vulnerability of the lagoon environment in a multi-risk assessment context and to draw up an overall adaptation plan to climate change for the Venice lagoon and its Metropolitan City (Maragno et al., 2021, 2022). Intervention and maintenance procedures for architectural elements and historical materials have also been developed, which allow for the definition and implementation of periodic monitoring of the historical, artistic, and cultural heritage present in the lagoon.

[2] http://venezia2021.corila.it/tematiche/tematica-5/.

Fig. 6.2 The impacts of extreme events in the lagoon system (November 2019). *Photo credit* Iuav photo service and images

6.1.1 Research Line 5.3. Adaptation Plan to Climate Change and Implementation of Intervention Strategies for the Protection of Architectural Heritage

In recent years we have witnessed a growing occurrence of extreme climatic events, which have particularly affected coastal Italian cities. These events have caused huge losses in social, economic, environmental, and tangible and intangible cultural heritage terms. Line 5.3, in parallel with the construction of the climate change scenarios defined in "Line 5.1. Climate change scenarios for Venice and its lagoon" and with the support of the analyses on impacts, vulnerability, and risk carried out in "Line 5.2. Impacts, vulnerabilities and risks induced by climate change", has as its main objective that of structuring methodologies and practices capable of capturing the changes induced both by climate change and by the implementation of the MOSE, and consequently develop actions aimed at adaptation, to the mitigation of impacts and the prevention of risks. Line 5.3, following the approaches and indications deriving from Line 5.1 and Line 5.2, defines an adaptation plan relating to the lagoon area and the first interface strip with the territory of the Metropolitan City of Venice, building for these areas, specific intervention strategies for the conservation of cultural and territorial heritage to the point of providing indications for increasing the resilience of the Venice system. The activities envisaged by Line 5.3. have been defined according to an interdisciplinary perspective that involves the disciplines of "project", from urban and territorial planning to construction techniques, to restoration, to technical and environmental physics, to petrography. The project seeks to provide an overall vision of the "Lagoon system" including the environmental systems as well as the built environment including the interface with the lagoon basin, in the perspective of the impacts of climate change. Furthermore, the research bases all proposals on a shared set of data and analyses.

The legitimation of the directions of this line of research derives from the common and shared basis of study and, above all, from the constant presence of the stakeholders and decision-makers involved in the decision-making process (a specific axis of the research supported the involvement aimed at evaluating the requests and preferences, despite a very complex socio-health situation which has hindered any joint initiative). The return of knowledge to the various actors (Municipalities, protection bodies, private citizens) is a necessary point of arrival for research to provide methods and tools for sustainable and long-lasting maintenance and management of the Lagoon system in all its components. The need to study the state of conservation of the architectural and landscape heritage in a systematic, integrated, and widespread way, with particular attention to the most sensitive and vulnerable elements, is essential to evaluate the future direct and indirect impacts of climate change.

6.1.2 The Plan for Venice and Its Lagoon: Structure and Objectives

This contribution relating to the climate adaptation plan for the lagoon border (11 municipalities involved), represents the final document of the Venezia2021 project, collecting and systematizing the material and studies produced over the years, and is structured in three macro-parts: the cognitive framework, the planning framework, and a final summary on the determining factors for systemic adaptation and integration (see Fig. 6.3). The cognitive framework, as we will see in the dedicated chapters, provides a wealth of information on the territory under study, from the main actors involved to the regulatory references, from the socioeconomic systems to the territorial vulnerabilities. The project framework, based on the studies of the cognitive framework, proposes a programmatic and strategic design, defining objectives, actions, and measures for adaptation to climate change. Finally, the final summary highlights the determining factors for planning territorial adaptation to climate change and for the related systemic integration within existing tools/systems.

The scale of the cognitive framework is large and embraces the entire lagoon border, focusing on 11 municipalities. The design framework, however, starting from large-scale considerations, goes into more detail to suggest and indicate adaptation options and measures at an urban and architectural level. As we will see in the next chapters, information, and studies on the entire "lagoon system" (actors, regulations, planning tools, socioeconomic systems, vulnerability analysis) will be reported

Fig. 6.3 Structure of the plan: 3 pillars of the adaptation plan

within the cognitive framework, which will allow us to support the strategic choices of the framework planning. The latter, however, is broken down into 11 municipal cards, in which strategies, objectives, actions, and options are specified.

This plan is optional and, through the territorial study (ecosystems, agricultural systems, urban areas, construction, and architectural characteristics), directs the planning tools and the administrations involved within a strategic framework, equipped with mapping, objectives, adaptation options, and indicators. The plan, in addition to providing a cognitive and planning framework, also allows local administrations to be disseminated and made aware of the issue of adaptation and to convert the territorial planning approach from sectoral-individual to holistic-integral-shared.

6.2 Towards Updated and Multi-Level Knowledge: The Cognitive Framework of the Lagoon System in Comparison with Climate Changes

The definition of the cognitive framework through the systematization of information and data is the first contribution of this document. The first phase of identification of the area under study took place through consideration of the boundaries of the last lagoon boundary in force, decreed in 1990 by the Ministry of Public Works. This delimitation includes the port inlets of San Niccolò, Malamocco, Chioggia, and the island of Sant'Erasmo. It was deemed appropriate to add a further buffer area of 1 km inland (mainland), to consider the territorial system as a whole and more correctly identify the possible impacts attributable to climate change. This delimitation is important both for the development of an inter-municipal adaptation strategy and for the need for a governance system capable of effectively responding to the needs of each administrative reality. In this perspective, (11) municipalities have been identified, for a total area of 1126.86 km^2, and a total population of 490 thousand people. The municipalities selected for the study are: Venice, Campagna Lupia, Cavallino-Treporti, Chioggia, Dolo, Jesolo, Marcon, Mira, Musile di Piave, Quarto D'altino, San Donà di Piave.

The main set of information consists of the vulnerability analysis of the lagoon (line 5.2), which was followed by reconnaissance and study of the current plans, to identify any measures already foreseen concerning the impacts due to climate change and capable of supporting the adaptation and mainstreaming of measures. Given the nature of the Venetian context, it was appropriate to extend this phase of study to include existing European regulations and projects, to exhaustively identify the totality of the overall forecasts of the lagoon basin.

In a multidisciplinary way, we worked to combine different types of data and information—environmental, climatic, socio-economic, and governance—necessary for the process of defining this Venice Lagoon Adaptation Plan. After the preliminary phases of gathering data and information on environmental-climatic, socio-economic, and relational aspects of governance, the investigation took place through various tools, previously made available by municipal, provincial, and regional public administrations, as well as by associations category, Chamber of Commerce of Venice and Rovigo and other sources, including territorial information systems and geo-portals.

At the same time, an examination of the scientific literature relating to the different types of most used adaptation measures was started, preparatory to the drafting of the project framework (including an abacus of actions, measures, and policies, already partly presented, and shared with the stakeholders involved) (Bezzi et al., 2015; Bockel, 2009; Keskitalo, 2010; Laukkonen et al., 2009; Magni, 2019; Musco et al., 2018; Padova Resiliente, 2016).

The construction of the cognitive framework is aimed at proposing contextualized adaptation measures and encouraging the mainstreaming process, starting from the following mappings:

- Climatic and geographical context;
- Governance of the internal structure of the organization and stakeholders involved at municipal and supra-municipal levels;
- Study of critical areas (focus on the issue of flooding).

The first phase of obtaining the necessary data was transformed into tools for reading and interpreting the territory (mapping), useful for bringing out relationships and themes, as well as critical issues and emergencies. At the same time, an analysis was also carried out of the legislation and planning in force in the area (PNACC, 2017; SNACC, 2014), as well as an investigation into the socio-economic systems that characterize the territory and the individual entities.

6.2.1 The Climatic Context of the Geographical Structure of the Vast Area

The area covered by the research, including the territory of the Metropolitan City of Venice, is part of the Veneto Region and is largely occupied by the Venice Lagoon, one of the largest and most important coastal ecosystems in Europe and with a characteristic structure by continuous changes (UNEP, 2016).

The territory is crossed by the terminal branches of the Piave, Brenta-Bacchiglione, and Adige rivers, with a contribution of debris that is fundamental for the morphology of the lagoon, and a dense hydraulic network, which makes it extremely complex in its management. It has land located at altitudes higher than 0 m above sea level and land located at altitudes lower than 0 m above sea level in the southern area (Cona, Cavarzere, and Chioggia) and in the northeastern area (Sandonatese). The altimetric peculiarity is due to hydraulic reclamations which have removed large tracts of territory from marshy lands and lagoons, which are now cultivated and on which urban centers and industrial areas have developed (ISPRA, 2014).

In the coastal areas of Chioggia-Sottomarina, Ca' Roman—Pellestrina, Lido, Cavallino, and Jesolo the elevations of the countryside level rise above sea level with dune alignments that reach several meters above sea level. The entire area boasts a rich environmental heritage, with important SCI sites—Sites of Community Importance, SPA sites—Special Protection Areas, and SCI/SPA sites.

From a climatic point of view, as has happened in a good part of Europe, as in Veneto and the Metropolitan City of Venice, significant changes have occurred in the last 70 years: the Region is positioned in a transition zone between the continental area of central Europe and the Mediterranean, presenting peculiar characteristics that derive from the mitigating action of the Adriatic Sea and the influence of the Alpine chain. In particular, the territory of the metropolitan city of Venice has two large areas, one coastal and one inland, with different climatic characteristics (see Fig. 6.4).

The first is mitigated by the waters only in a limited way and a restricted portion, resulting in winter temperatures that are on average lower also due to the Bora winds that blow from the Northeast. The internal area, further away from the coast, has continental climatic characteristics, with summers hotter and harsher winters than the coastal area and reduced ventilation.

As regards the analysis of temperatures in Veneto, according to ARPAV data which refer to the regional temperatures of the period 1955–2004, the trend is increasing in the annual averages (+ 1.8 °C/50 years), but also in those seasonal, more evident for the maximum in summer and winter (+ 2.3 °C/50 years). Minimum temperatures also record a significant increase, especially in summer (+ 1.6 °C/50 years) and in spring (+ 1.0 °C/50 years).

The most recent data, collected starting from 1993 and up to 2017, highlight higher thermal values throughout the region (+ 1.3 °C/25 years). The analysis records a trend of increase in average temperatures, particularly in the autumn period. The consequences of this increase, such as the decrease in the extension of the Dolomite glaciers or the variations in the ripening times of some crops, are already evident.

Regarding rainfall, the trend in Veneto, in the years from 1955 to 2004, recorded a slight tendency towards a decrease in annual and winter events, in the Region there was an increase in autumn rainfall, with significant cases of increases compared to the values maximum annual short-term precipitation.

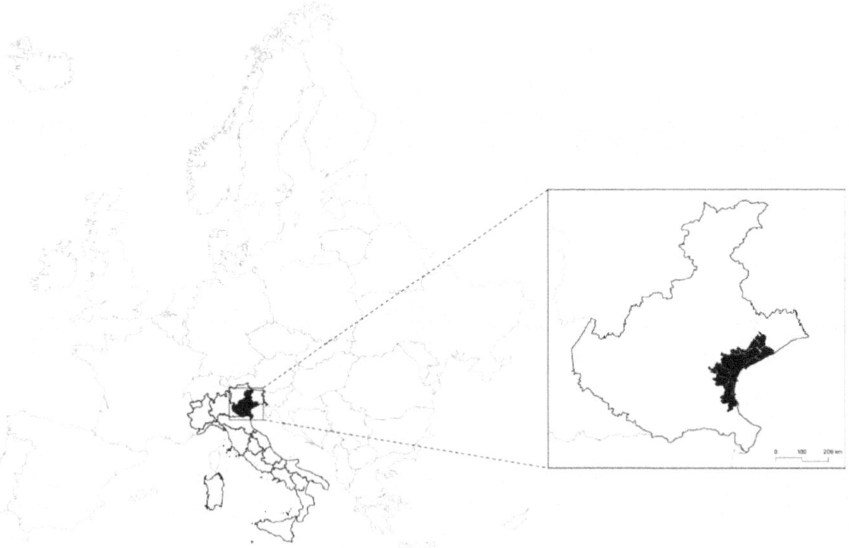

Fig. 6.4 Spatial framework of the municipalities in the lagoon basin. The map shows the location in Europe of the 11 municipalities of the study area

Nonetheless, compared to data relating to temperatures and precipitations, the statistics relating to extreme events are more limited, an intensification of their frequency also emerges in the Venice lagoon area, in the form of heat waves, an increase in the intensity of precipitation, strong winds and tornadoes.

As regards intense rainfall, in the decade 2003–2012, there was a greater number of events compared to the decade 1992–2001, especially for those of short duration, both in terms of increase in frequency and intensity. In 2007, a state of emergency was declared for part of the Veneto region following the floods of 26 September; which involved a large part of the territory of the Metropolitan City of Venice and the lagoon area, in particular, the Municipalities of Venice, Mestre, Marcon, Campagna Lupia, Mira. Events relating to strong winds, tornadoes, and small tornadoes occurred in June 2007 in Venice, in the San Giuliano area, in July 2010 in Pellestrina (VE), and in July 2015 in the municipalities of Dolo and Mira.

As regards the phenomenon of high water, the exceptional nature of the November 2019 event should be noted, with the reaching of + 182 cm in Venice, the second highest level since 1872, and 3 events greater than or equal to 150 cm in 6 days, all in November.

The Tide Forecasting and Reporting Center of the Municipality of Venice highlights how, throughout 2019, the sea level exceeded 110 cm 28 times, with a total stay of approximately 50 h in November alone. These numbers exceed the maximum values reached in the previous 150 years, in particular 2010 with 18 events in one year and 2012 in which a total of 24 h of permanence above 110 cm were recorded. Since

1875, the peaks, for which a rise in sea level of at least 110 cm has been recorded—which constitutes the lower threshold for defining an exceptionally high tide—have intensified in recent years up to a maximum of 14 days with 4 events, in which 130 cm were exceeded, in 2009 and 7 events, in which 120 cm were exceeded, in 2010. The historical series highlights how the frequency of high tides in this century is therefore progressively increasing, also about "normal" tides greater than or equal to + 60 cm (CMCC, 2021).

6.2.2 The Governance System and the Stakeholders Involved: Simplification and Systematization of a Management Matryoshka

Governance is a complex of variables linked to the definition of policies and its schematization considers the relationships between levels and components of government involved, the regulatory framework of planning, and the degree of community involvement that can be achieved.

For this reason, the governance analysis has set itself the objective of highlighting the sectors, specific tasks, and skills of each municipal structure about the issue of climate change and the territorial repercussions that this has in environmental, social, and economic terms, politicians.

In addition to institutional governance, we also tried to schematize the system of stakeholders present within the communities, that is, that complex and extremely diversified multiplicity of stakeholders, involved and capable of being involved in the development of adaptation processes. The relevance lies in the interest (of these actors) concerning the activities of the Administration(s) and in the ability to influence the achievement of the objectives of the process.

The study of the lagoon and peri-lagoon territory has made it possible to identify the socio-economic, productive, and political systems within which to search for the reference stakeholders for each municipality affected by the plan. The material and immaterial characteristics of the territories of the lagoon municipalities (described in the mapping of the cognitive framework) have made it possible to cross-reference data and information to identify the active networks that can be activated for an overall adaptation process.

The construction of the cognitive framework also lies in the identification and description of governance: the Public Administrations involved were analyzed through the reading of the municipal organizational charts and internal functions (internal public governance) and the system of relations between Municipalities and Bodies with territorial competences, inter-municipal entities, and other civil entities (external public governance). In inter-municipal and supra-municipal relations (large scale), territorial macro-challenges (such as climate change) are faced, overcoming municipal administrative limits/boundaries, and broadening the strategic vision concerning a larger territory. In this way, broad-based control and timely responses for local administrations can coexist.

Internal public governance is thus defined by sectors and offices that play a key role at a political and operational level in climate change and, specifically, the economic repercussions of its impacts. These offices were instrumental in the Venezia2021 project in drafting the deliverables and exchanging information, also preparatory to the development of this plan.

External public governance is often composed of private stakeholders with active or potential relationships with the Administrations. The survey made it possible to identify the categories of interlocutors who recognize the problem of climate change in a specific territory and who, even if guided by different value systems, can be involved in the process of defining and negotiating adaptation strategies, actions, and measures. These may be monitoring, planning, and governance bodies, environmental management and protection bodies, universities, and research centers, political bodies, operational bodies, professionals, trade associations and local associations (environmental in particular), or groups of unorganized citizens. These are therefore both public and private entities, operating at a regional, provincial, and municipal scale, which can have the role of influencing the activity of the Municipality, satisfying public needs, like those of the Administration, and completing the services which depend on the Public Body itself.

Systematizing information in a database facilitates the participatory process. The involvement of these and other strategic subjects has positive effects on the development of this plan, as it allows for greater incisiveness of the actions to be undertaken, avoiding the disconnect between indications and options of the plan and the reality of the territorial systems.

In the majority of municipalities, extensive collaboration was demonstrated through the availability and skills of officials and managers of the technical areas of the municipalities themselves. The tool used during the meetings was that of the open, discursive interview, characterized by considerable flexibility which allowed us to glean a greater amount of information, simply directing the reflections towards insights useful for the study, aimed at highlighting and clarifying the system of the relationships, which only partially emerged from the preliminary stakeholder reconnaissance phase. The outcome of the meetings therefore allowed the verification and co-definition of the administrative structures, their internal connections, but also of the system of external relations, in particular through the reporting of locally active subjects, such as trade associations and citizens' groups (which only direct knowledge of the territory can allow). Furthermore, the interview made it possible to obtain information on the organisation's perceptions of the risks of climate change; on the existence of rules or measures attributable to the adaptation contained in the municipal plans already adopted; on the socio-economic characteristics of the local productive fabric also linking them to the presence of monitoring, carried out following extreme climatic situations, to quantify the damage.

6.2.3 *The Analysis of Territorial Vulnerabilities: Between Historical-Structural Critical Issues and New Risk Perceptions*

Within the definition of the cognitive framework, the study of environmental and climatic aspects—in a territorial context characterized by calamitous events—is focused in particular on some impacts, which better than others represent elements of vulnerability and risk (EEA, 2016; Fritzsche et al., 2014; IPCC, 2014). This operation is carried out to provide fundamental information to be systematized with that described previously to understand which main socio-economic and territorial sectors are exposed to impacts.

The data used come from different information levels made available by the administrations, both at the municipal and supra-municipal scales. These are largely based on the different levels of sector planning which, at different scales and objectives, provide a knowledge base. Often this information forms part of the exploratory documents of the territory, which structure the cognitive framework on which the public administration bases its decisions. For these reasons, the detail is greatly affected by the lack of homogeneity of the processing scales (for example from the basin scale to the municipal scale). To complete these reconnaissance studies, further analysis[3] was conducted on urban flooding (Urban Flooding) and heat waves (Urban Heat Island).

In the specific Venetian territory, water management is historically crucial. The reclamation, management, and regulation of the water have allowed over time to increase the cultivable surfaces and agricultural production but have made the area bordering the lagoon particularly fragile and vulnerable to flooding of the river bodies typical of a delta area (too much water) and drought (little water), as can be seen with a progressive trend in recent years. The co-presence of particular weather conditions and high tides is capable of altering the normal flow of water at the mouth, causing the risk of flooding for a large part of the eastern territory of the Metropolitan City of Venice. The study of the climatic dynamics of recent years, together with the analysis of extreme climatic events that have occurred in the area, confirm the intensification in terms of frequency and intensity of extreme events of this type (peaks of large quantities of water and almost absolute).

The awareness of territorial fragility, revealed by the increase in unusual extreme phenomena, is leading to a radical change of approach in dealing with the problem and the multiple repercussions on the territory.

As regards the urban heat island (UHI) phenomenon, within the territory of the Metropolitan City of Venice, a geographical characterization can be seen. The northern part of the lagoon basin, characterized by medium-sized settlements interspersed with large rural areas and the flow of rivers, can present phenomena localized in urban centers, such as in the center of San Donà (Magni et al., 2020). Even the southern part, mainly agricultural and flat, presents slight heat island phenomena

[3] The analysis can be viewed in full in the documents of research line 5.2.1.

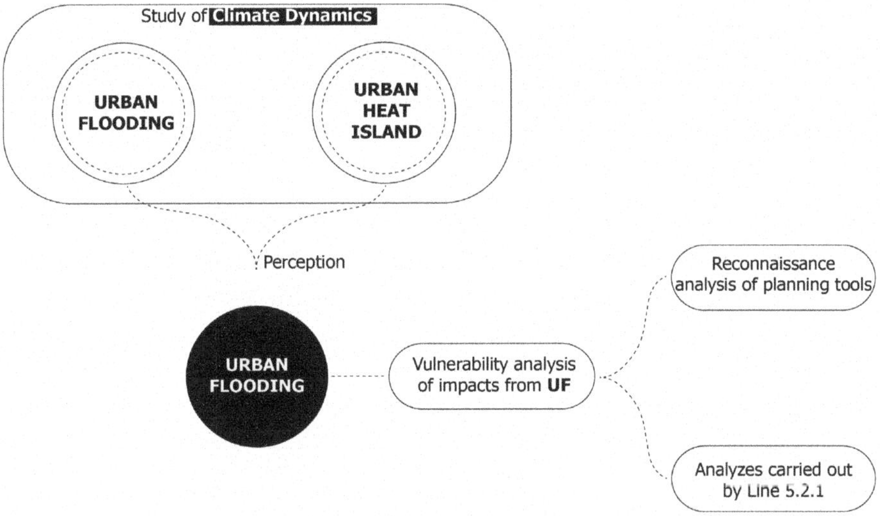

Fig. 6.5 Study of climate dynamics: risk perception. The figure shows the process of studying climate dynamics, in detail the first step consists of the study related to urban flooding (UF) and urban heat islands (UHI). Common perception drives the research to focus mainly on urban flooding and vulnerability analysis of impacts related to it

localized in the inhabited centers. The coastal municipalities, despite having a high building density, can count on the mitigating effect linked to the winds and breezes, inherent in the coastal system. This was also confirmed by the perceptions of the municipal bodies involved, who do not perceive this phenomenon (UHI) as relevant compared to the peculiarities of their socio-economic contexts (see Fig. 6.5).

On the contrary, the study conducted on urban flooding phenomena (UF) for each municipality was taken into greater consideration based on a common perception of the criticality (extreme weather events, flooding …). Considering the analyses carried out, confirmed by meetings with municipal administrations, the recent tangible disasters in the territories studied play a decisive role in the perception of the phenomena.

For these reasons, particular attention was paid to the analysis of the vulnerability of impacts from UF. For each municipality, the identification of the most vulnerable areas occurred through a comparison between the documents present in current plans and regulations and the analyses produced by research line 5.2.1:

- Reconnaissance analysis of planning tools;
- Analyzes carried out by Line 5.2.1 as part of Venice 2021.

The reconnaissance analysis part of the planning tools (see previous chapters) has placed its focus on those which in terms of theme, objective, and purpose are most similar to the theme of adaptation to climate change.

The planning and management of water resources, specifically, is fragmented into different local planning tools:

the basin scale in which the hydrogeological cycle is analyzed, the areas subject to hydrogeological instability and hydraulic danger;

the areas subject to continuous and mechanical management of the water managed by the reclamation consortia;
the inter-municipal or Municipal Territorial Planning Plan identifies in greater detail the floodable areas with various return times and the perimeter of flooding in particular catastrophic events;
the Water Plan of each municipality, promptly, regulates and regulates changes, transformations, and adaptations of the municipal territory.

The availability of these tools varies among the municipalities involved, making it necessary to adapt the individual calculations to the planning context of the municipalities. In summary, the various planning and regulatory tools examined are:

- Hydrogeological Management Plan (PAI)
- General Management Plan of the Reclamation Consortium
- Intermunicipal Territory Planning Plan (PATI)
- Municipal Territorial Planning Plan (PAT)
- Municipal water plan (PAc).

The vulnerability study consists of a framework of general and specific mapping for each municipality involved. The results of this investigation are very interesting and functional to the design choices of this plan, as well as an information basis for further considerations and research.

6.3 Innovations for a Large-Area Adaptation Tool: Key Factors for Systemic Integration

A clear and transparent policy, together with strong support from all stakeholders, is essential to put the climate issue at the top of the political agenda and thus to promote adaptation processes. Likewise, a solid scientific knowledge base plays a very important role throughout the entire process, requiring the participation of researchers and experts from the early stages. Overall, a new form of governance for climate action is needed based on the involvement of all actors, long-term commitment and vision, transparent and inclusive processes, and effective communication to the whole of society. Some factors are decisive for an adaptation process: vertical and horizontal integration; the involvement of interested parties; and the sources of financing (see Fig. 6.6).

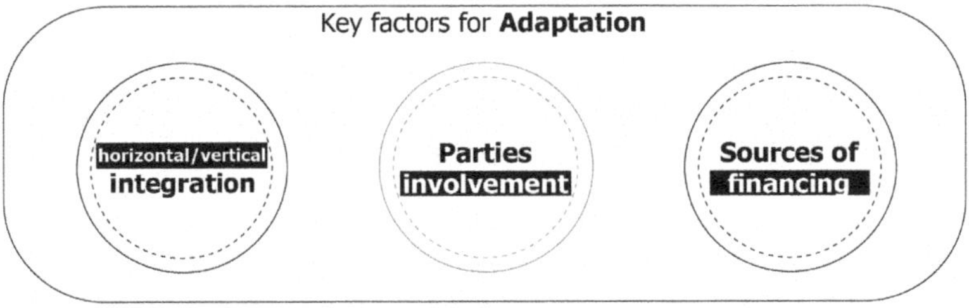

Fig. 6.6 Key factors for climate adaptation. These factors are decisive for an innovative adaptation process

6.3.1 Operational Verticality and Synergistic Horizontality for Integral Adaptation

Adaptation is a transversal and multi-sectoral process. Lack of integration is often a major obstacle. The concept of vertical integration highlights the importance of improving coherence between strategies and plans that deal with different spatial scales. The strategic choices adopted at national and subnational levels guide the measures implemented at the local level, while the different local actions must become part of a coherent framework that contributes to common objectives (Rauken et al., 2015). It must be developed hand in hand with horizontal integration, looking at the different sectors involved in the adaptation process through a common lens. Horizontal integration is essential for designing and implementing adaptation measures that are beneficial to more than one sector (as well as creating synergies), not generating problems for a specific sector and providing multiple benefits. However, the involvement of government institutions alone is not sufficient to make progress towards a new form of governance for climate action. The integration of disciplines is equally essential. Climate change resilience and adaptation actions are complex challenges that require the contribution of experts from different sectors and disciplines. The concept of integration is closely linked to that of stakeholder involvement. Given its cross-cutting nature, a wide range of stakeholders should be invited to contribute to the cogeneration of knowledge and co-definition of the key components of an adaptation plan (key vulnerabilities, vision, objectives, measures, tools, monitoring mechanisms, etc.).

6.3.2 Involvement and Interaction with Interested Parties

Stakeholder involvement is fundamental in all phases of adaptation. Involvement is specific to the context concerned and is important for: sharing data, and information on climate change risks, prioritizing impacts; co-creating the vision and defining related goals; sharing experiences on adaptation options, to contribute to the selection of specific measures; monitoring and evaluation planning. Stakeholders can also be active actors in the implementation of some of the envisaged measures, helping to monitor the implementation of the adaptation plan and disseminate the results of the process. Stakeholder mapping and analysis aims to identify who should be involved in the adaptation process, which is useful for understanding the interests and positions of stakeholders and their relationships with each other. The selection of interested parties should be based on transparent criteria and aim to ensure a balanced participation of the different categories (delegated administrations, departments or divisions of the same organization, sectoral authorities, academia, research institutes, NGOs, other interest groups, and representatives of sectors of activity). Defining roles and responsibilities for each stakeholder from the beginning is essential to increasing the success of the engagement.

Along with prioritization, stakeholders need to be regularly informed about the use of the contributions they provide. This can improve trust and create a sense of ownership during the process. Indeed, adaptation should be seen as a co-designed path and the resulting strategy or plan should be presented as a co-developed outcome for all stakeholders involved. Stakeholder engagement should be supported by communication and awareness-raising activities, which are essential to ensure more widespread public support for the process, to further strengthen public participation, and to raise awareness of the challenges of climate change (clarifying terminology and a shared dialectic; model the messages transmitted in a user-oriented way, communicate, and also share hypotheses, uncertainties, as well as examples and good practices).

An inclusive and integrated approach also recognizes the importance of cross-border cooperation (of two or more territorial entities sharing a common border, be they municipalities, regions, or nations). Cross-border cooperation aims, first, to improve coherence and synergies between strategies and adaptation plans, leading to the common management of a "shared territory", as in the case of the lagoon basin. This form of cooperation is very useful for the development and exchange of specific regional knowledge, tools, and good practices, contributing to the resolution of critical issues and the implementation of shared measures. However, jointly solving regional problems is a long process, requiring time, dedicated resources, and a long-term vision on a regional scale.

6.3.3 The Adaptation Process Between Local Self-financing and Multi-scalar Economic Opportunities

Finally, the timely and detailed identification of funding sources is essential to fuel the entire process and ensure that the necessary resources are available, both during the design phase and during the implementation phase (see Fig. 6.7). Funding opportunities and mechanisms can include a range of EU funding programs, such as LIFE and Horizon Europe which help support adaptation at sub-national and local levels. Added to these are the European Structural and Investment Funds (ESIF) which

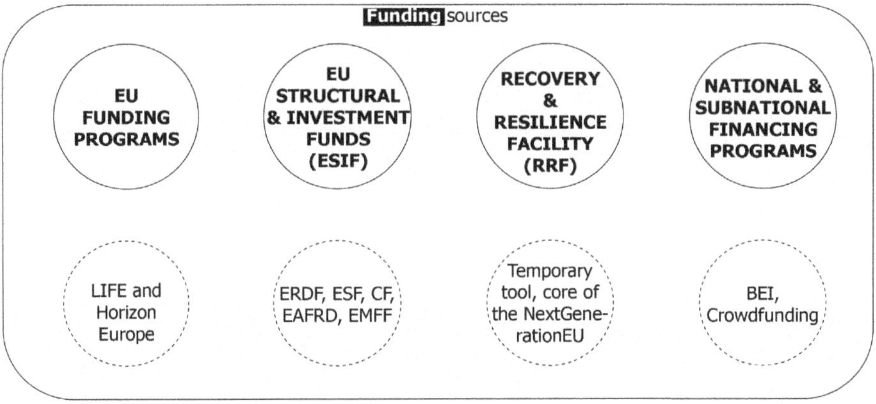

Fig. 6.7 Founding sources. Detailed identification of funding sources is essential to fuel the entire process and ensure that the necessary resources are available

include the European Regional Development Fund (ERDF), the European Social Fund (ESF), the Cohesion Fund (CF), the European Agricultural and Development Fund Rural (EAFRD), the European Maritime and Fisheries Fund (EMFF). Furthermore, to mitigate the economic and social impact of the Coronavirus pandemic, the Recovery and Resilience Facility is the EU financial instrument aimed at making European economies and societies more sustainable, resilient, and better prepared for green and digital transitions. At a different level, national and subnational financing programs, initiatives promoted by the private sector, loans from financial institutions (such as the European Investment Bank), or crowdfunding, support greater resilience, or a more general green transition.

6.4 The Project Framework for the Territories of the Venice Lagoon: Strategies, Objectives, Actions, and Measures

6.4.1 Adaptation Strategies and Objectives

Adaptation to climate change begins with a city's ability to adequately respond to environmental shocks and stresses. The Adaptation Plan proposes an analysis of the main shock and stress factors linked to climate change which affect, as anticipated in the cognitive framework, the study area included within the boundaries of the last lagoon boundary in force, decreed in 1990 by the Ministry of Works Public (see Fig. 6.8). 11 municipalities have been identified: Venice, Campagna Lupia, Cavallino-Treporti, Chioggia, Dolo, Jesolo, Marcon, Mira, Musile di Piave, Quarto D'altino, San Donà di Piave.

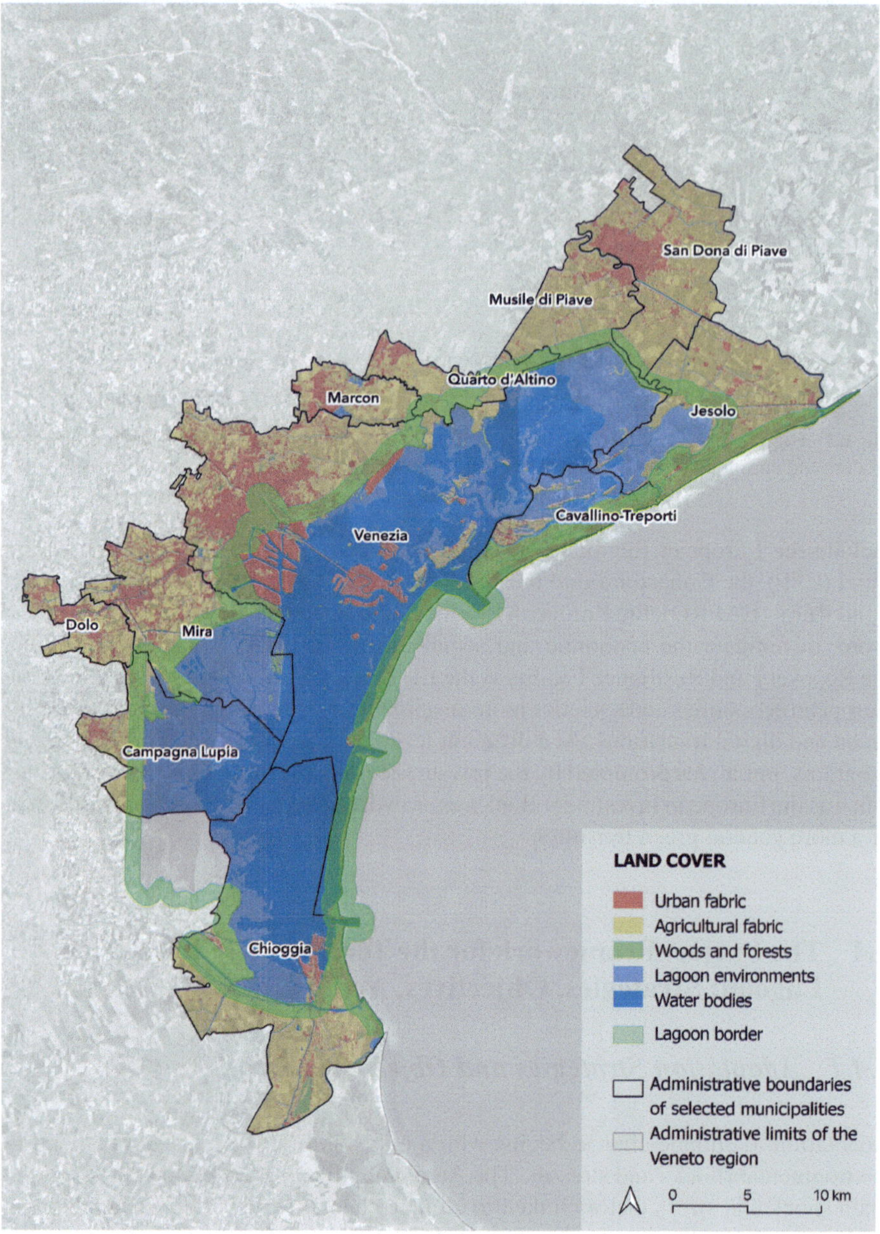

Fig. 6.8 Territorial framework of the lagoon basin municipalities. The map shows the land use (2020) of the study area, which includes the municipalities falling within the boundaries of the last current lagoon perimeter, decreed in 1990 by the Ministry of Public Works

Flooding events are linked to the occurrence of more intense rainfall in terms of quantity and frequency in correspondence with the hydrogeological configuration present on a larger scale. The safety measures carried out locally become the key to understanding the vulnerability of the different portions of the area throughout the territory. The phenomenon of water runoff from urban areas is one of the main issues linked to the risk of flooding (Jabareen, 2013). The problems arise from the degree of impermeability, where in natural environments rainwater is slowly washed away and filtered from and through the ground, while in urban environments impervious surfaces favor rapid washing away towards the receiving bodies. Urban sewage systems, in times of intense rainfall, receive water full of sediments and solid material, which in addition to reducing the volume of runoff, accumulates a large quantity of pollutants in the water. The occurrence of extreme meteorological phenomena increases the flow of water towards the receiving bodies, causing temporary flooding in the urban fabric and polluting the aquifers of the drainage basin (Arnbjerg-Nielsen et al., 2013). Climate change, therefore, places urban drainage systems in a condition of inefficiency during extreme rainfall events and the problem takes on a priority character as extreme rainfall events are destined to intensify (Lerer et al., 2015). To reduce urban flooding phenomena and protect the quality of surface and groundwater, it is necessary to limit runoff and reduce the polluting loads deriving from the urban run-off phenomenon. The Urban Run-off study, conducted by research line 5.2, returns the chronological study of the post-event safety interventions. In this sense, the presence of an inherent fragility in the territory is underlined regardless of the severity of the phenomenon that occurred. The level of information acquired refers to the reports received by the Fire Brigade since 2007. The information, originally collected promptly, recording the place and date of intervention, was summarized using a geo-database, through the provision of a regular hexagonal mesh with a 600 m side. The result shows a count of interventions distributed over each portion of the mesh (see Fig. 6.9).

The main objective of the adaptation plan is to structure methodologies and practices capable of capturing the changes induced by climate change, and consequently identifying actions aimed at adaptation, impact mitigation, and risk prevention of the entire lagoon area and the first interface strip with the territory of the Metropolitan City of Venice, building specific intervention strategies for these areas. The return of knowledge to the various actors will be a necessary point of arrival of the research to provide methods and tools for sustainable and long-lasting maintenance and management of the Lagoon system in all its components (Table 6.1; Fig. 6.10).

6.4.2 Adaptation Actions and Measures

The analysis of the lagoon and the metropolitan city of Venice which took place in the first part of this work and the delineation of a cognitive framework made it possible to facilitate the definition of adaptation measures concerning the specific socio-economic and morphological vocations of the territory. Adaptation must necessarily

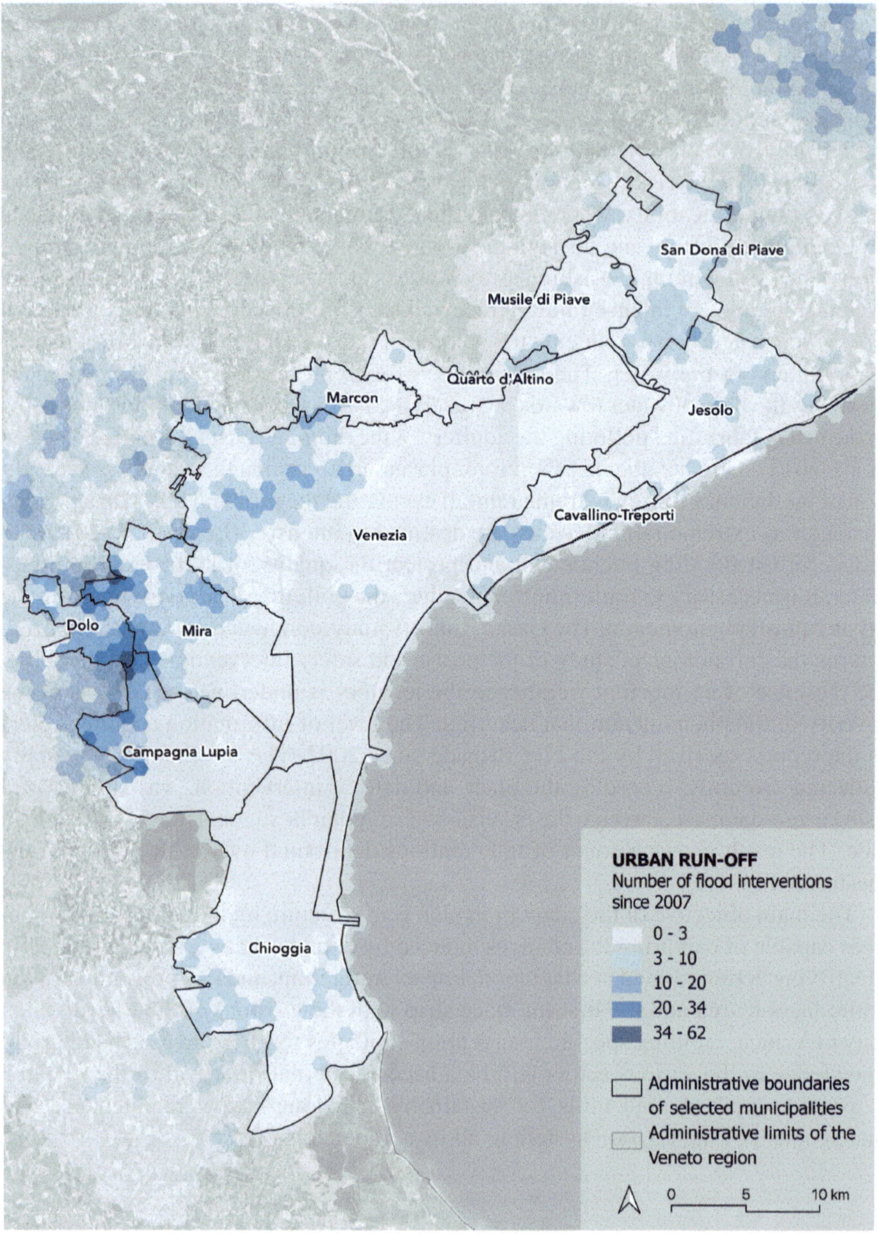

Fig. 6.9 Number of interventions due to flooding since 2007 in the lagoon basin municipalities. *Data source* Venezia2021 "Line 5.2.1—Study of environmental-climate vulnerabilities", Scientific Line Responsibles for IUAV: Denis Maragno

Table 6.1 Summary data of the Lagoon basin municipalities

Municipalities	Surface	Inhabitants	Population (density/km^2)
Campagna Lupia	87.59	7243	82.69
Cavallino-Treporti	44	12,897	290
Chioggia	185	49,744	270
Dolo	24.08	15,188	630
Jesolo	95	26,419	280
Marcon	25.58	17,536	690
Mira	99.14	38,487	388.21
Musile di Piave	44.87	11,473	260
Quartino D'altino	28.33	8.110	290
San Donà di piave	78.73	41,856	530
Venezia	414.57	260,897	630
	1,126.86	**489,850**	**434**

have as its starting point the needs of the territory in which climate changes will produce their effects.

The abacus proposes a series of measures that can be identified and selected based on the specific need required, such as the socio-economic implications or the impact it is called upon to respond to (see Fig. 6.11). This will be able to facilitate municipal technicians and public administrators in adopting measures that can adapt their territories to the pressing need to respond adaptively to the meteorological and climatic changes that are increasingly affecting cities and territories.

This abacus can be queried based on the specific needs of the territory, as the zoning classification to which the specific measure can be developed is also present. There are also other suggestions, such as: limits/advantages that the specific measure presents, whether it is a temporary or fixed measure, and the SDGs to which they respond. These specifications and those explained subsequently will be able to better guide administrators and technicians in identifying the measures most suited to the needs that the territories present. The abacus intends to spatialize the measurements identified in a place depending on:

Socio-economic implications

- *Sensitive family units*: includes all those measures that support family units which, due to density, age, and quantity, are particularly sensitive to potential shocks and stress.
- *Insufficient average disposable income per capita*: includes measures that act to support and relieve the costs borne by those who have an insufficient average disposable income per capita.
- *Absolute poverty*: measures that contribute to alleviating situations of poverty (such as food insufficiency, energy poverty, etc.).

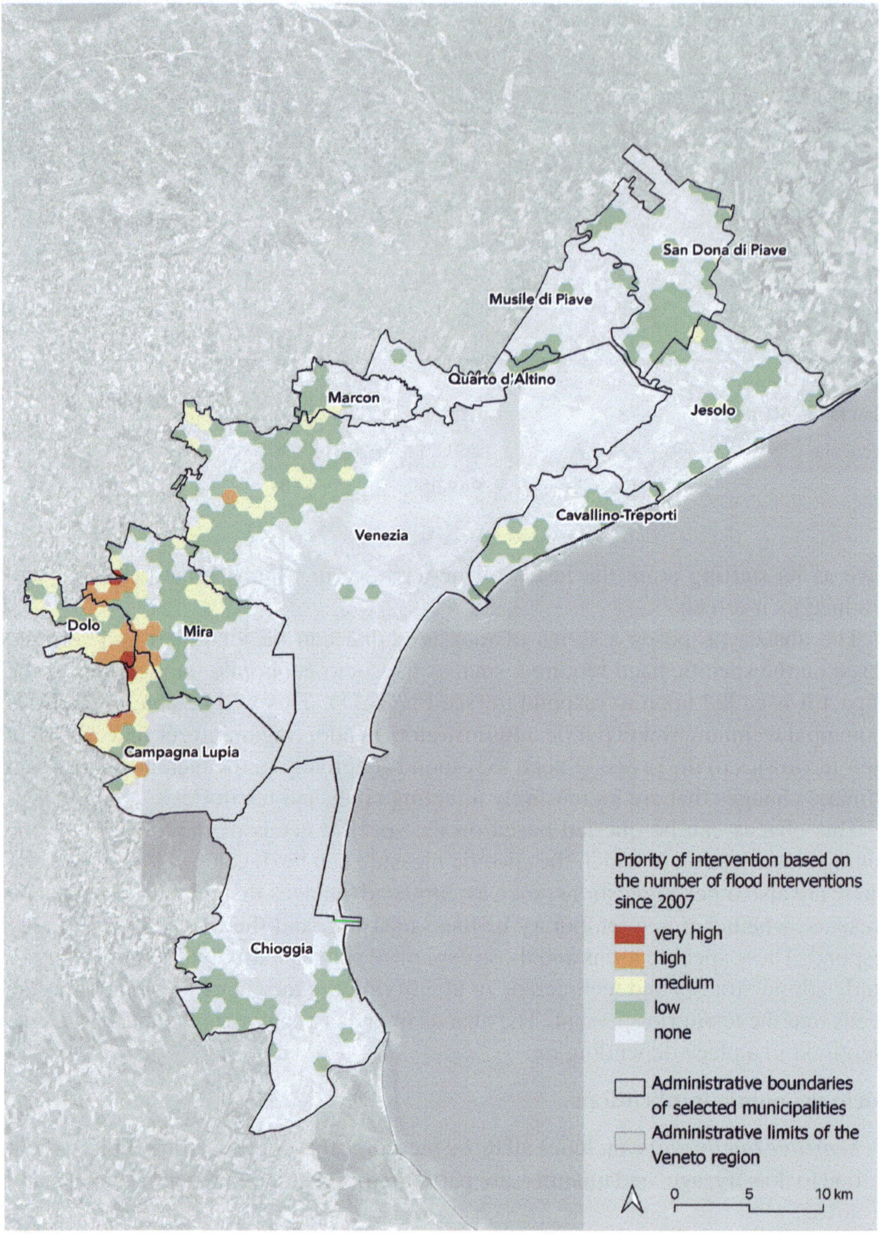

Fig. 6.10 Priority of intervention based on the number of interventions due to flooding since 2007* of the lagoon basin municipalities. **Data source* Venezia2021 "Line 5.2.1—Study of environmental-climate vulnerabilities", Scientific Line Responsibles for IUAV: Denis Maragno

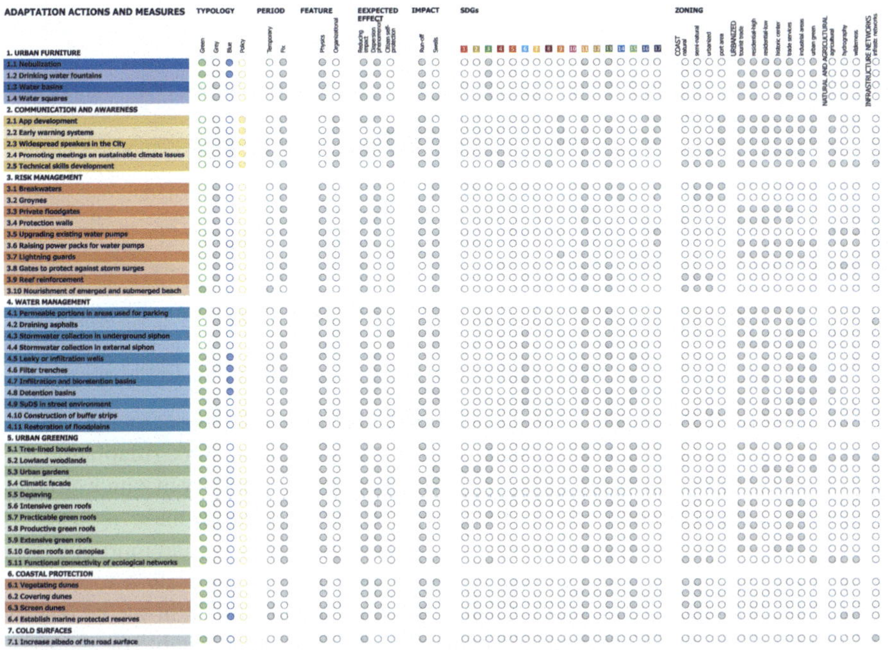

Fig. 6.11 Summary diagrams of the actions and adaptation measures. The abacus presents a range of adaptation measures that can be selected based on the specific need required, such as the socio-economic implications, the impact being addressed, the limitations/benefits that the specific measure presents, whether it is a temporary or fixed measure, and the SDGs it addresses. In addition, there is also the zoning classification based on which the specific measure can be developed

- *Low quality of housing*: measures that contribute to improving the physical-technical and aesthetic qualities of the home.
- *Quality of Public Space*: measures that contribute to improving the aesthetic and usability quality, but also the psycho-physical well-being of public space.

Impact to which it responds

- *Run-off/Urban flooding*: a phenomenon linked to the risk of flooding whose problematic nature arises from the degree of impermeability of the soil (storm surges).

Expected effect

- *Reducing impact.*
- *Dispersion of the phenomenon*: this is a spatial intervention capable of letting an event vent without having a serious effect on the continuity of urban life.
- *Citizen self-protection*: measures capable of suggesting and stimulating inhabitants to protect themselves and their material assets.

Typology

- *Green*: measures that use vegetal elements such as trees, lawns, hedges, parks, fields, forests, etc.
- *Grey*: Engineering solutions that use elements such as buildings, roads, and other urban constructions.
- *Blue*: measures that make use of water elements such as rivers, canals, ponds, wetlands, floodplains, water treatment plants, etc.
- *Policy*: strategic tools, economic incentives, public policies, etc.

Characteristic

- *Physics*: interventions that can be implemented with tangible modifications (of the building, of the urban form, etc.).
- *Organizational*: interventions that do not involve modifications to the space.

Socio-economic exposure translates, especially in urban environments, into a lower response capacity and resilience to shock and stress situations. Adaptation, by its nature, must necessarily refer to the needs localized in the territory in which climate changes will produce their effects, for these reasons, in order to make it easier to define adaptation measures with respect to the specific vocations and morphologies of each area, from the abacus of actions and measures for adaptation to climate change, specific measures suitable for the local level are identified, in such a way as to direct the technicians of the Municipalities to adopt the most suitable measure for the prerogative of the territory.

Adaptation must necessarily refer to the needs localized in the territory in which climate changes will produce their effects, for these reasons, in order to make it easier to define adaptation measures with respect to the specific vocations and morphologies of each area, from the Abacus of actions and measures for adaptation to climate change, specific measures suitable for the local level are identified, in such a way as to direct the technicians of the Municipalities to adopt the most suitable measure for the prerogative of the territory.

Therefore, following the study of the actions and measures of adaptation to climate change, we moved on to the identification of the areas suitable for the inclusion of these, and therefore to the zoning based on the specific morphological characteristics of the lagoon basin and the system of the Metropolitan City of Venice (see Fig. 6.12).

First, the basic intervention priority of the studies conducted by research line 5.2 was defined, which outlines the most fragile areas of the territory. The territory subjected to greatest vulnerability was divided into classes (5) representing the priority of intervention (none, low, medium, high, very high) based on the number of interventions due to flooding since 2007 on a regular hexagonal grid with a side of 600 m. The result is therefore based on the count of interventions distributed on each portion of the mesh in relation to the intervention priority (the higher the number of interventions due to flooding since 2007 in the lagoon basin municipalities, the higher the intervention priority will be).

The methodology subsequently used to identify the areas suitable for the inclusion of actions and measures to adapt to climate change made use of the Land Cover map referring to the year 2020 and to the third level of detail of the same paper.

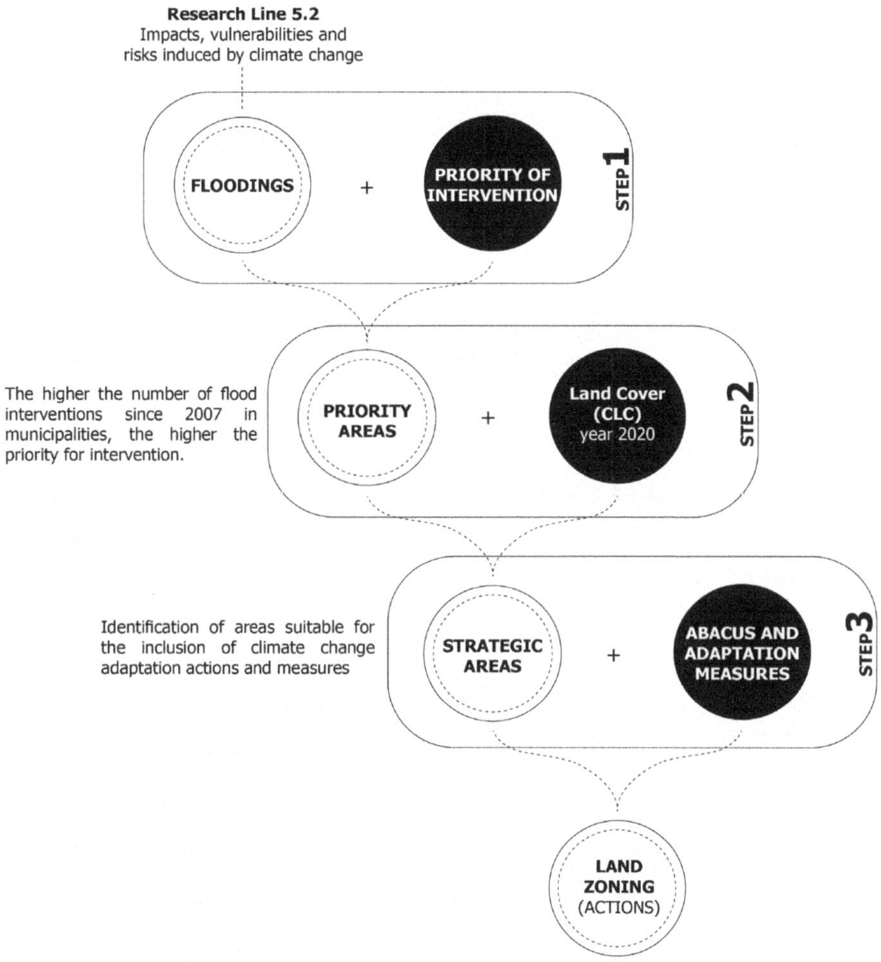

Fig. 6.12 Methodological scheme for the zoning of the territory. The methodology is divided into 3 steps. After defining the intervention priority based on the studies conducted by research line 5.2, which outlines the most fragile areas in the territory, the territory was divided into intervention priority classes based on the number of flood interventions since 2007. The methodology subsequently made use of the land cover map, and each level of detail was associated with a macroclass of actions to obtain the land zoning

A macro-class of actions was associated with each level of detail, based on both the level of priority and the prerogatives of the land, to obtain the zoning of the territory (see Fig. 6.13). The macro-classes of shares identified are:

- Type 1: street furniture
- Type 2: communication and awareness raising
- Type 3: risk management

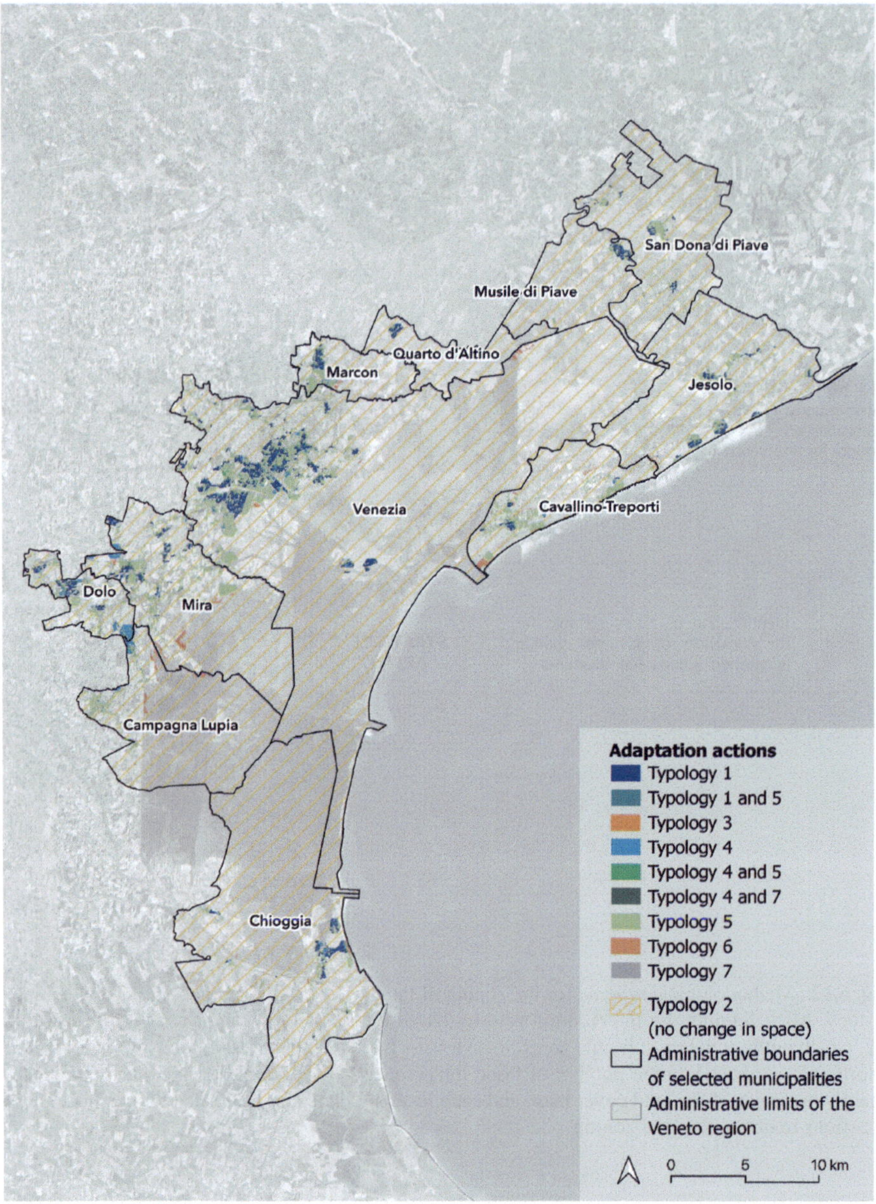

Fig. 6.13 Zoning relating to the identification of areas suitable for the inclusion of actions and measures to adapt to climate change. *Source* Federico K., personal elaboration for the drafting of the plan

- Type 4: water management
- Type 5: urban greening
- Type 6: coastal protection
- Type 7: cold surfaces

As regards typology 2: communication and awareness-raising, being actions of an organizational nature, therefore interventions which do not involve modifications of the space, no inclusions in the cartography have been foreseen.

6.5 Planning Adaptation in the Complexity of the Lagoon: Reflections and Perspectives

The research path that led to the drafting of the Adaptation Plan moved from a consideration/presupposition made explicit from the beginning: that urban planning and the dynamics of territorial governance must deal with the now inevitable consequences due to change climate. This relationship inevitably orients the change that local administrations that insist on and relate to the Venice Lagoon are now called upon to face toward the quality and improvement of the population's living conditions. Climatic imperatives intervene in this complexity by adding tension, threatening the balance, and increasing the vulnerability of this "system" already widely subjected to stress. The plan outlines in a clear and direct manner what the future involvement of urban planning processes must be, both on a single local scale and on an integrated lagoon-based system, the latter levels which local government initiatives have so far neglected (or left to voluntary and punctual actions) the relationship between climate and territorial planning. The initiatives implemented so far, despite confirming a stance on the part of some subjects wanting to undertake new paths, have not however led to adequate political responses, both in qualitative terms (types of instruments and policies) and quantitative (extension of involvement to the population or to the main local and supra-local economic systems).

Among the results highlighted by these first initiatives of the administrations of the lagoon basin, there is certainly the recognition of the need to overcome the specificity of partial planning, exclusively aimed at energy consumption, often without a real relationship with mandatory urban planning. The main reasons for this can be traced back to a lack of public and shared awareness of climate variability and its territorial repercussions, a late response to climate disasters due to a lack of capacity and resources, and a lack of public policies and urban and environmental planning regulations designed to manage climate change. However, if we look beyond these limits, we can then recognize the potential that resides within the territory of the Metropolitan City of Venice: the latter in fact, thanks to its broad thematic experience and its ever-increasing role in supporting local processes of climate planning, has the ability to contribute to reducing the causes of climate change (mitigation) and to effectively protect itself from expected local impacts (adaptation).

There are several aspects in which the adaptation plan addresses the challenges of climate change in the Venice Lagoon and highlights the importance of integration, coordination, and adaptability in spatial planning. In particular, the adaptation plan must be integrated into any pre-existing administrative plan or strategy to make climate change a central issue in spatial governance. It is important to coordinate climate initiatives with other environmental strategies already in place to avoid overlapping or conflicting responsibilities and to consider the various temporal and administrative scales in spatial planning, ensuring a coherent approach that balances local needs with those on a broader scale and turning current environmental challenges into development opportunities.

The Adaptation Plan designed within the research path "Venice 2021—Scientific research program for a regulated lagoon", together with the experiences carried out by the territorial bodies that relate at different scales with the Venice Lagoon (municipal administrations, consortia reclamation bodies, basin bodies, metropolitan cities, etc.) provides fertile ground from which to take inspiration to try to outline a roadmap for the future of the lagoon territory, imagining this tool from different aspects.

The plan as a catalyst for mainstreaming processes of adaptation

For climate change mitigation and adaptation to become structuring and structural development processes for the lagoon area, they must be incorporated or added into every plan or strategy. This will have to affect all the instruments of the administrations belonging to the lagoon area, thus making the issue of climate change mainstream within the territorial governance process. The strategic and operational proposals presented by the plan both for the entire lagoon area and for each local government thus propose new policies and actions that fit into an existing framework. These proposals sometimes initiate new lines of implementation and, very often, intervene with directions on existing plans or programs: variations to building codes, variations to traffic plans, variations to energy plans, etc.

These strategic indications therefore offer local government bodies a first topic on which to discuss to undertake (in case of complete absence) or strengthen (in case of paths already started) an ever-greater consideration of the climate issue within the planning dynamics territorial.

The plan as an example of thematic integration and operational interdisciplinarity

The Adaptation Plan, like many other plans and policies for climate protection, fits within the existing framework of management and planning policies of public administrations present in the lagoon area. Often this type of plans—remaining in their nature of voluntary instruments—are part of a consolidated context made up of multiple initiatives that touch on the topic of energy and environmental sustainability directly. If these processes (often very heterogeneous) are each carried out independently, they tend to weaken, generating an inefficient use of internal resources within local administrations, if not even entering into open conflict (where, for example, an overlap of competencies is created). Precisely to avoid these externalities, the Adaptation Plan considers it a priority to coordinate climate initiatives with the framework of other environmental protection strategies/plans that are already in place (energy programs, sustainability programs, etc.), both at local and macro-regional.

The plan as a good inter-scalar planning practice

There is no doubt that the ordinary and voluntary instruments of territorial governance and local climate and sustainability policies are directly dependent on the institutional and political conditions of governance at the higher level. Many of the initiatives built by local governments, however, are developed without an integrated territorial planning framework, aiming toward short-term feasibility, especially those for which it is relatively simple to make decisions and have direct control. On the contrary, long-term perspectives, such as land use and transport planning, find it difficult to integrate into ordinary planning tools. For the adaptation plan, the issue of scale, both in temporal terms (as short-term and/or long-term planning of actions) and above all in territorial-administrative terms, represents one of the aspects to which adequate attention must be paid in the future. If some policies or measures, especially those regarding the technological quality of systems and buildings, have a predominantly a-territorial character, urban planning choices regarding land use, density, and mobility systems, key variables for energy efficiency and territorial resilience cannot be placed indifferently under the control of a single local administrative entity. For this reason, the achievement of these objectives in the lagoon context required, during the drafting and will require during future implementation, an approach capable of always maintaining a strong link between the scales of intervention, allowing individual local administrations to govern their specific emergencies and critical issues, without losing the strategic coherence and balance guaranteed by the large-scale strategic system.

The plan as an exercise in methodological and operational coherence

As happens with any research path that aims to address only weakly explored themes, some results prove to be acquired, while others require more time and greater in-depth analysis to become an integral part of public discourse and operational land management processes. New challenges in many cases require new approaches to arrive at new solutions. From the operational panorama and the cognitive heritage investigated within the context of the Venice lagoon, it emerges that the territorial planning process is certainly enriching its design, lexical, and operational vocabulary. An evolution that acts on the continuity of knowledge that has traditionally influenced the governance practices of the lagoon and which today is taken up innovatively. It is, therefore, possible to state that the increasingly urgent issues regarding climate change are leading urban and territorial planning to a partial update of the disciplinary apparatus by taking advantage of opportunities such as the research path "Venice 2021—Scientific research program for a regulated lagoon". The originality of the approach used to formulate the Adaptation Plan lies precisely in the conversion of environmental risk into opportunities for the transformation and development of the Vast area.

In other words, it appears that today's environmental crisis offers the opportunity to take a different point of view, a more careful and far-sighted look at the processes and related consequences that concern urban and territorial transformations while respecting those balances that make the lagoon "an environment of man".

From the experiences investigated by the research, in addition to the possibility of expanding the concept of climate planning, it was also possible to reflect critically on the use of emergency planning practices (5.3.1.2) and decision support practices (M 5.3.1.3), both for research purposes as well as for territorial planning practice. Learning from practices means, by inductive process, extrapolating key concepts to reformulate methodological approaches, and recognizing the potential for innovation inherent in these processes. In turn, these theoretical-methodological advances allow them to be used again to direct and support the innovation of other practices, thus qualifying the planning process as circular, flexible, and adaptive.

Acknowledgements Thanks are due to the working group of research line 5.3 with Scientific Responsible Prof. Anna Saetta and in particular to the members of line 5.3.1 who contributed directly to the work carried out and described in the chapter composed by Giulia Lucertini, Filippo Magni, Gianmarco Di Giustino, Katia Federico and Giovanni Carraretto. Thanks are also due to the components of research line 5.2, with Scientific Responsible Prof. Andrea Critto (UniVE), for the materials useful to build the basis on which to carry out the plan and in particular Prof. Denis Maragno and Gianfranco Pozzer (Responsible for IUAV of the research line 5.2.1).

References

Arnbjerg-Nielsen, K., Willems, P., Olsson, J., Beecham, S., Pathirana, A., Bülow Gregersen, I., Madsen, H., & Nguyen, V. T. V. (2013). Impacts of climate change on rainfall extremes and urban drainage systems: A review. *Water Science and Technology*. https://doi.org/10.2166/wst.2013.251.c

Bezzi, A., Boccali, C., Calligaris, C., Colucci, R. R., Cucchi, F., Finocchiaro, F., et al. (2015). Impatti dei cambiamenti climatici sul territorio fisico regionale. Studio sullo stato di fatto concernente la conoscenza d'insieme del territorio fisico regionale per la valutazione degli impatti dovuti ai cambiamenti climatici. Retrieved from: http://www.regione.fvg.it/rafvg/export/sites/default/RAFVG/ambiente-territorio/geologia/FOGLIA22/allegati/Impatti_dei_cambiamenti_climatici_sul_territorio_fisico_regionale.pdf

Bockel, L. (2009). How to mainstream climate change adaptation and mitigation into agriculture policies, FAO. Retrieved from: http://www.fao.org/fileadmin/templates/ex_act/pdf/ppt/Climate_Change_and_Agricultural_Policies_ppt.pdf

Comune di Padova. (2016). Padova Resiliente: Linee Guida per la costruzione del Piano di Adattamento al cambiamento climatico, Padova.

Euro-Mediterranean Center on Climate Change—CMCC. (2021). Scenari climatici per l'Italia. Retrieved from: www.cmcc.it/it/scenari-climatici-per-litalia

European Environment Agency. (2016). Climate change, impacts and vulnerability in Europe 2016. Retrieved from: https://www.eea.europa.eu/publications/climate-change-impacts-and-vulnerability-2016

Fritzsche, K., Schneiderbauer, S., Bubeck, P., Kienberger, S., Buth, M., Zebisch, M., & Kahlenborn, W. (2014). The vulnerability sourcebook: Concept and guidelines for standardised vulnerability assessments; Deutsche Gesellschaft für Internationale Zusammenarbeit (GIZ) GmbH: Bonn, North Rhine-Westphalia, Germany. Retrieved from: https://www.adelphi.de/de/publikation/vulnerability-sourcebook-concept-and-guidelines-standardised-vulnerability-assessments

Gaudioso D., Giordano F., & Taurino, E. (Eds.) (2014). Qualità dell'ambiente Urbano X Rapporto Edizione 2014 Focus su Le Città e la Sfida dei Cambiamenti Climatici, ISPRA—Istituto Superiore per la protezione e la ricerca ambientale. Retrieved from: https://www.isprambiente.gov.it/files/pubblicazioni/statoambiente/

IPCC. (2014). AR5 climate change 2014: Impacts, adaptation, and vulnerability; chapter 14.

Jabareen, Y. (2013). Planning the resilient city: Concepts and strategies for coping with climate change and environmental risk. *Cities, 31*, 220–229. https://doi.org/10.1016/j.cities.2012.05.004

Keskitalo, E. C. H. (2010). Introduction–adaptation to climate change in Europe: Theoretical framework and study design. In *Developing adaptation policy and practice in Europe: Multi-level governance of climate change* (pp. 1–38). Springer International Publishing AG, Dordrecht. https://doi.org/10.1007/978-90-481-9325-7_1

Laukkonen, J., Blanco, P. K., Lenhart, J., Keiner, M., Cavric, B., & Kinuthia-Njenga, C. (2009). Combining climate change adaptation and mitigation measures at the local level. *Habitat International, 33*(3), 287–292. https://doi.org/10.1016/j.habitatint.2008.10.003

Lerer, S. M., Arnbjerg-Nielsen, K., & Mikkelsen, P. S. (2015) A mapping of tools for informing water sensitive urban design planning decisions—questions, aspects and context sensitivity. Water, 7(3), 993–1012. https://doi.org/10.3390/w7030993

Magni, F., Musco, F., Litt, G., & Carraretto, G. (2020). The Mainstreaming of NBS in the SECAP of San Donà di Piave: The LIFE master adapt methodology. *Sustainability, 12*(23), 10080. https://doi.org/10.3390/su122310080

Magni, F. (2019). *Climate proof planning: L'adattamento in Italia tra Sperimentazioni e Innovazioni* (pp. 1–245). Franco Angeli. ISBN 9788891797209.

Maragno, D., dall'Omo, C.F., Pozzer, G., & Musco, F. (2021). Multi-risk climate mapping for the adaptation of the Venice Metropolitan Area. *Sustainability, 13*, 1334.https://doi.org/10.3390/su13031334

Maragno, D., Litt, G., Ferretto, L., & Gerla, F. (2022). Abaco per la transizione climatica. Primo catalogo per pianificare l'adattamento nell'Alto Adriatico. Anteferma Edizioni. ISBN 9791259530219.

Ministero dell'Ambiente e della Tutela del Territorio e del Mare. (2014). SNACC—Strategia Nazionale di Adattamento ai Cambiamenti Climatici. Retrieved from: https://www.minambiente.it/notizie/strategia-nazionale-di-adattamento-ai-cambiamenti-climatici-0

Ministero dell'Ambiente e della Tutela del Territorio e del Mare (2017). PNACC—Piano Nazionale di Adattamento ai Cambiamenti Climatici. Retrieved from: https://www.minambiente.it/sites/default/files/archivio/allegati/clima/pnacc.pdf

Musco, F. (Eds.) (2016). Counteracting urban heat island effects in a global climate change scenario. Springer International Publishing AG. ISBN 9783319104249.

Musco, F., Maragno, D., & Litt, G. (2018). Abaco di azioni di adattamento ai cambiamenti climatici. Verso paesaggi dell'abitare e del lavorare a prova di clima: adattamento ai cambiamenti climatici e miglioramento del comfort degli spazi pubblici e delle aree produttive nel territorio peri-urbano milanese. Fondazione Cariplo.

Rauken, T., Mydske, P. K., & Winsvold, M. (2015). Mainstreaming climate change adaptation at the local level. *Local Environment, 20*(4), 408–423. https://doi.org/10.1080/13549839.2014.880412

Sistema Nazionale per la Protezione dell'Ambiente—SNPA. (2019). Rapporto "Consumo di Suolo, Dinamiche Territoriali e Servizi Ecosistemici". Retrieved from: https://www.snpambiente.it/2019/09/17/consumo-di-suolo-dinamiche-territoriali-e-servizi-ecosistemici-edizione-2019/

UNEP/MAP. (2016). Regional Climate Change Adaptation Framework for the Mediterranean Marine and Coastal Areas, (DEPI)/MED IG.22/28., Decision IG.22/6, 19th Meeting of the Contracting Parties to the Convention for the Protection of the Marine Environment and the Coastal Region of the Mediterranean.

Chapter 7
Adaptation of Coastal Infrastructures to Climate Change: Two Examples of Touristic Facilities' Adaptation in Western Mediterranean

Jurgi Areizaga and Oliver Canosa

Abstract The Mediterranean region is a hotspot for highly interconnected climate risks with key economic sectors highly vulnerable to climate hazards (agriculture, fisheries, forestry, tourism). This reinforces the socioeconomic vulnerability of the population living in the region (Ali et al. in Climate change 2022: impacts, adaptation and vulnerability. Contribution of working group II to the sixth assessment report of the intergovernmental panel on climate change. Cambridge University Press, Cambridge, UK, pp 2233–2272, 2022). Understanding climate impacts on coastal areas is a key element for adaptation, urban planning, and economic policy. Legally, in the context of Spain, the promoters of infrastructure projects on the coastal zone are required to submit a Coastal Dynamics Study to the competent authority (National Coastal Authority). These studies must include an assessment of the potential impacts of climate change on the proposed projects, as well as potential adaptation solutions. In this paper, two case studies are presented, on one side, the study of a real estate asset (restaurant) vulnerable to coastal erosion in Tarragona that was affected by storm Celia in 2022 where authorities required a coastal dynamics study to understand future climate risks. On the other side, non-permanent tourist-related and nautical facilities in Formentera Island. The paper presents the processes followed to assess the climate risk and the results obtained to understand the needed adaptation solutions.

Keywords Climate change adaptation · Tourism · Beach · Estate assets · Spain

---

J. Areizaga · O. Canosa (✉)
TecnoAmbiente, Cádiz, Spain
e-mail: oliver.canosa@ext.tecnoambiente.com

M. Granceri Bradaschia et al. (eds.), *Climate Change Adaptation, Flood Risk, and Beyond*, Planning for Climate Proof Cities,
https://doi.org/10.1007/978-3-031-65463-3_7

7.1 Introduction: Climate Change in the Mediterranean Coastal Areas

Coastal areas are particularly vulnerable to the impacts of climate change, including sea level rise, increased frequency and intensity of extreme weather events, and ocean acidification. Sea level rise is projected to continue in the twenty-first century, with projected regional relative sea level rise within 20% of the global mean along two-thirds of the global coastline. Due to relative sea level rise, extreme sea level events that occurred once per century in the recent past are projected to occur at least annually at more than half of all tide gauge locations by 2100. This will have significant impacts on coastal ecosystems, people, and infrastructure (Intergovernmental Panel on Climate Change, 2021).

The coastal zone is currently home to around 896 million people, nearly 11% of the 2020 global population, projected to reach more than one billion by 2050 across all five SSPs. Specifically in the Mediterranean region, "densities in coastal areas have continued to increase at unsustainable rates over the last decade. Over 1965–2015, urban pressures increased in 75% of Mediterranean countries" as mentioned in the State of the Environment and Development in the Mediterranean Report from the United Nations Environment Programme in 2020. The coastal population in the Mediterranean is now close to 150 million people, doubling during the tourist season (UfM, 2016). The concentration of population in coastal zones is heaviest in the western Mediterranean, the western shore of the Adriatic Sea, the eastern shore of the Aegean-Levantine region, and the Nile Delta. Overall, the concentration of population in the coastal zone is higher in the southern Mediterranean countries.

Changes in risk distribution result from changes in the degree of the hazard, the population exposed, and the degree of vulnerability of people, assets, or ecosystems. Coastal flooding events affect many of the highly populated regions of the world where large percentages of the population are exposed. The IPCC has identified near-term projected increase of population exposed to 100-year flooding events depicted as the increase from the year 2020–2040 (due to sea level rise and population change), based on the intermediate GHG emissions scenario (SSP2-4.5) and current adaptation measures (Intergovernmental Panel on Climate Change, 2021).

As the scale and pace of sea level rise accelerates beyond 2050, long-term adjustments may in some locations be beyond the limits of current adaptation options. Higher rates of sea level rise demand earlier and stronger responses and reduce the lifetime of measures. The magnitude, rate, timing of threshold exceedances, and the long-term commitment of sea level rise depend on emissions, with higher emissions leading to greater and faster rates of sea level rise.

Climate change risks to cities, settlements, and key infrastructure will rise sharply in the mid- and long-term with further global warming, especially in places already exposed to high temperatures, along coastlines, or with high vulnerabilities. Climate change has contributed to desertification and exacerbated land degradation, particularly in low lying coastal areas, river deltas, drylands, etc. Nearly 50% of coastal wetlands have been lost over the last 100 years, as a result of the combined effects of localised human pressures, sea level rise, warming, and extreme climate events (IPCC, 2022). In the Mediterranean we can see a mix of increasing desertification, rebound in wildfires, profound droughts and flash flood events that are generally impacting coastal communities. Additionally, the overexploitation of aquifers is leading to saline intrusion that could be a source of additional socioeconomic impacts.

Sea level rise poses a distinctive and severe adaptation challenge as it implies both dealing with slow onset changes and increases in the frequency and magnitude of extreme sea level events. Such adaptation challenges would occur much earlier under high rates of sea level rise. Responses to ongoing sea level rise and land subsidence include protection, accommodation, advance, and planned relocation. These responses are more effective if combined and/or sequenced, planned well ahead, aligned with sociocultural values, and underpinned by inclusive community engagement processes. Ecosystem-based solutions such as wetlands provide co-benefits for the environment and climate mitigation, and reduce costs for flood defences, but have site-specific physical limits, at least above 1.5 °C of global warming and lose effectiveness at high rates of sea level rise beyond 0.5–1 cm/yr (Hoegh-Guldberg et al., 2018).

Climate change modelling efforts conclude that coastal areas will be significantly affected by climate change, with sea level rise, increased frequency and intensity of extreme weather events, and ocean acidification posing significant risks to coastal ecosystems, people, and infrastructure (Griggs & Reguero, 2021). Effective adaptation measures, including protection, accommodation, advance, and planned relocation, are necessary to mitigate these risks, and ecosystem-based solutions such as wetlands can provide co-benefits for the environment and climate mitigation. However, the effectiveness of these measures will depend on the rate of sea level rise and the degree of community engagement and planning.

Intense droughts, torrential rains, violent tides, climatic discomfort, changes in agricultural patterns and disturbances to tourist seasons are the main effects of climate change in the Mediterranean (Romero & Olcina, 2021). As mentioned in the First Mediterranean Assessment Report (MAR1) (Cramer et al., 2020) these impacts are no longer expected in the long run, countries of the region are already feeling the impact on a variety of variables. The rapid and significant impacts of climate change in the Mediterranean Basin underscore the urgent need for proactive adaptation measures. Drawing upon examples from other regions that have implemented successful adaptation solutions can provide valuable insights and inform the development of effective solutions tailored to the unique challenges of the Mediterranean Basin. The aim of this paper is to present the need for evaluating the coastal risk of blue economy activities, adapting the complexity of the studies and adaptation solutions identified. As climate change poses a risk to coastal infrastructure, adaptation must be streamlined in permitting or operational activities to avoid economic damages.

7.2 Legal Framework for Climate Coastal Risk Analysis of Economic Activities

Coastal areas are highly valuable and of great strategic importance for the future of coastal states. This is particularly true in the case of Spain, which is a predominantly coastal country mostly surrounded by sea. The country also has geographical, social and administrative peculiarities that are key to understanding the coastal management system implemented. The two cases analysed in this chapter serve as a good example for highlighting the differences and the specificities of each case in terms of complexity and adaptation needs. The first one is a touristic real estate asset that given its current situation and future projections it will face more frequent structural damages or situations where it won't be able to open and normally function unless it implements adequate adaptation measures. Secondly, the chapter presents the case of non-permanent structures needed to assess climate risk and adaptation needs in the short term during the permitting phase.

From a political and administrative point of view, Spain has a central government and subsequent regional governments or so-called Autonomous Communities with different jurisdictional powers. From an administrative point of view, below the autonomous governments, each of them is formed by provinces. Of the seventeen autonomous communities, ten are at least partly coastal, including more than 25 coastal provinces and around 487 coastal towns.

In terms of geological and natural systems, Spanish coastal areas present a high diversity of structures. Coastal areas could be considered transition zones between marine and terrestrial systems setting ecological boundaries or buffer zones characterised by intense energy and material exchanges. In Spain we can see two types of coastal areas, cliffs resulting from erosion and beaches, sandbanks or coastal wetlands as a result of sedimentation dynamics.

The differences in coastal areas have raised the need to treat them accordingly from an environmental perspective but also a socioeconomic one. Dating back to Roman Law, coastal zones have been considered Public Domains. In Spain, the public domain of coastal zones was part of legal documents (*Las Siete Partidas del Rey Alfonso X "El Sabio"*) published during the XIII century. Such is the importance of the concept of Public Domain that the main Spanish legislation piece, the 1978s Constitution, states in its article 132.1 that the Law must regulate the Public Domain, on the basis that such areas may not be subject to embargo, divestment, or prescription (article 132). The law shall regulate the legal regime of property in the public domain and communal property, based on the principles that such property may not be subject to embargo, divestment, or prescription. The areas deemed to belong to the public domain shall be determined by law and shall, in any event, include the maritime terrestrial zone, beaches, territorial waters and the natural resources of the economic zone and the continental shelf. The importance of the Maritime-Terrestrial Public Domain is clearly relevant for our purposes here since it is the only category which is singled out for individual mention and is defined in the Constitution (art. 132.2), probably to eliminate the previous confusion and opposition to public domain being exercised over this important area (Losada et al., 2019).

Ten years after the Constitution, Spain's Coast Law (Law 22/1988) defined and demarcated the areas making up the maritime-terrestrial public domain. The Spanish Coastal Law 22/1988 (Ley de Costas 1988) defines a public domain area along the coast and a further zone where special restrictions apply to private ownership. The aim is to make the whole length of the coastline accessible to the public and to defend the coast against erosion and excessive urbanisation.

Inside the coastal law, specific requirements have been incorporated in order to reflect the risks of climate change on economic activities and the potential externalities it might have. Any project developed on the coastal area occupying maritime-terrestrial public domain must present a Coastal Dynamics Study, which, according to the article 91 of the general regulations of the coast, should contain:

2. Must predict adaptation of Works to the environment and, if necessary, its influence on the coast and the possible regression effects (article 44.2 of the Law 22/1988, July 28th, modified by Law 2/2013).

Additionally, projects must contain an assessment of possible effects of climate change on the area where the works will be placed.

This legal requirement is based on a holistic approach of coastal management, where every element or activity placed or developed on coastal areas must include a comprehensive analysis of coastal dynamics, including effects of climate change, in order to properly define its design and technical as well as environmental requirements. This approach also allows the promoters in early phases of their activity, to identify adaptation needs and define specific actions to avoid being impacted by climate change in the short or long-term.

In the context of coastal management, public administrations differentiate between the adaptation planning of temporary and fixed activities or infrastructure as the impacts of climate hazards have different consequences.

- For temporary activities, acute climate risk analysis is key to ensure the operational and economic feasibility of light infrastructure in terms of location and engineering needs. Additionally, proper coastal dynamics studies add an additional layer of occupational health and safety in terms of extreme weather events and its impacts on both employees and clients.
- For fixed infrastructure or activities that will occupy public terrain in the next decades, acute and chronic climate risk analysis is needed. Chronic risks are included in fixed infrastructure assessments as the lifespan of the assets will see important modifications of climate variables such as sea-level rise, storm surge, temperatures or drought. In case of future climate risks, the company will be required to implement adaptation measures to ensure the operability of the activity to avoid negative socioeconomic or environmental externalities.

This paper presents two coastal climate risk case studies, the first one being a permanent activity and the second one a temporary structure on maritime—terrestrial public domain. As the coastal law 22/1988 requires the climate risk assessment, they had to develop their corresponding coastal dynamics studies.

7.3 Case Studies

Both study cases are located in the western Mediterranean, in the Spanish Mediterranean coast, the first one on the coast of Catalonia, and the second, in the island of Formentera as shown in Fig. 7.1.

The first of our cases is placed in Altafulla, a coastal destination in Tarragona province, on the eastern coast of Spain and one hour south of Barcelona. The infrastructure in need of developing a climate risk assessment was Voramar, a restaurant located on the coastline, a frontline real estate asset at Altafulla's beach.

This infrastructure suffered serious structural damages caused by storm Celia in March 2022 (from 14th to 22nd). In Fig. 7.2 it can be appreciated the situation before the storm, and in Figs. 7.3 and 7.4, the affectations after the storm. Visually, one can clearly see the sand erosion caused by the storm and the difference with the coastline. The decrease in sediment material below the front-line of the structure caused important impacts on walls, roofs, facilities, and entrance.

These types of mediterranean storms have the potential to impact a large number of coastal constructions that were built before the entry into force of Law 2/2013 and Law 22/1988, of protection and sustainable use of coastal areas. For constructions built before 1988, no limitations on the minimum distance to the water were established on a general basis even though some local councils established urban restrictions. After the entry into force of the Coastal Law 22/1988 and the subsequent amendment made in Law 2/2013, a minimum of 100 m between the water and the construction cannot be occupied as it is considered a protection zone easement.

Law 2/2013 incorporated regulations to tackle the impacts of climate on coastal areas. A system for coastal zones at high risk was incorporated, limiting the activities allowed in such areas. Management actions to promote protection, preservation or restoration were also included in the 2013 law. It established the requirement to evaluate possible effects of climate change into projects, assets or activities in coastal

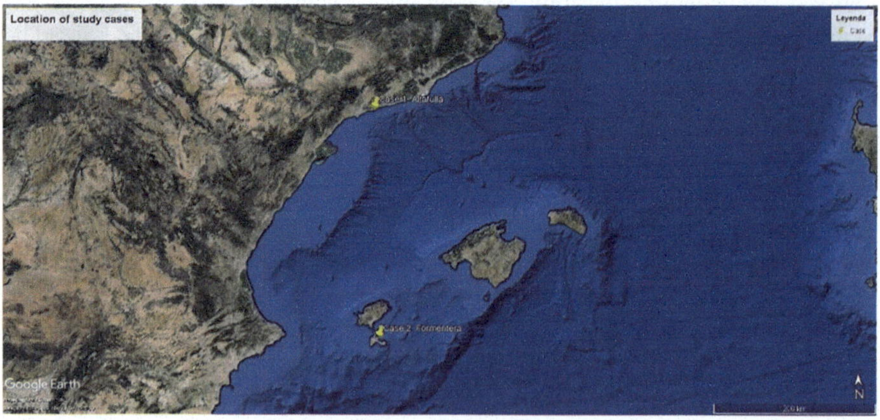

Fig. 7.1 Location of study cases (map made by the authors)

Fig. 7.2 Situation of the restaurant before the storm (picture made by the authors)

Fig. 7.3 Situation of the restaurant after the storm (picture made by the authors)

Fig. 7.4 Detail of damages after the storm (picture made by the authors)

areas, including provisions that projects could be cancelled if they are considered at risk of sea level rise. Alongside these legal measures, the Act required the Ministry of Ecological Transition to develop a strategy for coastal adaptation to climate change impacts. This was the first step towards the adoption of the Climate Adaptation Strategy of the Spanish Coast (Ministerio de Agricultura y Pesca, Alimentación y Medio Ambiente, 2016).

The second of our case studies is the request of the Island Authority of the Balearic Islands (Consell Insular de les Illes Balears) to renew the authorization for 4 years, to occupy part of the coastal public domain beaches of the island with many temporary (watchtowers, deckchairs, beach umbrellas…) and permanent (pavilions, terraces, beach cabins and sailing schools) facilities for touristic activities.

On a general basis, beach facilities and infrastructure are of foremost importance for the regional government. Such activities are directly linked with the most prominent economic sector of the Balearic Islands economy, tourism. At the same time, it is also relevant to highlight the fact that the regulatory procedure to analyse potential climate affectations to coastal activities was required for short term facilities. In Figs. 7.5 and 7.6 we provide the location of these elements.

7.4 Methodology

7.4.1 Aim and Objectives

Even if in both cases it was necessary to carry out a coastal dynamics study in order to fulfil a legal requirement, the aim of each study was considerably different given the zones considered, the type of asset and the risk time horizon. In the first case, the

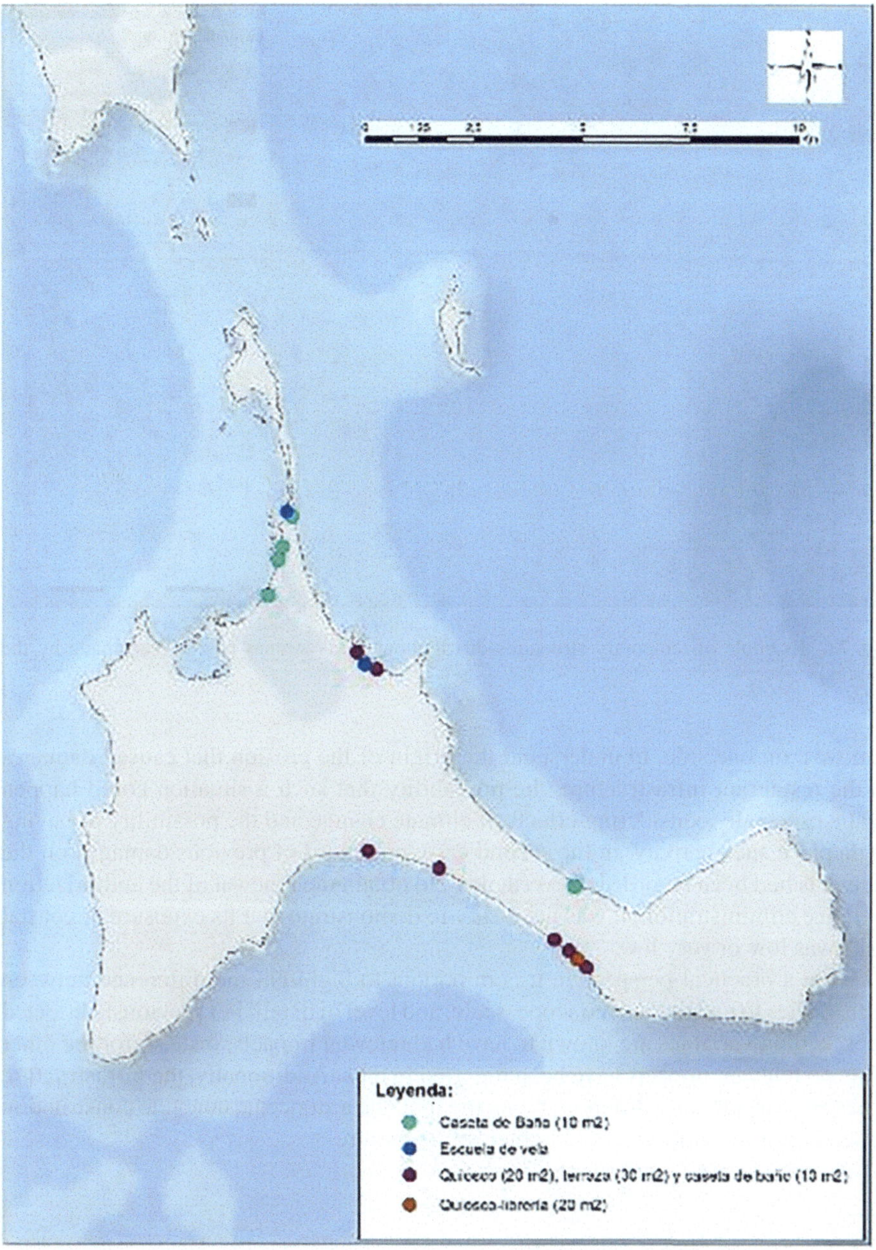

Fig. 7.5 Location of different permanent structures (map made by the authors)

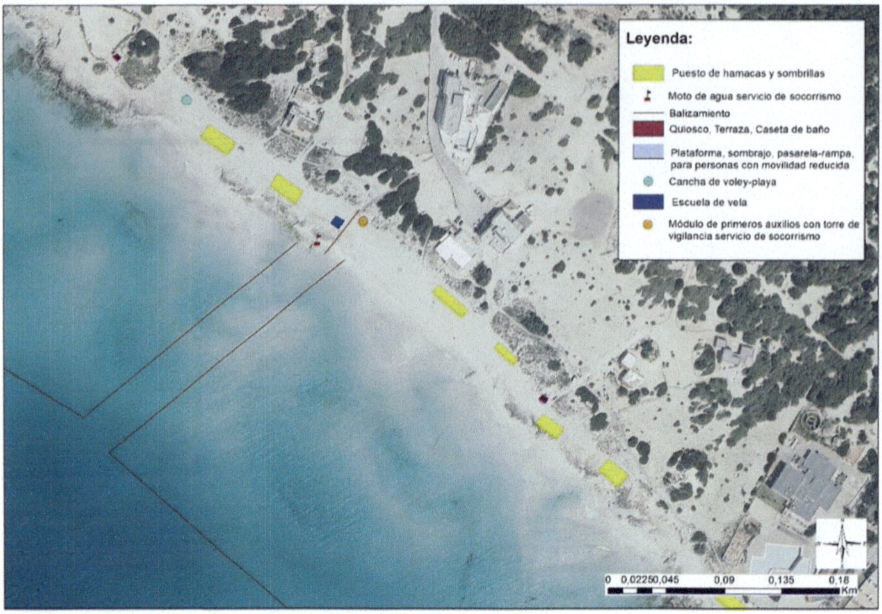

Fig. 7.6 Example of temporary structures distribution at Els Arenas beach (map made by the authors)

aim was, on one side, to understand the origin of the erosion that caused damages to the restaurant infrastructure, the probability that such a situation could happen again, especially considering effects of climate change and the possibility of having to displace such activity. In the second case, no record of previous damages on the elements had been recorded. Nevertheless, to obtain the renewal of the authorization from the administration, it was necessary to demonstrate that its exposure to coastal risks was low or very low.

From a practical perspective it is important to highlight the difference between both studies to understand the scope, scale, and level of detail. For previously affected sites, certain variables are known to have had relevant impacts, instead, for the other case, no relevant impacts have been reported to date. Additionally, the infrastructure types are considerably different, being the first one a stone and concrete construction instead of more temporary and lighter infrastructure.

7.4.2 Analytical Approach

Article 93 of the Regulations that develop the Coastal Law (RD 876/2014) sets the general aspects that a coastal dynamics study of a project placed in the coastal area must include. Considering the conditions of the coastal area in both study cases, in both case studies, the analysed aspects, among those indicated by the Coastal Regulations were the following:

I. Offshore marine climate (wave, wind and sea level statistics) and propagation to the coast
II. Coastal currents and transport patterns
III. Coastline evolution
IV. Beach profile evolution during storms
V. Coastal flooding
VI. Effects of Climate Change.

7.4.2.1 Marine Climate Analysis

The first step towards evaluating the coastal risk of the studies sites was to acquire proper offshore wave climate information to start all the modelling from a strong basis. This information was obtained from the public dataset available at the Spanish Ports Authority (Puertos del Estado) website (https://www.puertos.es/es-es/oceanografia/Paginas/portus.aspx). The dataset provides historical information for several ocean variables from 1959 until now.

Based on the data analysis conducted, a representative marine climate conditions profile was generated. In each case this profile was defined by wave height, wave period and wave direction. Such statistical information was numerically modelled using the OLUCA-SP[1] model (Falqués et al., 2006), which forms part of the Spanish Coastal Modelling System (SMC) (González et al., 2004) software package to propagate the conditions to the points of study. Once propagated to the coast, the whole register of wave data was reproduced on a variety of points in front of the study area.

7.4.2.2 Breaking Zone Currents, Coastal Transport and Coastline Evolution

Near the coast, waves are the main driving force for coastal dynamics, and breaking zone currents are its core element. For simulation of breaking zone currents, COPLA module was used, included also in the SMC package, which calculates currents inside the breaking zone. Using it, current patterns were obtained for different wave conditions, which explain changes in short term beach dynamics.

Using the wave register reproduced at different points along the coast, mean wave flux energy direction was also obtained (FME) as it is a relevant indicator for coastal sediment transport trends (Verón & Bertola, 2014).

In order to determine coastal transport, the sand transport formula proposed by CERC (Formula 7.1) (USACE, 1984) was applied at different points along the coast, to determine areas with erosive and accumulative trends. These estimations have

[1] Spectral wave propagation model part (OLUCA—SP) of the Spanish Coastal Modeling System (SMC).

been compared to coastline evolution (through aerial photographs) and record of past coastal engineering works (beach restoration, construction of jetties, harbours and seawalls). With the application of the CERC formula and the bibliographic study, confirmations on erosive or accumulative trends were identified for the different spots analysed.

CERC formula:

$$Q_L = \frac{K_\rho \sqrt{g}}{16\sqrt{\gamma}(\rho_s - \rho)(1 - \rho)} H_b^{5/2} \sin(2\alpha_b) \tag{7.1}$$

K Coefficient that depends on the size of the sediment
H_b Breaking wave height
α_b angle of the breaking wave with respect to the normal to the coast
γ Breakage index
ρ Density of water
ρ_s Density of solid particles
p Porosity of the material.

In addition to historical, long term coastline evolution analysis, short term evolution after big storms has also been studied, in order to determine which particular conditions have bigger effects. This analysis was done using aerial photographs of the coast provided by the GIS viewer of the Catalan geographical service.

7.4.2.3 Beach Profile

Regarding beach profile stability, both stability in the long term and profile evolution during storms have been calculated. For long term stability, closure depth has been defined using Hallermeier's (1981) formula. Closure depth corresponds to the most landward depth seaward of which there is no significant change in bottom elevation during a given time interval.

$$h_{in} = 2.28H_{12h/y} - 68.5\left(\frac{H^2_{12h/y}}{gT^2_{12h/y}}\right)$$

where

$H_{s12/y}$ significant wave height only exceeded 12 h a year.
$T_{12h/y}$ mean wave period associated with H_{s12}.
g acceleration of gravity.

Evolution during storms has been simulated using PETRA model (included in SMC package), which simulates evolution of beach profile during a specific storm defined by its duration, wave height, wave period and storm surge.

7.4.2.4 Coastal Flooding

Coastal flooding level has been calculated using both Iole probabilistic model and the analytical equations (Holman, 1986).

$$\frac{R_{u2\%}}{H_s} = 0.822 \cdot I_r + 0.2$$

where

$R_{u2\%}$ run-up only exceeded 2% of time
H_s significant wave height
Ir Iribarren's number, which has the following expression.

$$I_r = \frac{\tan \tan \beta}{\sqrt{\frac{H}{L_0}}}$$

where

Tan β beach slope
H offshore wave height
L_0 offshore wavelength.

Iole model (Tomás et al., 2015), based on wave and sea level databases generated within the framework of the C3E project (https://c3e.ihcantabria.com/), and the two-dimensional numerical model IH-2VOF (Torres-Freyermuth et al., 2007), which solves the Navier–Stokes equations, is used to simulate coastal flooding processes. Thus, from the total data set for the entire record of levels and waves, for each profile, the 183 time periods (3 per year) with the highest run-up are selected. Subsequently, the POT extreme regime is adjusted with all the sea states in each profile. As a result of the modelling with the IH-2VOF model, wave run-up and overtopping results were obtained to understand how coastal infrastructure would be impacted by waves.

Climate change effects have been analysed by using the C3E viewer. C3E viewer, https://c3e.ihcantabria.com/, is the tool for visualising the results of climate change projections, which allows obtaining the values of various parameters of waves, sea level and water temperature for the current climate (mean value between 1985 and 2005) as well as for two projections of 2 periods (2026–2045) and 2081–2100, under two climate scenarios, R.C.P. 4.5 and R.C.P. 8.5. The viewer allows to obtain sea level, storm surge, wave height, wave period and wave direction values for the exposed projections and periods, at many points in front of the coast of Spain. Using the values given by this viewer effects over already estimated and propagated results has been assessed.

7.5 Results and Discussion

In the case of the Voramar restaurant, wave data were collected from Tarragona buoy, placed offshore, on the south of the study area. Wave data have been propagated to the coast with OLUCA-SP (Fig. 7.7), and FME (Flujo Medio de Energía, or, in English, Mean Energy Flux) has been obtained at different points in front of the restaurant (Fig. 7.8).

FME is perpendicular to the coastline at 2 m depth, but at 5 m depth, FME is crosswise, from the East, so, transport at that depth will happen to the West.

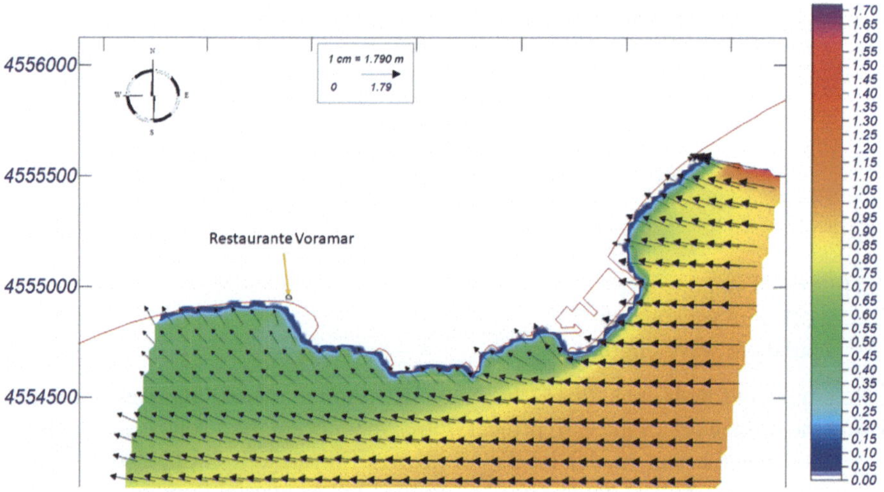

Fig. 7.7 Wave vectors and height for waves coming from E (Hs: 1 m; Tp: 5 s) (map made by the authors)

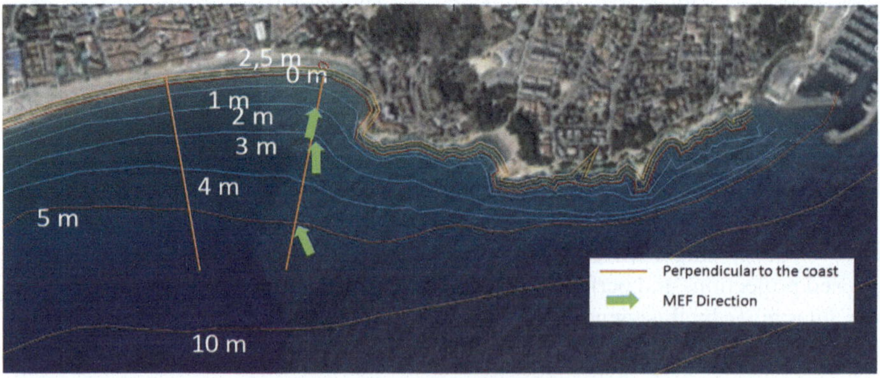

Fig. 7.8 Mean energy flux direction in front of the study area (map made by the authors)

Estimated net coastal transport capacity, at 5 m depth, is 25.148 m^3/year, a consistent result with the transport capacity estimated by LIM (Maritime Engineering Laboratory, Libro Verde de la Zona Costera de Cataluña, 2010). The study carried out by LIM about historic coastline evolution between 1995 and 2004 showed a clear erosive pattern in the Eastern end of the beach (in front of the restaurant, about 5 m/year) as well as an accumulation on the western end. Sedimentary balance is negative in the Eastern end, and positive in the western end between 1995 and 2004. The analysis of aerial photographs between 1945 and 2022 (https://betaportal.icgc.cat/comparador-gificador/#16/41.8620/3.1532) showed a mainly stable coastline before 1993, that changed due to a beach restoration related to the Torredembarra harbour on the east in 1994. In 2009 a new beach restoration was carried out. Since then, beach width gradually diminishes, accelerated from 2016, reaching a maximum in March 2022, causing damages to the restaurant. Even if this trend is clear, damages on the restaurant did not arrive in a gradual way, but in an acute process, during a short storm.

Analysing most relevant storms from 2017 to 2022, we saw that all, except that of October 2019, which came from the E, come from ENE (offshore). The storm of March 2022, which has the second lowest Hs and lowest Sea Water level rise, provoked the biggest beach width reduction, probably because it was the one that lasted longer (8 days) and had the highest wave period.

Simulation of profile evolution during the March 2022 storm shows that sand movements are significant until 8 m depth. The model underestimates erosion on the beach front, but properly reflects sand accumulation in the submerged area. Simulated maximum setback of the coastline coincides with the effects recorded after the storm, when the base of the restaurant lost its sand.

Coastal flooding calculations using Iole model indicate, for 10 and 100-year return, that the restaurant will be flooded, as its base is at 3 m, and flooding levels reach 4 and 4.5 m, respectively.

Flooding results with Holman's equation (1986) (considering run-up and maximum sea water level including astronomical and meteorological tide), indicate that, for a 50 year-return period, the restaurant would be flooded.

Climate change projections estimates do not indicate an increase in Hs nor in Tp, but yes in MSL (around 0.15 cm in 2045 and around 0.56 m in 2100). This could lower the return period as it would require slightly less wave height to inflict the same damage.

Transport pattern, coastline evolution and historical response to storms clearly show a coherent functioning of the area where the restaurant Voramar is located:

- Coastal transport tends to the west, but before 1994, the coastline was mainly stable.
- Since the construction of the Torredembarra harbour in 1994, a constant erosive trend appeared, with acute effects during long lasting and long period storms, that move sediment seawards and then transport it to the west, causing a net loss of sand in the section.

- Climate change will increase the risk and frequency of flooding, mainly due to sea level rise.

This situation provokes that the restaurant will more frequently face structural damages or situations where it won't be able to open and normally function, unless it implements adequate adaptation measures.

Given that the case in Formentera focuses on short-term temporary activities, the overall analysis was kept simpler to not over-complicate. In this case, analysed elements are small and mobile, and located on the beach just for the summer period, even though some of them stayed installed during the whole year but with no active usage for most of the time. It can be considered that the main risks for these installations are coastal flooding and beach profile erosion.

Coastal flooding area was estimated using the viewer of climate change impacts on Balearic Islands coast developed by the regional administration (https://ideib.caib.es/impactes_costa_canvi_climatic/). The GIS visor showed flooding under both the RCP 4.5 and 8.5 climate projections, in mean and extreme conditions, in 2050 and 2100. For these scenarios, many of the facilities on beaches will be flooded, but as most of the infrastructure or assets are mobile, location of facilities will have to be defined every year on the basis of updated projections.

Beach profile evolution was also analysed using PETRA (GIOC, 2001). The coastline setback was estimated for the different beaches where facilities are present and none of them are considered to be affected by setbacks. Given this result, it was considered that no relevant adaptation actions were expected for the administration to grant the 4 years legal authorization.

Beach facilities and services locations were analysed all along Formentera beaches, in order to assess its exposure to risk caused by waves or coastal flooding during the authorisation period (4 years). As a result, the analysis obtained the following points:

- Due to the short term of the climate risk analysis—four years—, climate change effects were not considered to be significant as potential chronic impacts would not be visible.
- Present coastal flooding and coastline setbacks do not represent a risk for analysed facilities. Despite this, the location of all facilities has to be periodically assessed on the basis of updated projections, as sea level rise might put those elements in the future.

7.6 Conclusions

Analysed cases show two different situations of coastal facilities and infrastructures on which intensity of climate change effects is different, and strategies and actions to adapt to them are also different. We could summarise as follows. The case of Voramar Restaurant, in Altafulla, is a recurrent case of a facility or infrastructure that was built in the beach front without proper analysis of future coastal risks. Current legislation in Spain would avoid the inconvenience and socioeconomic impact of starting a business on coastal areas vulnerable to climate change. This reflects the importance of requiring coastal and climate risk analysis in specific locations.

In the Altafulla example, the estate was constructed in the seawards' limits of vegetated ancient dunes, in a coastal strip on dynamic equilibrium, that, due to construction of Torredembarra harbour, lost this equilibrium and began an erosive trend in front of the restaurant. This trend, which is not intense, has acute episodes during long lasting storms, which take sediment seawards and then transport it to the west causing a net loss of sand, and an intensification of the erosive process. With all information presented two conclusions can be extracted; (i) understanding the previous ecological and geological status of the area occupied could give an idea of the future dynamics of it. Additionally, (ii) environmental externalities from surrounding infrastructure developments must be taken into account to avoid generating downstream impacts. This case shows how the development of a harbour one kilometre north destabilised the coastal dynamics of the Altafulla beach leading to infrastructure damage to the infrastructure and low long-term feasibility if no adaptation measures are implemented.

Projections on climate change do not indicate an increase in wave height nor period, but, even if they are not quantified, it seems that extreme storms will be more frequent, intensifying the problem. Sea level rise will also increase, so damages recorded in 2022 will be more and more frequent and intense due to climate change. Due to the permanent character of the facility, it is not possible to move or displace it, only two options are possible: increase of its defence (construction of a seawall to protect from erosion and flooding) or dismantling. For the moment, reinforcement of defence will be executed, but the long-term sustainability of this option has to be properly assessed as investments in adaptation might need to increase with time and return from that investment could not be enough. As a conclusion, administrations have an important role to play in setting the criteria for adaptation or relocation of future impacted assets. As far as adaptation finance must be considered an option to avoid climate risks and socioeconomic impacts, for assets or activities with a high statistical probability of being severely impacted on a regular basis, relocation planning should be prioritised. The case of beach facilities in Formentera shows the importance of considering climate change and coastal dynamics effects even on assets and infrastructure of temporary beach facilities. It is important to highlight the need of simplifying procedures but maintain the need of studying such topics to access the permit to occupy public land. With this procedure business owners are increasingly aware of potential impacts and raises awareness on the topic importance.

Even though in this specific case, no climate impacts were identified in the short term (4 years period), projections up to 2050 indicate that some of the beaches won't be able to carry these facilities. This highlights the need of updating such studies on a regular basis in order to use improved modelling results and real-time data to re-assess physical climate risks.

As a general conclusion, we want to highlight the relevance of the Spanish legal framework for managing coastal areas, which makes compulsory to study the whole functioning of coastal area, as well as the effects of climate change, in order to achieve an integrated coastal management based on knowledge about coastal systems and be able to adapt our actions to changes in the environment. Without proper administrative procedures, adaptation to climate change will be difficult to implement and require private sector actors more focused on short-term returns. This holistic approach to coastal analysis is essential for proper coastal management. Presented case studies raise awareness on the importance of evaluating coastal risk of permanent and non-permanent infrastructure to adapt the tourism sector to the effects of climate change. Other regions of the world can be inspired by the cases presented and include new requirements to current permitting procedures or increase the interest in evaluating present infrastructure.

References

Cramer, W., Guiot, J., & Katarzyna, M. (2020). *UNEP/MAP MedECC climate and environmental change in the Mediterranean Basin—Current situation and risks for the future, first Mediterranean assessment report.*

Falqués, A., Medellín, G., Medina, R., González, M., & Ávila, M. (2006). Testing high angle waves instability on a low energy beach. *Proceedings of the Coastal Engineering Conference*. https://doi.org/10.1142/9789812709554_0225

GIOC. (2001). "PETRA: Modelo de evolución del perfil transversal de la playa"; Versión: 3.0. (2001). Manual de Referencia. Universidad de Cantabria, España. Grupo de Ingeniería Oceanográfica y de Costas.

González, M., Santamaría, M., Osorio, A., & Lomonaco, P. (2004). Sistema de Modelado Costero Español (SMC). https://doi.org/10.13140/RG.2.1.2105.0405

Griggs, G., & Reguero, B. (2021). Coastal adaptation to climate change and sea level rise. *Water, 13*(16), 2151.

Hallermeier, R. J. (1981). A profile zonation for seasonal sand beaches from wave climate. *Coastal Engineering, 4*, 253–277.

Hoegh-Guldberg, O., Jacob, D., Taylor, M., Bindi, M., Brown, S., Camilloni, I., Diedhiou, A., Djalante, R., Ebi, K. L., Engelbrecht, F., Guiot, J., Hijioka, Y., Mehrotra, S., Payne, A., Seneviratne, S. I., Thomas, A., Warren, R., & Zhou, G. (2018). Impacts of 1.5 °C global warming on natural and human systems. In V. Masson-Delmotte, P. Zhai, H. -O. Pörtner, D. Roberts, J. Skea, P. R. Shukla, A. Pirani, W. Moufouma-Okia, C. Péan, R. Pidcock, S. Connors, J. B. R. Matthews, Y. Chen, X. Zhou, M. I. Gomis, E. Lonnoy, T. Maycock, M. Tignor, & T. Waterfield (Eds.), *Global Warming of 1.5 °C. An IPCC Special Report on the Impacts of Global Warming of 1.5 °C Above Pre-industrial Levels and Related Global Greenhouse Gas Emission Pathways, in the Context of Strengthening the Global Response to the Threat of Climate Change, Sustainable Development, and Efforts to Eradicate Poverty* (pp. 175–312). Cambridge University Press, Cambridge, UK and New York, NY, USA. https://doi.org/10.1017/9781009157940.005

Holman, R. A. (1986). Extreme value statistics for wave run-up on a natural beach. *Coastal Engineering, 9*(6), 527–544.

IPCC. (2021). Climate Change 2021: The Physical Science Basis.

IPCC. (2022), In H.- O. Pörtner, D. C. Roberts, M. Tignor, E. S. Poloczanska, K. Mintenbeck, A. Alegría, M. Craig, S. Langsdorf, S. Löschke, V. Möller, A. Okem & B. Rama (Eds.), *Climate Change 2022: Impacts, Adaptation, And Vulnerability. Contribution of Working Group II to the Sixth Assessment Report of the Intergovernmental Panel on Climate Change* (3056 p.). Cambridge University Press. Cambridge University Press, Cambridge, UK and New York, NY, USA. https://doi.org/10.1017/9781009325844

Ley 22/1988, de 28 de julio, de Costas.

LIM. (2010). Libro Verde de la Zona Costera de Cataluña.

Losada, I., Toimil, A., Muñoz, A. P., Garcia-Fletcher, A., & Diaz-Simal, P. (2019). A planning strategy for the adaptation of coastal areas to climate change: The Spanish case. *Ocean & Coastal Management, 182*, 104983. ISSN 0964-5691.https://doi.org/10.1016/j.ocecoaman.2019.104983

Ministerio de Agricultura y Pesca, Alimentación y Medio Ambiente. (2016). Estrategia de adaptación al cambio climático de la costa española.

Romero, J., & Olcina, J., (2021). Cambio climático en el Mediterráneo. Procesos, riesgos y políticas. Ed. Tirant lo Blanch.

Tomas, A., Mendez, F., Medina, R. Jaime F., Higuera, P., Lara, J., Ortiz, M. D., & Alvarez de Eulate, M. (2015). A methodology to estimate wave-induced coastal flooding hazard maps in Spain. *Journal of Flood Risk Management, 9*.https://doi.org/10.1111/jfr3.12198

Torres-Freyermuth, A., Losada, I. J., & Lara, J. L. (2007). Modeling of surf zone processes on a natural beach using Reynolds-Averaged Navier-Stokes equations. *Journal of Geophysical Research. Oceans. AGU. (American Geophysical Union), 112*, C09014.

Union for Mediterranean (UfM). (2016). Untapping the potential of the blue economy in the Mediterranean region. https://ufmsecretariat.org/untapping-the-potential-of-blue-economy-in-the-mediterranean-region/

USACE. (1984). CERC. Shore Protection Manual (SPM). U.S. Army Coastal Engineering Research Center.

Verón, M., & Bertola, G. (2014). Aplicación del método de flujo de energía en el litoral de la provincia de Buenos Aires, Argentina. *Latin American Journal of Sedimentology and Basin Analysis, 21*, 17–23.

Chapter 8
Enabling Adaptation to Climate Change Through Sustainable Energy and Climate Action Plans: Three Case Studies from Italy

Matteo Marchesi, Massimiliano Granceri Bradaschia, Roberto Menghi, Matteo Riccardi, and Claudia Verna

Abstract Adapting to climate change is a fundamental component within the Sustainable Energy and Climate Action Plans (SECAPs). In a context of increasing climate uncertainty, cities and communities must prepare to address the current and future impacts of climate change. SECAPs tackle adaptation through a series of specific measures and actions. First and foremost, a climate characterization of the examined territory is carried out, considering both the current situation and future short- and medium-term climate scenarios. Based on these assessments, a vulnerability analysis is conducted to identify specific climate-related risks that can influence the reference area. These risks may include extreme events such as floods, droughts, heatwaves, and more intense storms. The SECAPs then encompasses a range of actions aimed at reducing vulnerability and improving community resilience in the face of such risks. Adapting to climate change within a SECAP is a volunteer and dynamic process that requires long-term planning and ongoing evaluation of the adopted strategies. Cooperation among various stakeholders, including government entities, local institutions, communities, and the private sector, is essential to ensure effective adaptation and sustainable management of climate risks. The article will focus on SECAP development case studies about the following local governments: Reno Galliera Union, Municipality of Vieste, Municipality of Cervia.

Keywords Urban planning · Covenant of Mayors · Italy · Climate change adaptation · Monitoring · Indicators

---

M. Marchesi (✉) · R. Menghi · M. Riccardi · C. Verna
NIER SpA, Bologna, Italy
e-mail: m.marchesi@nier.it

M. Granceri Bradaschia
IUAV University of Venice, Venice, Italy

M. Granceri Bradaschia et al. (eds.), *Climate Change Adaptation, Flood Risk, and Beyond*, Planning for Climate Proof Cities,
https://doi.org/10.1007/978-3-031-65463-3_8

8.1 Introduction

The objective of this paper is to describe the outcomes achieved from the application of the Sustainable Energy and Climate Action Plan (SECAP) to three different contexts and the related workflow. The research work led to the identification of some of the main adaptation actions that can be considered to provide useful insights to practitioners and policy makers on adaptation strategies in different contexts of the Italian territory. The case studies are the reduction of surface runoff in Reno Galliera Union, coastal erosion and the resulting risk of flooding from the sea in the Municipality of Vieste, risk analysis of flooding due to sea-level rise in the Municipality of Cervia.

The case studies concern the Italian territory since Italy is in an area identified as particularly vulnerable to climate change. In detail, the latest reports on climate change show how Italy is particularly sensitive to the increased temperatures and reduction of water security under different aspects, from flood risk to drought, with repercussions in terms of safety of populations and exposed assets. These trends will have different impacts on specific territories (i.e., coastal areas, urban areas, mountainous areas, etc.), biodiversity and ecosystems, as well as the many economic sectors involved (e.g., agriculture, building, manufacturing, etc.) (Ali et al., 2022). In addition, differently from published works, the case studies cover small and medium-sized Italian municipalities as they represent the majority of municipalities and where the greater part of the population lives (> 68.5%) (ISTAT, 2020).

The paper is structured as follows: the following chapter of Sect. 8.1 contextualises the issue of climate change and the main tools adopted at community level to address it, Sect. 8.2 describes the method and workflow applied in designing the Sustainable Energy Action Plan while Sect. 8.3 shows the results of the application to three case studies differing in geographical location, municipality size and population involved. Finally, the main conclusions are highlighted in the last section of the document.

8.2 Background

It is well known that the fight against climate change requires concerted efforts from the global community on two main sides. While reducing and mitigating carbon emissions is crucial to slow down global warming, the effects of climate change are already happening in the form of an increasing incidence of extreme weather events and other similar trends, necessitating adequate adaptation measures. There are many intersections between mitigation and adaptation actions (Fig. 8.1); however, the two streams of climate action have so far evolved largely independently from each other, and the interactions between them are still underestimated (Chapman & Stiff, 2018).

Instead, tackling climate change requires an integrated and holistic approach that encompasses both mitigation and adaptation actions.

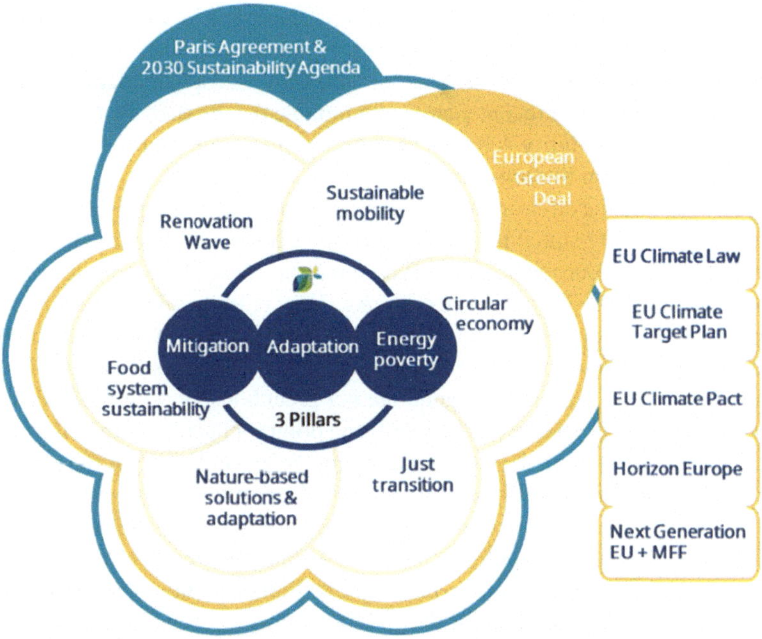

Fig. 8.1 Covenant of Mayor classification activities. *Source* EU Commission

The European Environment Agency defines climate change mitigation as a set of actions aimed at making the impacts of climate change less severe by preventing or reducing greenhouse gas (GHG) emissions into the atmosphere. Mitigation is achieved by reducing the sources of these gases (e.g., by increasing the share of renewable energies or creating a cleaner mobility system) or enhancing their sinks (e.g., through expanding forests). In short, mitigation is a human intervention that reduces the sources of greenhouse gas emissions and/or strengthens the absorption wells.

These actions are crucial to slow down global warming and limit the future impacts of climate change. However, even if emissions were drastically reduced today, the effects of climate change would still be present due to past emissions and delays in the response of the climate system.

This is where adaptation to climate change comes into play. According to the European Environment Agency, adaptation means anticipating the adverse effects of climate change and taking appropriate measures to prevent or minimise the damages they can cause or seize the opportunities that may arise. Examples of adaptation measures include large-scale infrastructure modifications, such as building defences to protect against sea-level rise, and behavioural changes, such as reducing individual food waste. Essentially, adaptation can be understood as the process of adjusting to the current and future effects of climate change by adopting actions aimed at reducing the vulnerability and improving the resilience of communities, infrastructure, and ecosystems to the impacts of climate change. Adaptation is essential because the effects of climate change are already visible.

To achieve the goal of combating climate change, European adaptation policies have progressed quickly in recent years, with notable results such as the Paris Agreement, the 2030 Sustainability Agenda, the introduction of the European Green Deal, the new EU Adaptation Strategy, and the Climate Law of 2021.

The protection of the environment is now a fundamental principle of the European Community, as already stated in Article 37 of the Charter of Fundamental Rights of the European Union (European Union Proclamation, 2000). However, it is thanks to the communication titled *The European Green Deal* (European Commission, 2019) that the European Commission launched a new strategy aimed at transforming the Union into a prosperous society, equipped with a modern, resource-efficient, and competitive economy that will not generate net greenhouse gas emissions by 2050. The Green Deal has effectively marked a turning point for the planning of the community strategy for sustainable development, laying the groundwork for the publication of important regulations and directives aimed at decoupling economic growth from the improper use of natural resources and greenhouse gas emissions.

For example, consider the definition of EU Sustainable Finance Action Plan (EU Commission, 2018), which has led to the classification of economic activities considered environmentally sustainable in the EU Taxonomy (European Union, 2020), to support the transition to a low-impact economy not only with policy regulations but also from a financial point of view. As a matter of fact, the first two goals of Taxonomy Regulation concern mitigation and adaptation to climate change and are further elaborated in the Climate Delegated Act (EUD Regulation, 2021a, 2021b), which establishes under what conditions an activity is considered capable of contributing to the taxonomic objectives related to the climate, through Technical Screening Criteria.

It is undeniable, however, that climate change has several severe implications for other environmental issues. In fact, the European Climate Law (EU Regulation, 2021a, 2021b), which establishes the framework for achieving climate neutrality, reflects the concerns of the IPBES (Intergovernmental Science-Policy Platform on Biodiversity and Ecosystem Services) to emphasise that climate change is the third most significant driver of global biodiversity loss.

The climate issue, far from being a confined topic, is part of the UN Agenda 2030 (UN General Assembly, 2015), one of the main attempts to address the dimensions of sustainable development in a synergistic way, highlighting the mutual influences between environmental concerns, economic growth, and the well-being of people considered in its entirety. The SDG 13, through its five targets (United Nation), promotes the fight against climate change, integrating a strong component of awareness, education, and enhancement of global skills, both from a human and institutional perspective, in terms of mitigation and adaptation.

On the other hand, the new EU Strategy on Adaptation to Climate Change (EU Commission, 2021) also highlights the close relationship between various risks associated with climate change—from the health of the planet and people to risks for infrastructure, supply chains, agriculture, and some of the major human activities. It calls for maintaining a high level of monitoring, starting with specific data. In particular: «Data on climate-related risk and losses are crucial to improve the accuracy of climate risk assessment. Any new investment and policy decision should be climate-informed and future-proof, from households renovating their homes, to SMEs setting up business in a vulnerable area, larger businesses managing supply chains, banks agreeing new loans, or cities planning zoning developments».

It is true that, in line with the scientific findings of the IPCC (Intergovernmental Panel on Climate Change) and the IPBES (Intergovernmental Science-Policy Platform on Biodiversity and Ecosystem Services), achieving the goal of climate neutrality by 2050 requires not only reducing greenhouse gas emissions, but also increasing absorption. This is why Europe is committed to reducing Union emissions by at least 55% compared to 1990 levels by 2030.

Not less important is addressing adaptation adequately. As stated in the European Climate Law: «Adaptation is an essential element of the long-term global response to climate change. The adverse effects of climate change could potentially exceed the adaptive capacity of the Member States. Member States and the Union should therefore enhance their adaptation capacity, strengthen resilience, and reduce vulnerability to climate change, as provided for in Article 7 of the Paris Agreement, and maximise co-benefits deriving from other policies and regulations» (EU Regulation, 2021a, 2021b).

Notably, the 2023 edition of the Global Risks Report, published by the World Economic Forum, ranks the failure of measures to increase adaptation to climate change as the second most significant global risk over a ten-year time frame. For this reason, it can be even more crucial to direct efforts towards greater resilience at the local level, because "the local level is the bedrock of adaptation" (European Commission, 2021). In this context, the Covenant of Mayors for Climate and Energy comes into play (EU Commission).

Launched in 2008, it brings together thousands of local governments eager to ensure a better future for their citizens. By participating in the initiative, they voluntarily commit to implementing the EU's climate and energy objectives. The initiative not only represented the first bottom-up movement for climate and energy, but its success quickly exceeded expectations: since its launch in 2008, the initiative has mobilised over 10.000 local authorities in Europe, representing roughly one third of the EU population.

Local authorities that join the Covenant of Mayors—Europe commit to acting at the pace set by science, in a joint effort to keep the global temperature increase below 1.5 °C—the updated target of the Paris Agreement. Covenant signatories commit to (1) reducing greenhouse gas emissions in their territory, (2) increasing resilience and preparing for the negative impacts of climate change, and (3) addressing energy poverty as a key action to ensure a just transition.

The success of this initiative and its ongoing evolution do not remove the complexity and problems of the Sustainable Energy and Climate Action Plan (SECAP) implementation (Delponte et al., 2017).

The scientific literature shows that works focusing on European initiatives to combat the effects of climate change and on SECAP are not very abundant (Abarca-Alvarez et al., 2019). Most of them being focused on mitigation criteria, such as, for example, the correlation between reducing emissions by saving energy in buildings and increasing air quality (Monforti-Ferrario et al., 2018), the evaluation of emission reduction policies of a hundred cities with more than 100,000 inhabitants (Croci et al., 2017), and the analysis of the results of a multi-municipality energy approach (Kamenders et al., 2017).

Other works have focused on the development of methodologies and tools to support the planning, implementation, and monitoring process of SECAPs. They range from the application of the International Standard ISO 50001 by municipalities to facilitate the implementation of their sustainable energy action plans (Dzene et al., 2015), to the development of monitoring methodologies to support the Public Administration (Cinocca et al., 2018) and tools for mapping the actual state of energy resources in order to develop sustainable energy policies (Gagliano et al., 2015). Case studies and SECAP applications are limited and mostly involve large metropolitan cities, e.g., Milan, Palermo, Rome and Genoa (Berghi, 2016; Kamenders et al., 2017).

At the same time, the ongoing increase in the signatories of the Pact and the fact that the majority of these signatories (67.38% of the whole CoM Signatories in April 2021) fall under the so-called 'small' municipalities (i.e., those with fewer than 10,000 inhabitants), is an encouraging fact (Scorza-Santopietro, 2021).

8.3 SECAP's Methodological Approach

The members of Covenant of Mayors for Climate and Energy—Europe initiative commit to submitting an action plan within two years of formal accession to the initiative. This action plan is a key implementation tool for Covenant signatories. It defines mitigation and adaptation objectives, and it is based on a Baseline Emission Inventory and a Risk and Vulnerability Assessment, which provide an analysis of the situation at a given time. They serve as a basis for defining a comprehensive set of actions that signatories intend to undertake to achieve their objectives, as well as to address energy poverty.

In practical terms, the SECAP is structured into four different steps:

- Territorial characterization of the examined area.
- Analysis of vulnerabilities and risks for the relevant territory.
- Reconstruction of energy balances and emissions inventory.
- Proposal of actions for climate change mitigation and adaptation.

8.3.1 Territorial Characterization of the Examined Area

Territorial characterization within the SECAP is a process aimed at understanding the characteristics, resources, and potentials of the examined territory.

Territorial characterization should be based on data and information collected from various sources, including local administrations, environmental agencies, research institutes, national statistics, remote sensing tools, and specific studies conducted on the territory. The goal is to develop a comprehensive understanding of the territory's characteristics to guide the development of a concrete action plan, tailored to local needs for the transition to a sustainable and low-carbon economy.

In the context of SECAPs, territorial characterization generally involves two subjects: the first aims to provide an overview of the territory, while the second focuses on the socio-economic context. In the first phase, a land use analysis is conducted, providing information on urban, rural, and natural areas present in the territory. Vegetation coverage, including the presence of forests or green areas, is an important factor for assessing the territory's capacity to absorb carbon emissions and mitigate the effects of climate change. Information on traffic and mobility in the territory is also collected during this first phase.

Territorial characterization also considers socio-economic aspects such as resident population, employment, dominant economic activities, and social dynamics. This provides a complete picture of territorial dynamics and the socio-economic implications of actions aimed at promoting sustainable energy.

Socio-economic and land use characterization of a territory are two of the main input data considered at the very beginning of the following phase: vulnerabilities and risks analysis.

8.3.2 Analysis of Vulnerabilities and Risks for Relevant Territories

Risk analysis helps to deal with potential threats and vulnerabilities associated with energy and climate change in territory. The objective of risk analysis is to identify and understand the potential negative impacts and consequences of actions or extreme weather events on the territory, as well as to identify opportunities to mitigate such risks.

The most important components for assessing and managing climate change-related risks are hazard analysis, exposure analysis, and vulnerability analysis (Field, 2012).

Risk exists only when a hazard source, a target system (or vulnerable receptor) that can experience negative consequences, and exposure, meaning the possibility of contact between a hazard and the receptor, coexist in well-defined areas and time intervals. The vulnerability component, defined as the propensity or predisposition of a system to be negatively altered, encompasses sensitivity, which refers to the susceptibility to damage, and adaptive capacity. Adaptive capacity expresses the ability of a system (nation, community, group) to adjust its characteristics to present and/or future climatic conditions and reduce the level of vulnerability, in relation to specific dynamic contexts of a biophysical, social, economic, technological, and political nature.

The proposed risk and vulnerability analysis is conducted based on the bi-dimensional risk index presented and described in the National Plan for Adaptation to Climate Change (MASE, 2018). The index combines "potential impact" (determined by hazard, exposure, and sensitivity) with adaptive capacity. Thus, the proposed index includes the three fundamental components in risk assessment and management, but vulnerability is differentiated into sensitivity and adaptive capacity.

In this conceptual framework, vulnerability is partly quantified by territorial indicators (spatially distributed) that assess not only exposure, but also the greater or lesser propensity for damage and sensitivity (e.g., physical, geological, and land use characteristics). The other part is described by some adaptive capacity components, measured by indicators at the provincial level, regarding the greater or lesser ability of the territory to address climate change.

While in the broad sense, vulnerability encompasses both sensitivity and adaptive capacity according to the theoretical definition of the IPCC, in practical terms, sensitivity expresses the susceptibility to damage for each hazard separately, while adaptive capacity reflects the ability to deal with overall damage, i.e., the sum of the damages resulting from all considered hazards (Fig. 8.2). For this reason, in calculating the risk index, sensitivity and adaptive capacity cannot be treated together (MASE, 2023).

Therefore, the index reported in Fig. 8.3, developed and detailed in the PNACC, in accordance with the IPCC guidelines, is configured as a bi-dimensional index (Potential Impact and adaptive capacity) presented at the provincial level, and it is an efficient tool for nationwide investigation.

It should be noted that, while the bi-dimensional index, based on parameters mainly derived from national or European databases, certainly provides a useful contextual analysis for a national plan, it should be further explored to identify and guide the implementation of specific adaptation actions at the local level (see Sect. 8.3.4). In other words, the bi-dimensional index proposed by the PNACC should

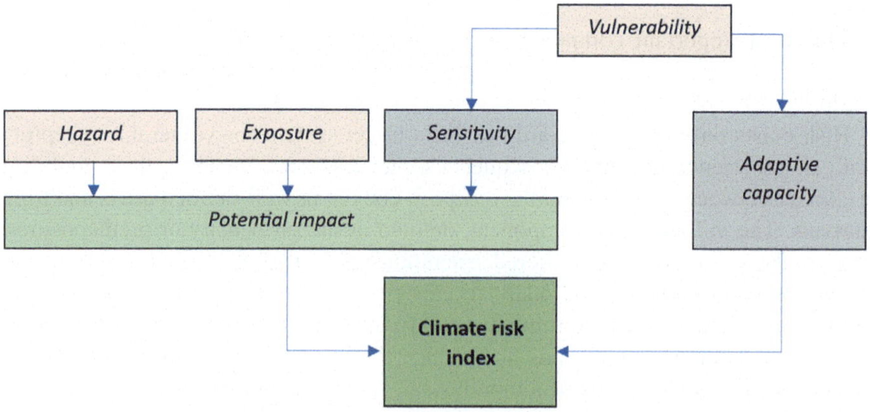

Fig. 8.2 Climate risk index definition approach (made by the authors)

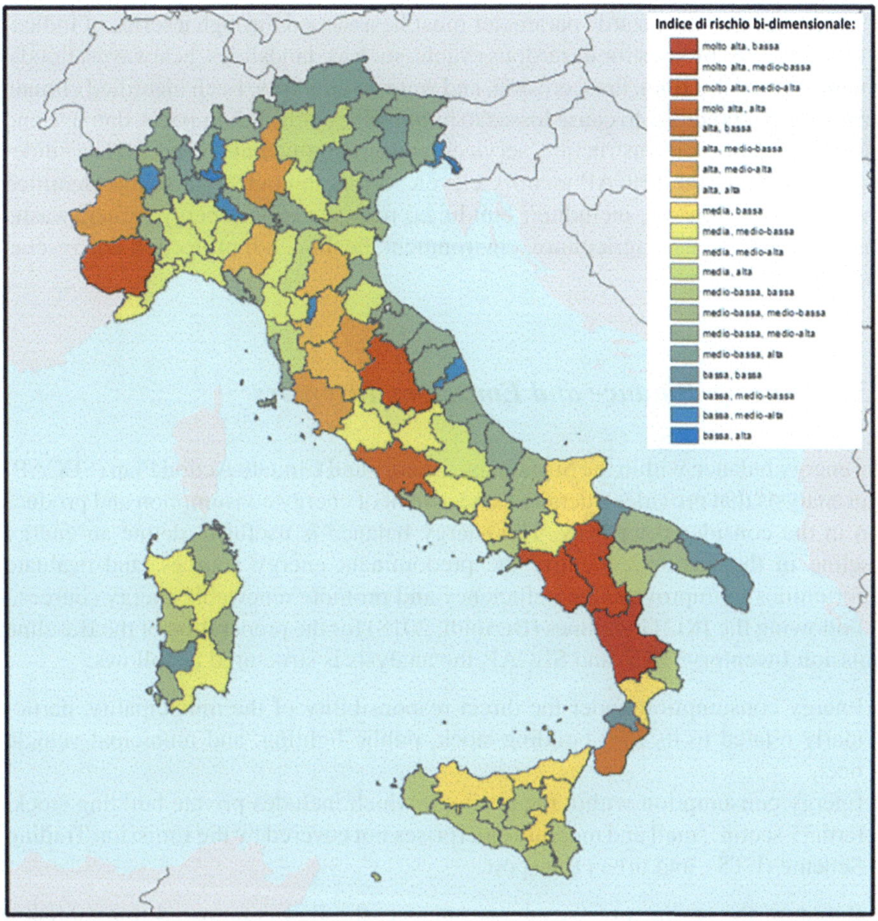

Fig. 8.3 Vulnerability index of Italian districts according to PNACC. *Source* PNACC 2022

be supplemented with the findings from SECAP phase 1 (territorial characterization of an area) to obtain a comprehensive map of risks and vulnerabilities in the studied territory.

In particular, the "hazard" parameter must be assessed through a series of indicators that serve as proxies for hazardous events, such as: landslides, heatwaves, floods, droughts, frost, flooding, fires, erosion, and water availability. Each identified climate hazard has the potential to cause losses to human lives or health impacts, damage and losses to properties, infrastructure, services, and environmental resources. The guidelines for drafting the SECAP identify eleven sectors to contextualise the identified environmental impacts, including: buildings, transportation, energy, water, waste, land planning and use, agriculture, environment, health, civil protection and rescue, and tourism.

8.3.3 Energy Balance and Emissions Inventory

The energy balance within the Sustainable Energy and Climate Action Plan (SECAP) is an analysis that provides a detailed assessment of energy consumption and production in the considered territory. The energy balance is useful to define an energy baseline in the territory, identify the predominant energy sources, and evaluate opportunities to improve energy efficiency and promote renewable energy sources.

Following the JRC Guidelines (Bertoldi, 2018) for the preparation of the Baseline Emission Inventory (BEI) and SECAP, the analysis is structured as follows:

- Energy consumption under the direct responsibility of the municipality, particularly related to its own building stock, public lighting, and municipal vehicle fleet.
- Energy consumption within the territory, which includes private building stock, tertiary sector, small and medium enterprises not covered by the Emission Trading Scheme (ETS), and urban transport.

Based on the findings of the energy balance, the BEI is prepared to establish a snapshot of the municipal energy situation in the reference year, to compare it with the current state. The BEI thus serves as the starting point for the SECAP, from which objectives can be defined, an appropriate Action Plan and following monitoring can be developed. The BEI quantifies the CO_2 emitted in the local authority's territory during the reference year in assessing the reduction commitments by 2030. To show the emissive trend of the territory, SECAP signatories must analyse a Monitoring Emission Inventory (MEI), starting from energy consumptions of subsequent years.

8.3.4 Proposal of Actions for Climate Change Adaptation

Adaptation strategies in the SECAP include a series of actions aimed at adapting to the impacts of climate change. Some of these actions can deal with reducing greenhouse gas emissions as well (i.e. forestation). Each action is associated with a specific SECAP sector. Since adaptation and mitigation strategies can vary depending on territory characteristics, available resources, and specific needs of local communities, it is difficult to standardise action proposals.

It is important to develop a customised SECAP that considers the specific challenges and opportunities of the considered territory, as identified in the territorial characterization and emissions inventory phases.

8.4 Case Studies

Three illustrative SECAPs adaptation case studies are presented below (Fig. 8.4). They emphasise adaptation actions, particularly in the context of floods, as the main theme of this discussion. Specifically:

- **Reno Galliera Union**: Special attention will be given to actions related to the reduction of surface runoff.
- **Municipality of Cervia**: Adaptation actions are linked to coastal erosion and the resulting risk of flooding from the sea.
- **Municipality of Vieste**: Adaptation actions are related to the possibility of flooding due to sea-level rise.

In addition to these, some of the main proposed actions related to other relevant issues for each territorial context will be reported. The summary of all adaptation actions, for each case study and sector of interest, is provided in Table 8.1. Every adaptation measure is associated with a Key Performance Indicator (KPI) which either is or can be used for monitoring the application of action itself in its territory.

8.4.1 Reno Galliera Union

The Reno Galliera Union includes a group of 8 municipalities within the metropolitan city of Bologna, Emilia Romagna (Italy). The union encompasses an area of 295.56 km^2 with approximately 74.000 residents (ISTAT). The analysis was conducted following the guidelines outlined in Chap. 2. The starting point was to conduct a detailed territorial analysis to identify the risks and vulnerabilities specific to the area under examination, as well as to compile the BEI. The results of this assessment demonstrate that heatwaves and intense precipitation are the climate hazards with the highest level of risk. Furthermore, it is noted that in a short to medium-term scenario, there is an expected slight increase in the intensity of all specified climate hazards. As for the sectors to focus on, they have been identified as water management, buildings, agriculture and forestry, and territorial planning.

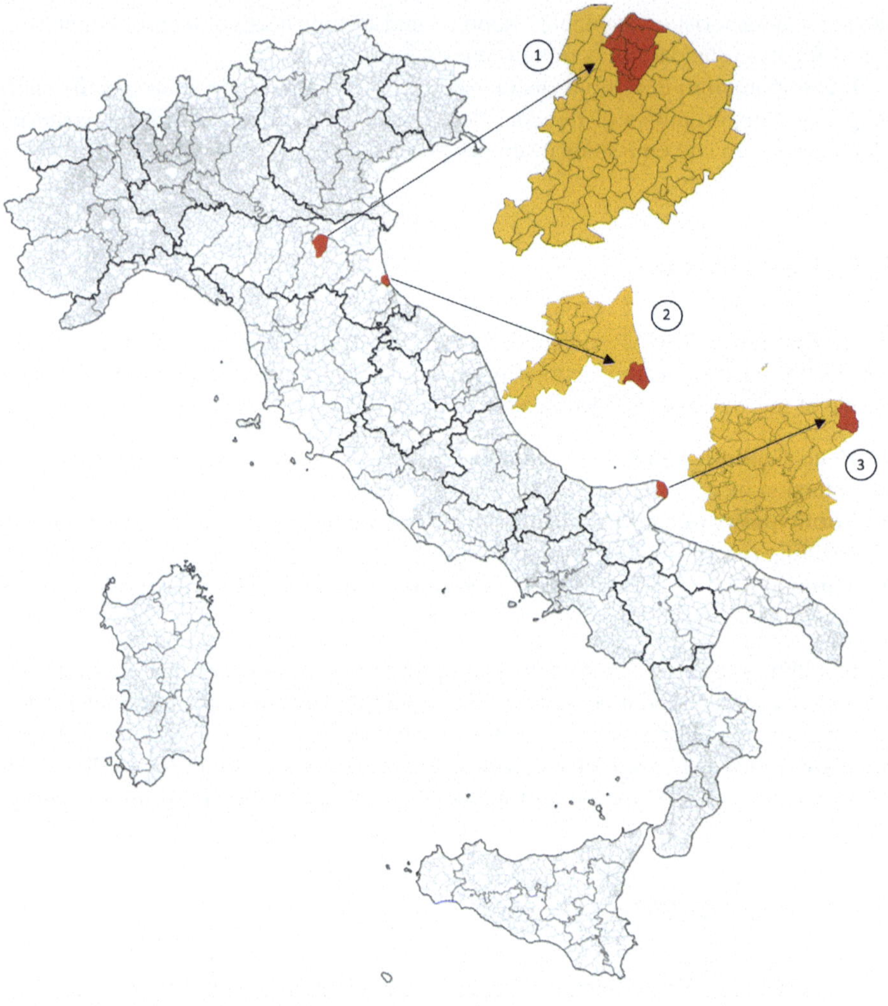

Fig. 8.4 Localization (areas coloured in red) of the case studies of adaptation to climate change within the Italian territory. *Note* 1: Reno Galliera Union, 2: Cervia and 3: Vieste (made by the authors)

8.4.1.1 Agricultural Land

The agricultural sector is largely affected by direct and indirect impacts of climate change. Rising average temperatures and heat waves create stressful conditions for crops, a situation that can be exacerbated by water shortages due to drought. All these factors disrupt the normal course of crop cultivation. Additionally, crop yields can be significantly influenced by the progressive increase in extreme weather events (such as storms and hailstorms) that can damage the harvest.

Adaptation Measures

To facilitate adaptation to climate change in the agricultural sector, first of all, it is essential to conduct a strategic assessment regarding the choice of plant species to cultivate, taking into consideration the evolution of the local territory. Furthermore, the responsible management of water resources is highly relevant, aiming to promote techniques that reduce waste. To mitigate extreme events, where possible, protective structures and crop protection systems should be implemented (such as nets for hail protection or anti-insect nets). In addition to introducing specific agricultural practices, the optimal management of irrigation systems, the selection of crop species, and the expansion of organic and local farming (km) are equally important. Supporting closed-loop farming enterprises and promoting private urban gardens are

Table 8.1 Adaptation measures listed by hazard, with each associated KPI, for each case study

Case study	Hazard	Adaptation measures	KPI [SI units]
Reno Galliera Union	Increase of extreme precipitation events, which often lead to damaging floods and challenging drainage situations	Reduction of surface runoff implementing: 1. Construction of drainage ditches alongside roads and parking lots 2. Installation of infiltration wells 3. Use of high-permeability pavements 4. De-pavement measures	Surface change in green and blue infrastructure [%] Surface change in waterproof pavement [%]
		Cleaning and maintenance of the main and secondary hydraulic networks (rivers, streams, canals, etc.)	Number and type of infrastructure on which adaptation interventions have been carried out
		Maintenance, improvement, or expansion, (including infrastructure upgrades) of the sewage system	Number and type of infrastructure on which adaptation interventions have been carried out

(continued)

Table 8.1 (continued)

Case study	Hazard	Adaptation measures	KPI [SI units]
Municipality of Cervia	Coastal erosion with the connected risk of sea flooding	Protection and restoration of the coastal dune: 1. Placement of sand in front of the current dune face 2. Installation of windbreak fences 3. Cover with dry plant residues spread on the ground to promote the deposition and capture of seeds and the development of new vegetation 4. Keep access to the dune area closed to prevent internal trampling 5. Annual monitoring of the dune-beach system	Dune slope [°]. Reference values: 21°/25° Before the intervention: 12°/15° Presence of halo-nitrophic pioneer vegetation in the arenile area (community interest 1210), vegetation of embryonic and mobile dunes (community interest 2110), therophytic vegetation of dune mosaics (community interest 2230), vegetation of the grey dunes (community interest 2130), hygrophile vegetation back of the dune (community interest 6420), shrough vegetation back of the dune (community interest 2260) Georeferenced length, width and height of the dune [m]
Municipality of Vieste	Increased intensity of storm and the long-term risk of sea-level rise	Infrastructure improvements in the areas facing the sea and raising road levels	Number and type of infrastructure on which adaptation interventions have been carried out Coastal retreat [m]
		Demolition of unauthorised structures (8 interventions) and the development of paths and/or natural engineering works (7 interventions)	Number and type of infrastructure on which adaptation interventions have been carried out

also significant. The communicative strategy is also important to make these solutions effective; this requires adequate information and training for those involved, the introduction of new management criteria and practices by industry associations.

8.4.1.2 Residential Land

Urban areas are highly sensitive to the effects of climate change due to the concentration of population and exposed infrastructure. Furthermore, climate phenomena are increasing in frequency and intensity, resulting in growing impacts on people's health and comfort, with a decrease in overall quality of life.

Adaptation Measures

The microclimate inside buildings is influenced by both external building components and internal systems. Therefore, the proposed adaptation measures aim to encourage the implementation of cool roofs and pavements, horizontal green roofs (green roofs and rooftop gardens), and vertical greenery (green walls) in both private and public residential structures.

8.4.1.3 Water

Among the primary risks associated with climate change in the Reno Galliera Union territory is the increase in extreme precipitation events, which often leads to damaging floods and challenging drainage situations (surface runoff). Surface runoff is directly proportional to land urbanisation (land use) and inversely proportional to soil permeability. The municipalities of Reno Galliera Union are situated in an area with a medium to high land consumption and low permeability, hence they are in a high-risk area for surface runoff (Fig. 8.5).

Adaptation Measures

To address this dual issue (surface runoff and damaging floods), it is possible to implement two different types of interventions. The first is aimed at reducing the issue related to surface runoff, while the second focuses on reducing the risk of flooding. The surface runoff depends not only on the intensity of rainfall events, but also on the permeability of the soil and the capacity of the drainage network. In natural environments, rainwater is washed and slowly filtered by and through the soil. In urban environments, impermeable surfaces hinder the natural process of water infiltration, causing a rapid runoff toward final receiving systems. In case of extreme precipitations, excessive runoff and limited receptor capacity can lead to temporary flooding of urban spaces. Due to climate change, it is expected that extreme events will increase in intensity, putting additional pressure on urban drainage systems and exacerbating their inefficiency.

Regarding the surface runoff there are multiple possible adaptation actions, including: the construction of drainage ditches alongside roads and parking lots,

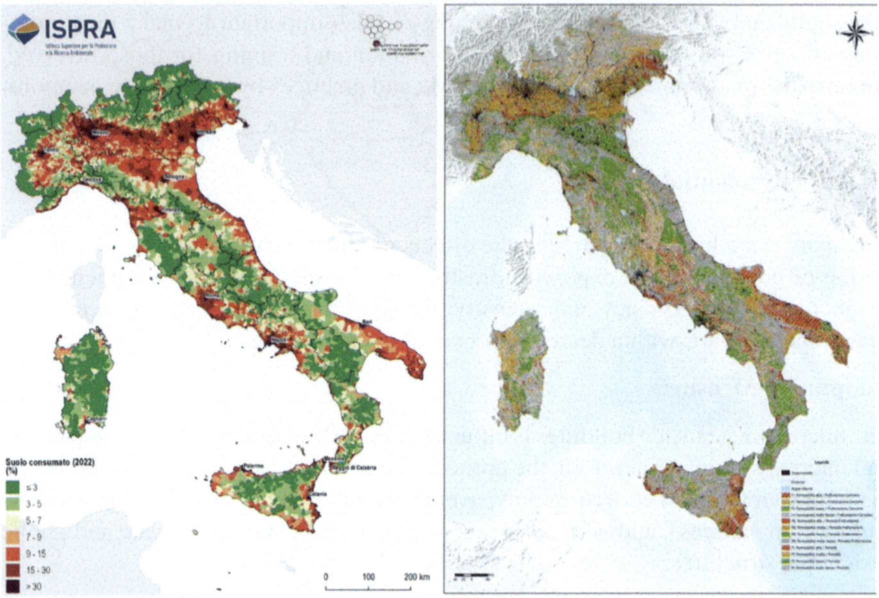

Fig. 8.5 Right: Italy land use (by municipalities) (SNPA, 2023); Left: Italy soil permeability (ISPRA, 2023)

the installation of infiltration wells, the use of high-permeability pavements as replacements for traditional pavements (on roads and in parking areas), de-paving and de-sealing measures, and, finally, incentives to increase the percentage of high-permeability areas (i.e., non-paved areas) within private properties.

Simultaneously with actions to reduce surface runoff, both in terms of quantity and drainage time, there is a desire to implement measures to reduce the hydraulic risk in the territory. To move in this direction, the proposal is to enhance collaboration with competent entities for the maintenance of the main and secondary hydraulic networks (rivers, streams, canals, etc.) and for the planning and execution of interventions aimed at reducing the hydraulic risk in the area (Fig. 8.6 represents an extreme event of flood). Furthermore, Reno Galliera Union intends to continue its collaboration with the sewage network management organisation (HERA Spa) to carry out maintenance and improvement or expansion, including infrastructure upgrades, of the sewage system. This collaboration aims to reduce critical situations related to the insufficient drainage capacity of the network and enhance its resilience. This will be done in conjunction with an analysis of the vulnerability of the overall knowledge framework of the General Urban Plan (PUG).

Fig. 8.6 Urban floods in Reno Galliera Union, May 2023 (picture made by the authors)

8.4.2 Municipality of Cervia

Cervia is an Italian municipality located in the province of Ravenna, Emilia-Romagna. The adaptation actions undertaken for the Municipality of Cervia, in addition to aiming at the drafting of a SECAP are part of the Adriadapt project, "a Resilience Information Platform for Adriatic Cities and Towns," funded within the framework of the cross-border cooperation program between Italy and Croatia. The Adriadapt project aimed to support the construction of local and regional resilience by developing the necessary knowledge base to identify and plan appropriate options for adapting to climate change.

8.4.2.1 Coastline

The analysis has outlined that the coastline near the dune area is characterised by strong erosion, most likely due to several factors, including high human activity, the presence of beachfront buildings and heavy use, various rigid elements disrupting coastal currents, cliffs and rocky outcrops, a significant decrease in the transport of sandy sediment from rivers, and improper management of vegetation and sand in the dune and beach area. Coastal erosion, which means loss of land along the coast due to the action of waves and tides, can lead to coastal flooding. When the coastline recedes due to erosion, coastal areas can become more vulnerable to flooding during storms or high tides. Waves can more easily penetrate inland, causing flooding in areas that were previously not subject to this type of event.

Currently, in Italy, a significant issue of coastal erosion is being observed. Approximately 60% of Italy's coastlines are affected by erosion, which can range in intensity from mild to severe. These coastal erosion phenomena pose a threat to the marine and coastal environment as well as the communities residing in these areas (Fig. 8.7). In particular, in the Emilia-Romagna region, located on the Adriatic coast of Italy, the coastal erosion rate is 31.4%, meaning that 31.4% of the region's coastline is undergoing erosion processes. This statistic is equivalent to a length of 44 km of coastline affected by erosion (ISPRA, 2018).

Adaptation Measures

Based on the considerations presented, a project for the protection and restoration of the dune has been planned. Considering the geomorphological challenges, the decision was made to create new conditions that favour the deposition and capture

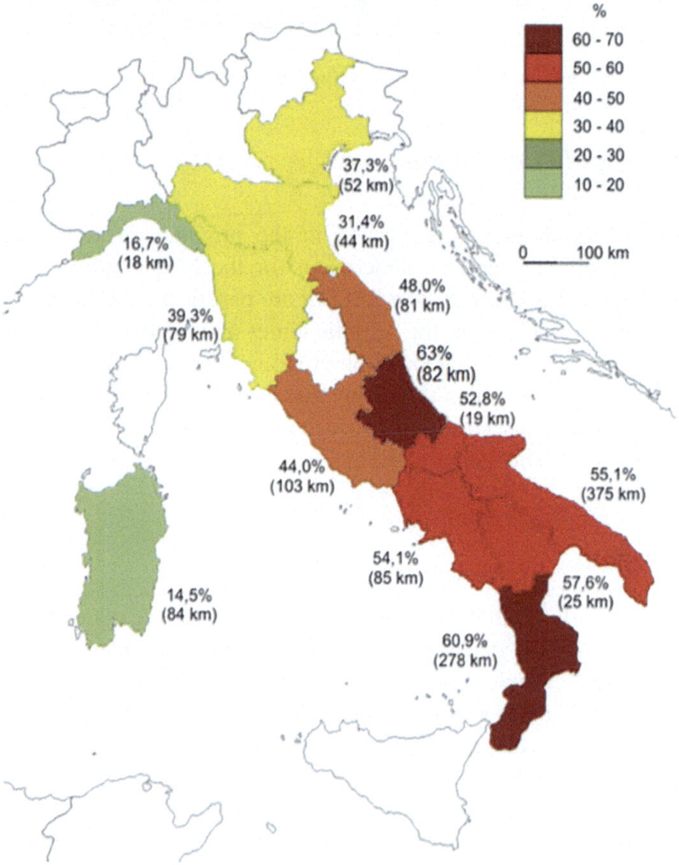

Fig. 8.7 Coastal erosion in Italy. *Source* ISPRA (2018)

Fig. 8.8 Left: dune-beach system in Cervia; Right: dune-beach ganivelles. *Source* Adriadapt Cervia project images by NIER Ingegneria

of sand carried by the wind, to achieve a dune body profile as close as possible to a natural profile, thus promoting the growth of native coastal vegetation (Fig. 8.8).

1. The first intervention to be carried out is the placement of sand in front of the current dune face, to provide the current face with a new slope of approximately 23°/25° and thereby create the formation of a new dune toe with dry and mobilizable sand to cover the existing face, primarily composed of very compact and non-mobilizable shell-like sand. It was observed to have a slope of about 12°/15° before the intervention.
2. The second intervention involved the installation of windbreak fences (ganivelles) to facilitate sand capture and deposition. Windbreak fences were placed with a defence line in front of the new dune toe on the seaside, a line on the current dune ridge, and multiple lines inside erosion channels to restore lateral continuity of the dune body. In total, the project envisaged the installation of 635 linear metres of windbreak fences (ganivelles), with the placement of 287 chestnut support posts spaced at 1-m intervals and one bracing stay per post on the seaside line, 165 chestnut support posts spaced at 1-m intervals and one bracing stay per post in the lines inside the blow-outs (erosion channels), and 92 chestnut support posts spaced at 2-m intervals and one bracing stay per two posts on the inner line on the dune crest.
3. The parts of the dune which was losing vegetation has been covered with dry plant residues spread on the ground, without excessive density, to promote the deposition and capture of seeds and the development of new vegetation.
4. It is of fundamental importance to not forbid access to the dune area, to prevent internal trampling, which promotes erosion, loss of vegetation, and the development of invasive and disruptive vegetation, all of which lead to an imbalance in the system and the onset of critical conditions.
5. Annual monitoring of the dune-beach system, according to the parameters listed in Table 8.1.

8.4.2.2 Spatial Planning

The conducted territorial analysis has highlighted an increase in average temperatures and heatwaves. To adapt to these climate hazards, proper territorial planning is crucial.

Adaptation Measures

The study carried out on the territory has emphasized the importance of sea breezes as local climate regulators. This is a phenomenon that must be considered in urban planning to avoid constructing built fronts along the coastline. Such constructions would have the undesired effect of obstructing the flow of air into the city, thereby increasing the likelihood of heatwaves.

8.4.3 Municipality of Vieste

Vieste is the easternmost municipality on the Gargano promontory and in the province of Foggia, Puglia (Italy) with a population of 13.459 inhabitants and an area of 169.19 km^2. Like the previous case, the analysis followed the same logical structure, highlighting that the most significant climate hazards (on which to focus the main adaptation actions) are droughts, fires, and heatwaves. Concerning a medium to long-term scenario, the risk related to storm surges and sea-level rise will increase. The sectors to concentrate on are the water sector, buildings, agriculture and forestry, civil protection and emergency services, and territorial planning.

8.4.3.1 Water

One of the main risks associated with climate change in the municipality's territory is the increased intensity of storm surges and the long-term risk of sea-level rise.

Adaptation Measures

The Municipality of Vieste intends to undertake adaptation and enhancement measures, including infrastructure improvements, in the areas facing the sea to ensure safety, including the possibility of raising road levels. Among the actions already planned by the Municipality are the revitalization of "Enrico Mattei" and "Europa" Promenades. Additionally, SECAP includes the completion of the safety measures for the "Pizzomunno" Cliff and the protection of the rocky outcrop in the "Reginella" area (Fig. 8.9). These actions will be combined with initiatives to restore sites with landscape and environmental value. The action, for example, will include the demolition of unauthorised structures (8 interventions) and the development of paths and/or natural engineering works (7 interventions).

Fig. 8.9 Costone Reginella in Vieste (picture made by the authors)

8.4.3.2 Civil Protection and Emergency Services

The forestry sector is significantly affected by both direct and indirect impacts of climate change. Particularly noteworthy are the number of wildfires in forested areas, mostly in protected regions.

Adaptation Measures

In this case, adaptation measures aim to enhance the maintenance of forested areas, such as clearing and deforestation of critical areas to create firebreaks. These actions will involve agreements with agricultural entrepreneurs. Additionally, the development of a municipal civil protection plan has been planned, including the management of the risk of forest fires.

8.4.3.3 Territorial Planning

Among the primary risks associated with climate change in the municipality's territory, there is an increase in the intensity of storm surges, and in a medium to long-term scenario, the risk of sea-level rise.

Adaptation Measures

Adaptation measures will include actions to adapt and enhance the infrastructure along the coast to secure the area, which may involve raising the road level as well. Each specific intervention can be monitored according to the parameters listed in Table 8.1.

8.5 Conclusions

Since climate changes are at the basis of the sociological and demographic evolution of the world population, it is necessary to investigate the causes that originate them and convey the main environmental changes on our planet. The European sustainability objectives frame and dedicate a space to the issues of climate change mitigation (implementing preventive policies with respect to greenhouse gas emissions into the atmosphere) and adaptation to extreme climatic events (starting from the assumption that the changes themselves are now inevitable and their consequences on the population and on the environment in which they occur).

In this sense, the Sustainable Energy and the Climate Action Plans (SECAPs) play a decisive role as one of the main programmatic tools of the Public Administrations for planning both the aspects.

Since a comprehensive understanding of the areas to which adaptation processes can be applied is required, it becomes essential to study local characteristics and convey them to policymakers and technical offices. This will enable them to make informed and conscious decisions regarding the priority actions to be implemented (Magni et al., 2020). At the same time, cities and municipalities are fundamental in pursuing sustainability, as they lead the local community in developing local policies and regulations that go beyond national targets; they encourage, for example, energy efficiency in public buildings and promote sustainable products and services through green public procurement; and they share information with the local population about experiences and examples to improve knowledge and awareness regarding climate change (D'Onofrio et al., 2023). This aspect enhances resilience at the local level, not only from an environmental perspective, but also in terms of social relationships, thanks to increased trust between institutions and the population.

With regard to adaptation to climate change, although there is no consolidated literature and case studies such as for the issue of mitigation, there are already several examples of how a territory, starting from the initiatives of its administrative policy, can implement practices that would have been seen as innovative in the past, but which are currently a necessary condition for guaranteeing a multi-year vision, and not just an emergency one, of the planning of the activities of a territory.

Despite some difficulties in applying the SECAPs and the related actions, for example the voluntary nature of the instrument, which is therefore not widely applied throughout the national territory, and the difficult coexistence with other territorial planning tools (for example urban planning), there is no evidence that the regulatory obligation has led, in the past, to greater awareness on the topic of adaptation to climate change. Therefore, it is believed that, in line with the emerging trends of the European taxonomy and consistently with the main initiatives on sustainability, both in the public and private sectors, the voluntary instrument, such as SECAPs, can be equated with regulatory obligations, not only for the rules that the municipalities decide to adopt, but also for the assumptions that the study behind SECAPs reports can originate.

References

Abarca-Alvarez, F. J., Navarro-Ligero, M. L., Valenzuela-Montes, L. M., & Campos-Sánchez, F. S. (2019). European strategies for adaptation to climate change with the mayors adapt initiative by self-organizing maps. *Applied Sciences, 9*(18), 3859.

Ali, E., Cramer, W., Carnicer, J., Georgopoulou, E., Hilmi, N. J. M., Le Cozannet, G., & Lionello, P. (2022). Cross-chapter paper 4: Mediterranean region. *Climate Change*, 2233–2272.

Berghi, S. (2016). Energy planning for metropolitan context: Potential and perspectives of Sustainable energy action plans (SEAPs) of three Italian big cities. *Energy Procedia, 101*, 1072–1078.

Bertoldi, P. (2018). Guidebook 'how to develop a sustainable energy and climate action plan (SECAP)'. Publication Office of the European Union.

Chapman, E., & Stiff, G. (2018). Guidebook 'how to develop a sustainable energy and climate action plan (SECAP)'. Compete4SECAP project.

Cinocca, A., Santini, F., & Cipollone, R. (2018). Monitoring methodologies and tools for the sustainable energy action plans to support the public administration. *Energy Procedia, 148*, 758–765.

Croci, E., Lucchitta, B., Janssens-Maenhout, G., Martelli, S., & Molteni, T. (2017). Urban CO_2 mitigation strategies under the Covenant of Mayors: An assessment of 124 European cities. *Journal of Cleaner Production, 169*, 161–177.

Delponte, I., Pittaluga, I., & Schenone, C. (2017). Monitoring and evaluation of sustainable energy action plan: Practice and perspective. *Energy Policy, 100*, 9–17.

D'Onofrio, R., Camaioni, C., & Mugnoz, S. (2023). Local climate adaptation and governance: The utility of joint SECAP plans for networks of small-medium Italian municipalities. *Sustainability, 2023*(15), 1.

Dzene, I., Polikarpova, I., Zogla, L., & Rosa, M. (2015). Application of ISO 50001 for implementation of sustainable energy action plans. *Energy Procedia, 72*, 111–118.

EU Commission. (2018). Communication from EU Commission: the Action Plan: Financing Sustainable Growth. https://eurlex.europa.eu/legal-content/EN/TXT/?uri=CELEX:52018DC0097

EU Commission. (2021). *Forging a climate-resilient Europe—The new EU strategy on adaptation to climate change.*

EU Commission. (n.d.). Why a covenant of mayors?|Covenant of Mayors—Europe. https://eu-mayors.ec.europa.eu/en/about

EU Regulation. (2021a). 1119 of the European Parliament and of the Council of 30 June 2021 establishing the framework for achieving climate neutrality and amending Regulations (EC) No 401/2009 and (EU) 2018/1999 ('European Climate Law'). *Official Journal of the European Union, 50*, 243.

EUD Regulation. (2021). 2139, Commission Delegated Regulation (EU) 2021/2139 of 4 June 2021 supplementing Regulation (EU) 2020/852 of the European Parliament and of the Council by establishing the technical screening criteria for determining the conditions under which an economic activity qualifies as contributing substantially to climate change mitigation or climate change adaptation and for determining whether that economic activity causes no significant harm to any of the other environmental objectives. *Official Journal of the European Union No L, 442*, 1–349.

European Commission. (2019). The European green deal.

European Commission. (2021). Forging a climate-resilient Europe—The new EU strategy on adaptation to climate change. Communication from the Commission to the European Parliament, the Council, the European economic and social committee and the Committee of the regions, Brussels, February 24, 2021.

European Union. (2020). Regulation (EU) 2020/852 of the European Parliament and of the Council of 18 June 2020 on the establishment of a framework to facilitate sustainable investment, and amending Regulation (EU) 2019/2088 (Text with EEA relevance). *Official Journal of the European Communities, 198*, 13–43.

European Union Proclamation. (2000). Charter of fundamental rights of the European Union. Communities, 100(364/8).

Field, C. B. (Ed.). (2012). *Managing the risks of extreme events and disasters to advance climate change adaptation: Special report of the intergovernmental panel on climate change.* Cambridge University Press.

Gagliano, A., Nocera, F., D'Amico, A., & Spataru, C. (2015). Geographical information system as support tool for sustainable energy action plan. *Energy Procedia, 83*, 310–319.

ISTAT, R. S. (2020). Informazioni statistiche per l'Agenda 2030 in Italia.

ISPRA (Istituto Superiore per la Protezione e la Ricerca Ambiente). (2018). Linee guida per la difesa della costa dei fenomeni di erosione e dagli effetti dei cambiamenti climatici.

ISPRA (Istituto Superiore per la Protezione e la Ricerca Ambiente). (2023). Atlante dei dati ambientali, Edizione 2023.

Kamenders, A., Rosa, M., & Kass, K. (2017). Low carbon municipalities. The impact of energy management on climate mitigation at local scale. *Energy Procedia, 128*, 172–178.

Magni, F., Musco, F., Litt, G., & Carraretto, G. (2020). The mainstreaming of NBS in the SECAP of San Donà di Piave: The LIFE master adapt methodology. *Sustainability, 2020*(12), 19.

MASE (Ministero dell'Ambiente e della Sicurezza Energetica). (2018). *Piano Nazionale di Adattamento ai Cambiamenti Climatici.*

MASE (Ministero dell'Ambiente e della Sicurezza Energetica). (2023). *Valutazione Ambientale Strategica del Piano Nazionale di Adattamento ai Cambiamenti Climatici.*

Monforti-Ferrario, F., Kona, A., Peduzzi, E., Pernigotti, D., & Pisoni, E. (2018). The impact on air quality of energy saving measures in the major cities signatories of the Covenant of Mayors initiative. *Environment International, 118*, 222–234.

Scorza, F., & Santopietro, L. (2021). A systemic perspective for the sustainable energy and climate action plan (SECAP). *European Planning Studies*, 15–16.https://doi.org/10.1080/09654313.2021.1954603

Sistema Nazionale Per La Protezione Dell'Ambiente (SNPA). Consumo di suolo, dinamiche territoriali e servizi ecosistemici, Edizione 2023, Ottobre, 11, 2023.

UN General Assembly. (2015). Transforming our world: the 2030 Agenda for Sustainable Development. UN Doc. A/RES/70/, September 25, 2015.

United Nation. (n.d.). Goal 13|Department of Economic and Social Affairs. https://sdgs.un.org/goals/goal13

Chapter 9
Systemic Approaches and Principles for Urban Stormwater Management

Vittore Negretto

Abstract In the frame of adaptation planning, there is a need to address climate change impacts in a systemic way that goes beyond sectoral logic. The management of stormwater in urban areas, mainly relying on sewer systems, requires exploiting every planning opportunity to compensate for both the increased intensity of heavy rainfalls and the high density of soil sealing. This chapter explores different approaches to stormwater management connected to the compensation of new development and retrofitting of the existing built environment, stretching from quantity-based approaches to ones also considering quality and stormwater reuse. Based on applied research to support planning processes in Padua, a medium-sized Italian city, the chapter then outlines indications and principles of intervention to allow a systematic mainstreaming of these approaches in the planning and design system, targeting those policymakers, technicians and practitioners that can steer the physical modification of urban space.

Keywords Spatial planning · Urban drainage · Climate change adaptation · Sustainable drainage · Policy integration

9.1 The Spatial Planning Challenge: Managing Risk Arising from Climate Change

"The climate crisis is exacerbating existing risks and creating new challenges for cities worldwide. Across the globe, communities grapple with the same pressing questions, such as, "How can we adapt to changing rainfall patterns, floods and storms made worse by the climate crisis?" These challenges are often combined with limited access to information. To understand and project evolving climate risks, cities need to bridge these knowledge gaps to enable measures that offer the best cost-to-benefit ratio and protect the most people" (UNEP, 2023).

V. Negretto (✉)
IUAV University of Venice, Venice, Italy
e-mail: vnegretto@iuav.it

M. Granceri Bradaschia et al. (eds.), *Climate Change Adaptation, Flood Risk, and Beyond*, Planning for Climate Proof Cities,
https://doi.org/10.1007/978-3-031-65463-3_9

Global rising temperatures are associated with widespread changes in weather patterns, causing extreme events to be more frequent and intense (IPCC, 2013, 2023). In recent years, these extreme events have caused a tragic loss of life and damages in the billions of euros. Figure 9.1 shows how some CC-related events are also seriously damaging the economies of the European Economic Area. According to the report published by the European Environment Agency, the economic losses caused by extreme events related to climate change from 1980 to 2019 account for a total of 446 billion euros for the European Economic Area member states. On average, this equals 11.1 billion euros per year, which is nearly 3 per cent of the gross domestic product of the countries analysed (EEA, 2021). Flooding, storms, severe droughts, heatwaves, and other events are considered catastrophic when they strike an exposed system, with high values and many lives (Carraro & Mazzai, 2015). Cities, with their densely populated and anthropic environment, are exposed to increasing threats connected to a changing climate and are highly vulnerable to shocks and stresses.

The combination of increasing urbanisation and climate change could raise even more the impacts of these events on the urban environment, exposing a larger urban area and more of the urban population. This scenario makes essential both mitigation measures to stem climate change and adaptation measures to prepare our urban environment to cope with the changes that are now inevitable. Effective mitigation requires a synergy of international and national policies, with each country taking action to reduce human activity's impacts on the climate and the environment. Unlike emission mitigation, adaptation and risk management from climate change impacts are distinctly local issues. This is due to several factors, as reported in the latest IPCC assessment report (2023), including:

- The change in intensity of hazards at the regional level. The different intensities of climate change according to different regions, which in some places reduce

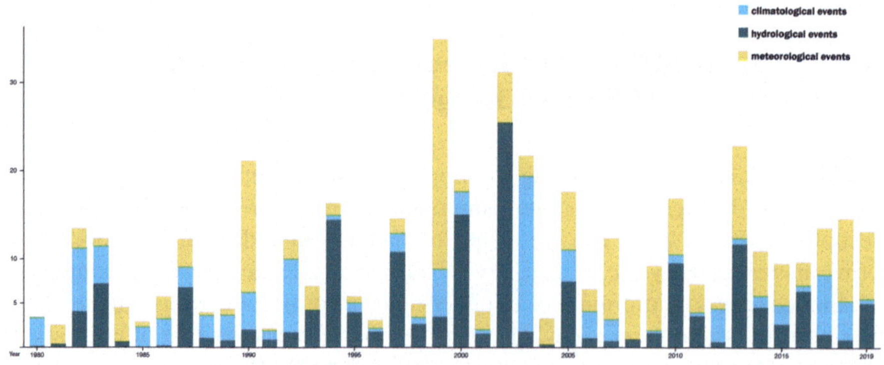

Fig. 9.1 Economic damage caused by weather and climate-related extreme events in Europe, in EUR millions, adjusted for 2019 value (1980–2019). This graph focuses on events of three categories: meteorological events (storms), hydrological events (floods and mass movements) and climatological events (heatwaves, cold waves, droughts, and forest fires). *Source* European Environment Agency

the intensity of some hazards and, in many others, exacerbate their intensity and frequency.

- The vulnerability of territories is a different response of territories to hazards. Some more vulnerable territories react unfavourably and experience substantial impacts from weather and climate events that other territories can manage without impacts, which is also connected to their adaptive capacity.
- And the value of the assets and populations exposed to the impact, as it varies significantly between different territories and within the same territory.

Given the inherently local character of risk, interventions to improve its management also have a dimension closely linked to specific territories and contexts. Urban and regional planning can, through different tools and approaches, intervene considerably in the dimensions of risk that shape vulnerability and exposure. Stormwater management systems are a relevant part of the urban infrastructure and strongly relate to urban planning. By collecting stormwater, they aim at reducing flooding, avoiding damages, and preventing potential environmental impacts (Butler & Davies, 2000). In the context of climate change and rapid urbanisation, with changes in rainfall patterns and land uses, the ability of these systems to comply with intense events may be severely affected (Ferreira Santos et al., 2019).

This chapter aims precisely to address the challenge of managing the increase in urban runoff that exceeds the capacity of infrastructure networks through the domain of urban planning practices and tools. The perspective in which this work stands is to help bridge the existing knowledge gap in line with the principle stated within the adaptation paragraph of the Paris Agreement (2016), which endorses the central role that science and local knowledge play in adapting a specific territory to the impacts of climate change.

It is, then, the purpose of this chapter to highlight the lessons learned from an experience of decision support applied research on reducing vulnerabilities to intense precipitations and managing stormwater. In this regard, the main objectives of this chapter are: (1) Highlight the hazards and vulnerabilities of the relation between rain and urban areas; (2) discussing and assessing different approaches to stormwater management; (3) identifying the policy opportunities for stormwater management; and (4) proposing a system of operational priorities for technical projects and policies.

9.1.1 Knowledge to Inform Policymakers

There has been global recognition of the need for effective communication between scientific research, policymakers, and urban planning professionals. The United Nations and the European Commission have advocated for the importance of knowledge transfer to ensure sustainable and resilient urban planning (Böhme & Gløersen, 2023; Henriques et al., 2020). This recognition is reflected in documents such as the United Nations' 2030 Agenda for Sustainable Development (UN GA, 2015) and the European Union's Horizon Europe research program (European Council, 2019), which include specific initiatives to promote the sharing of knowledge and skills from science towards the diverse stakeholders involved in urban planning.

Knowledge transfer in the field of urban planning is crucial for ensuring the effective dissemination of scientific knowledge to professionals and policymakers. Various international research agendas actively promote methods and programs to overcome this siloed logic, with a focus on knowledge transfer and capacity building as tools to equip policymakers and practitioners with the knowledge and tools to systematically address risks and impacts stemming from climate change (GCoM & Un-Habitat, 2022). The emphasis is on bridging the gap between global objectives based on scientific findings and subnational plans grounded in local needs. This gap is both spatial, which requires considering the specific local characteristics, and temporal. In fact, available projections of long-term climatic changes often clash with the short-term horizons of local policy mandates, typically spanning only a few years at most.

9.1.2 Operational Support to a Local Flooding Challenge

This chapter contributes to the discourse described above by providing valuable insights to systematically implement stormwater management principles in the practices of various public and private professionals engaged in urban spaces. It applies these concepts to an actual case study in which available science, data and planning practices are used to support a planning process and inform policymakers and practitioners. Within this applied research, the science and practice-based knowledge is translated into an operational and implementable dimension, leading to the adoption of the Sustainable Urban Drainage guidelines (Padua Municipality, 2022a). The chosen case study is Padova, a medium-sized Italian city in the Veneto region, and its territorial, technical, and planning context. Building on these characteristics, the chapter outlines the dimensions of local hazards, vulnerabilities, existing policies, and possible planning innovations that can be undertaken to reduce the risk of frequent urban flooding (Fig. 9.2).

The subsequent paragraphs highlight the relationship between local precipitation and urbanity, leading to the vulnerability to urban flooding and requiring planning intervention. Following this, various management approaches in force in different contexts are compared and analysed, spanning from traditional quantity-based approaches to more recent sustainable urban drainage approaches that embrace the recent concept of nature-based solutions (Fletcher et al., 2015). Applying this knowledge to the specific context of Padua results in the definition of a locally valid set of principles, integrating the existing policies and supporting the decision-making process to improve stormwater management continuously. The final discussion highlights the lessons learned and the recommendations that emerged in this process.

Fig. 9.2 Flooding-prone areas in Padua (made by the author)

9.2 Hazard and Vulnerability: Intense Precipitations and Urban Areas

9.2.1 Identifying the Target Weather Event

One of the local climate change hazards is cloudbursts, characterised by intense precipitation that pours significant amounts of rain over specific territories within minutes or at most hours. When such events overcome the capacity of an urban system to handle rainfall runoff, floods of varying magnitudes occur, often gaining prominence in the news. Urban flooding is thus the visible manifestation of an urban environment exposed to water quantities it cannot effectively manage.

It is not necessarily an exceptional storm the one that stresses the drainage system, especially in heavily impermeable territories with pre-existing vulnerabilities like the Padua's one. The frequency of flooding phenomena is such that intense precipitation events cannot be considered extraordinary. While the subsequent paragraph will delve into various aspects of the relationship between runoff and urban environments, this section focuses on identifying rains that exceed the response capacity of drainage systems, leading to urban flooding.

In the case of the Municipality of Padua, a review of major local newspapers highlights rainfall events causing flooding in that specific context. The starting point is the historical series of impacts, based on the effects of precipitation-territory relationships. From this empiric perspective, characteristics of rainfall events that challenge the local Paduan drainage system have emerged (Arpav, 2022). Prolonged periods of adverse weather (several consecutive days) with widespread and particularly persistent precipitation may occur. During these periods, intense phases alternate with less significant ones, but the rains causing the most damage are those with a large amount of precipitation concentrated in a short time. Precipitation events challenging the rainwater drainage system typically have durations of less than an hour (15–60 min). They concentrate in the summer months from June to September, especially in July and August, predominantly from late afternoon to early evening (Fig. 9.3).

The global warming expected by the IPCC (2023) involves an increase in global mean surface temperature and several climate changes in Europe, including an increase in the frequency and intensity of intense precipitation events for the region. These two values are used for defining return periods (RP), or probability values of a given intense event, which are the basis for sizing stormwater management infrastructure. Statistical calculations can determine its probability for a given event intensity (e.g., intense precipitation). For instance, a 20-year return period has a 5% chance of occurring in a given year, while a 5-year return period has a 20% chance. These values facilitate comparable intensity and probability thresholds across different territories and are the ones used in infrastructure sizing. Under the current changing climate, the time series on which return times are calculated appears to deviate from the current and future probabilities and intensities of intense precipitation events. Table 9.1 reports data from two events on two consecutive days with corresponding probabilities (RP). These values are not presented to statistically confirm high probability

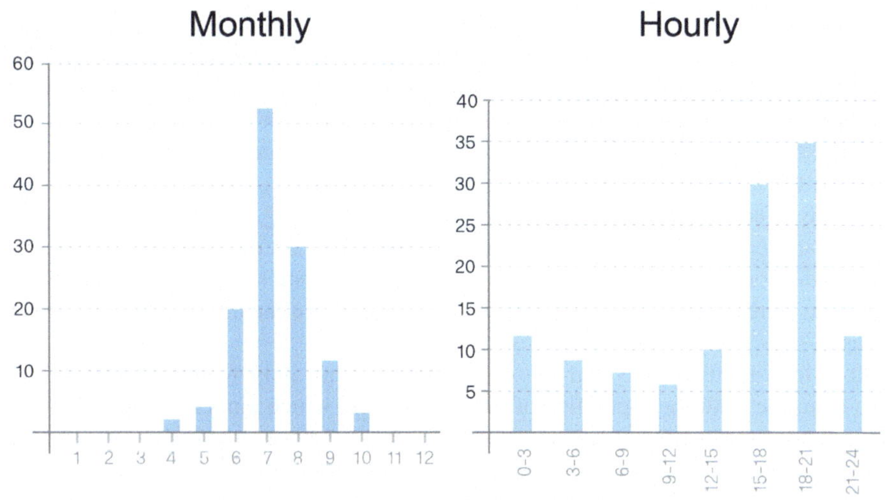

Fig. 9.3 Distribution of intense events with a 15-min duration and return time of at least 2 years for the stations of Campodarsego, Faedo, Legnaro, Mira, Montegalda, and Teolo during 2013–2022. *Source* ARPAV (2022)

Table 9.1 Rain quantities in mm and corresponding return period (RP) registered at the Padova-orto botanico meteorological station during two consecutive events (ARPAV, 2022)

Day	5 min	RP	10 min	RP	15 min	RP	30 min	RP
21/07/18	12.8	7	25	29	34	38	42.8	16
22/07/18	9.6	2	19.2	5	28.6	10	47.2	32

but to quantify the intensity and quantity of previously described precipitations. They also highlight the relationship with public infrastructure for managing and diverting flows, typically designed for return periods of 5 or 10 years. The result of this increasing mismatch is an intensification of impacts in urban settings that occur when intense precipitation events, added to dense soil sealing, generate more water than the capacity of traditional stormwater management systems.

9.2.2 Urban Soil Response to Precipitations

In the context of flood risk-based urban planning, it is essential to consider the multiple environmental pressures arising from urbanisation, with particular attention to soil sealing. This phenomenon is recognised as a critical factor in disrupting ecosystem balances, with direct and indirect repercussions on the hydrological cycle, carbon balance, local microclimate, biodiversity, and agro-environmental equilibrium (Brenna, 2012).

The sealing of urban soils has a well-documented series of effects on the hydrological cycle (Butler & Davies, 2000), complicating the management of intense precipitation events. The primary effects involve increased surface runoff due to reduced infiltration and evapotranspiration capacity, and a reduction in runoff times. Consequently, compared to permeable soils, impermeable soils generate a larger volume of water to be managed in a shorter time frame.

These effects expose urban areas to the risk of flooding, leading to a deterioration in surface water quality due to increased mobilisation of sediments and pollutants. Additionally, they alter the hydrological characteristics of water bodies, reduce available water resources, and contribute to a general decline in the quality of life in cities.

It is important to note that these effects result not only from soil sealing but also from the traditional practice of urban drainage, which aims to rapidly dispose of meteoric and wastewater (Chocat et al., 2001). Although initially effective in conveying runoff away from urban areas, this approach has often transferred problems downstream, especially when urbanisation has covered a significant part of territories and altered basin-level flows. To prevent the increase in river flooding risk at the watershed level whenever a new urbanisation or soil sealing occurs, different places have adopted regulations and policies to keep the amount of runoff "unaltered" and compensate for the contribution of new land-use transformations.

9.3 Stormwater Management Approaches

Italian approaches

Current practices and approaches applied in Italy, albeit with differences between different contexts, arise from this need to reduce watershed flooding and embrace the concept of invariance.

In Veneto, as in other Italian regions, the approach to offset the effects of urbanisation on the water cycle is still firmly tied to the principle of hydraulic invariance (Veneto region, 2012). This principle states that no land use transformation should contribute to an increase in the peak flow of the receiving water body. Hydraulic invariance requires quantifying and making available a volume (reservoir) to offset the effects of urbanisation, obliging those proposing a transformation to bear the burdens of resulting soil sealing. However, preparing such storage volumes is not aimed at retaining runoff on-site but rather at maintaining the formation of floods at the basin scale unchanged. The objective is to artificially maintain the maximum flow rate from the urbanised area within the limits of pre-urbanisation conditions. Hydraulic invariance, therefore, concerns the maximum flow rate that can flow downstream in a given time (flow rate), not the absolute quantity of runoff (volume). It follows that the urbanisation of, for example, an agricultural field, despite adhering to the principle of hydraulic invariance, will not eliminate impacts on the hydrological cycle; only the maximum outflow from urbanisation will remain "invariant".

Other impacts on the hydrological cycle need to be considered, such as the total quantity of runoff leaving the transformed area. In some Italian contexts, such as the Lombardy region (2017), the concept of invariance has been extended to include hydrological invariance. This means that the principle states that the volumes discharged into the receiving water body (sewer, watercourse, etc.) from transformed areas should not exceed those pre-existing the transformation. The additional runoff caused by the new transformation—and subsequent soil sealing—must, therefore, be managed locally through infiltration and reuse.

Lombardy, along with the Autonomous Province of Bolzano (2004), represents the most relevant Italian cases where the principle of hydraulic invariance is integrated, primarily focusing on controlling peak flows from transforming areas, with attention also to the total runoff quantity.

The combination of the two principles of hydraulic and hydrological invariance is more effective than the exclusive application of the former: in compensating for an intervention, attention is focused not only on maintaining the outflow within certain limits but also on ensuring that part of the runoff does not flow downstream. Adopting these two management principles as a guide for compensations restores an active role to soils both in flood mitigation and in support of all phases of the hydrological cycle. The positive effects also include increased soil infiltration, greater evapotranspiration for the benefit of microclimate and air quality, and the potential reuse of water as a resource (Table 9.2).

9.3.1 Anglo-Saxon Approaches

To encourage Italian approaches toward focusing on urban flooding phenomena, a comparison with some international planning practices highlights aspects that can be integrated into runoff management. Remarkably, the Anglo-Saxon context provides insights into both attention to issues that are not exclusively quantitative at a basin scale and a more flexible implementation method based on principles and standards.

The English and Welsh standards are two examples of new approaches to managing rainfall runoff that put the benefits at their core. These approaches are based on the idea of restoring the natural hydrological functions of the soil through distributed and local stormwater management, considering such water as a resource to be managed within urban areas (Ashley et al., 2013; Wong, 2007). These approaches aim to promote environmental sustainability, resilience, and adaptation to climate change in cities by proactively addressing challenges related to stormwater management and flood risk mitigation, both urban and riverine.

These benefits can be available and activated during intense precipitation and for ordinary and frequent ones. It is indeed necessary for attention not to be exclusively directed toward extreme events but also to consider the response of urban soils to modest precipitation. Alongside current underground infrastructure systems—based on channelling and diverting flows and sized for intense events—there is an opportunity to intervene on the surface with sustainable urban drainage techniques

Table 9.2 Comparative table between stormwater management approaches of hydraulic invariance, hydrogeological invariance and Anglo-Saxon approach and selected performance indicators

Category	Objective	Performance indicator	Hydraulic invariance	Hydrogeological invariance	Anglo-Saxon approach
Surface water	Reduction of hydraulic stress	Reduction of one-year event peak runoff rate [%]	x	x	x
		Reduction of one-year event volume runoff [%]	–	x	x
		Total interception of a 5 mm-event	–	–	x
	Support flood prevention	Reduction of 100-year event peak runoff rate [%]	x	x	x
		Reduction of 100-year event volume runoff rate [%]	–	x	x
	Avoid deterioration of surface water quality	Reduction in pollutants loading of first-flush [%]	x	x	x
		Reduction in pollutants loading of total runoff [%]	–	–	x
Urban liveability	Increase in micro-climate	Evaporation rate [%]	–	x	x
Biodiversity	Support for ecosystem	Increase in vegetated area (m^2)	–	x	x
		Increase in floristic diversity [number of species]	–	–	x

(continued)

Table 9.2 (continued)

Category	Objective	Performance indicator	Hydraulic invariance	Hydrogeological invariance	Anglo-Saxon approach
Water resources	Local groundwater level	Increase in runoff infiltration [%]	–	x	x
	Reduction of building freshwater consumption	Rainwater recycling and reuse for service water [%]	–	x	x

Based on Ferreira Santos et al. (2019)

that positively affect the hydrological cycle every day. Within these Anglo-Saxon contexts, for example, it is required that land transformations guarantee effective management of both intense and ordinary runoff events. For example, the English standards (DEFRA, 2015) adopt a broader strategy that limits runoff rates and runoff volumes. This dual strategy, implementable through storage, reuse, infiltration, and evapotranspiration processes, offers numerous advantages. It enables a more effective reduction of ecological and morphological impacts on receiving water bodies and reduces hydraulic risk downstream (Woods-Ballard et al., 2015).

English standards also mandate considering events with two different return periods (1 and 100 years) to design and size sustainable urban drainage systems (SuDS). This approach aims to improve the effectiveness of urban drainage systems even in ordinary situations. Drainage systems must be able to retain and manage on-site more intense and rare rainfall events (with return periods on the order of 50–100 years) and more frequent and less intense events (with rainfall depth of 5–10 mm). These minor events significantly contribute to the overall hydrological-hydraulic response of a basin and influence water quality. Furthermore, this provision implies that no single type of device is suitable for stormwater management, but rather the need to combine detention, infiltration, and evapotranspiration systems. Considering these indications in the design of urban drainage systems increases their effectiveness even in the face of frequent precipitation, resulting in perceptible benefits in ordinary conditions (Pasi et al., 2019).

In addition to these quantitative criteria, Welsh standards (Welsh Government, 2016) emphasise multifunctionality aspects. For example, standard S1 explicitly promotes the reuse of stormwater and soil infiltration. Standards S3, S4, and S5 require the design of multifunctional urban drainage systems that maximise benefits in terms of water quality management, improvement of urban areas' liveability, and support for ecosystems. The comparison with the Anglo-Saxon approach reveals a different mentality underlying surface runoff management, characterised by an awareness of the multifunctional potential of alternative drainage systems and the desire to restore the natural hydrological functions of the soil, considering the overall context of climate change.

9.3.2 *Performance Indicators of the Different Approaches*

Although with some regional differences, the various regulations considered can be grouped into three main approaches: hydraulic invariance, hydrogeological invariance, and the Anglo-Saxon approach to sustainable urban drainage. Table 9.2 aims to compare these three approaches to highlight the differences and points of contact. The different categories considered are used in the classification of Sustainable Urban Drainage Systems (Fletcher et al., 2015) and for stormwater systems in general (Ferreira Santos et al., 2019). The main objectives and some Performance Indicators (PI) are given for each category to indicate the result of applying the different approaches to a hypothetical green-field to urban area transformation. The indicators chosen to emphasise the relationship of these three approaches to surface waters, urban liveability, biodiversity, and water resources. From the various indicators in the literature, indicators that can represent, even partially, the benefits generated in addition to exclusive flood prevention were chosen. These include the goals of reducing hydraulic stress, avoiding deterioration of surface water quality, increasing micro-climate, supporting the ecosystem, recharging local groundwater level, and reducing building freshwater consumption.

Finally, beyond issues exclusively related to water management, there is an additional aspect that the Anglo-Saxon approach distinguishes from the Italian one. It concerns how prescriptions are provided: compared to the definition of precise rules and regulations, in the Anglo-Saxon context, standards are provided in the form of principles rather than rules. Based on Braithwaite (2002), in this framework, the distinction between rules and principles is that the former implies a specific prescription, while the latter has a more generic or unspecific prescription. Principles tend to be vaguer, have a broader range of applications, and find a more defined shape and boundary once applied in a specific context. The Anglo-Saxon standards analysed tend to offer general indications for intervention rather than precise rules or practices so that they can be relevant and applied in very different contexts—be they regulatory, geographical, or instrumental.

9.4 Exploiting Opportunities in the Padua Planning Framework

In order to improve the responsiveness of the territory of the city of Padua in the management of extreme weather/climate events (Fig. 9.2), a space emerged from the analysis between the regional regulations of hydraulic invariance and the interventions of new design and extraordinary maintenance of land-based elements such as roads, squares, parking lots, and street areas. This can be inspired by the Anglo-Saxon approaches of management through principles, to have a spillover effect of improving runoff management through the systematic action of different sectors and professionals, guided by a set of policies and principles that can make the most of every opportunity for land intervention, even if not explicitly for hydraulic purposes. In this perspective, the current framework of local policies and plans is first described, and then a proposal for principles specific to the Padua context is presented.

9.4.1 The Padua Policy Framework

There are provisions in the body of available planning documents that allow an approach to managing surface water before it reaches the centralised removal system.

In the various existing instruments that regulate and plan land use in the City of Padua, various aspects can support the implementation of urban drainage. The Land Use Plan (City of Padua, 2014, 2023) presents the strategic elements for coordinating urban transformability, respecting the identified hydraulic hazards. It provides requirements and obligations for managing hydraulic risk in urban transformations. There are also references to soil permeability (Art. 9) and its compensation in any urban/building transformation by applying invariance hydraulics (art. 8.1.1 and art. 8.1.2).

Instead, the Plan of Interventions takes on a prescriptive character: by regulating the use and functions of spaces, in fact, it provides for limiting impervious surfaces in favour of permeable ones and suggests the techniques and devices to be used for the containment of runoff in the construction of building works or in the design of open spaces. Within the Technical Implementation Standards, the regional approach for hydraulic management related to new land transformations, the hydraulic invariance, is followed. Specifically, this means that the peak flows generated by the transformed land must be unchanged from those before urbanisation. The "General Prescriptions" give the possibility to employ structural compensatory measures, typical of the traditional "grey" approach, and "green" measures such as sustainable urban drainage techniques and green infrastructure, as well as the digging of new ditches or the establishment of temporary flood zones in green areas.

With a vision geared toward design sustainability and improving the quality of public space, the Building Regulations promote the use of techniques for stormwater management and reuse, specific standards for green roofs and facades, and guidance on using materials for increasing the albedo of surfaces.

Finally, the SECAP and the Green Plan (City of Padua, 2021, 2022) represent two voluntary tools for systematising adaptation interventions through natural systems that contribute to reducing the city's hydraulic risk, improving urban comfort and the quality of public space. Key objectives include mitigating the impacts of climate change and preserving ecosystem services in the city.

The set of local policies and regulations of the City of Padua outlines an extension of the regional hydraulic invariance guidance regarding urban runoff management. Although the set of documents analysed has different levels of cogency, the local level embraces indications or prescriptions that follow the more advanced approach of hydrogeological invariance that is not mandatory in the Veneto Region. From the summary Table 9.3, the set of local policies and prescriptions also include indications on increasing vegetated areas over urbanised ones, improving the resulting urban microclimate, and reusing rainwater in buildings. However, with respect to hydrogeological invariance, there is no focus on reducing the total volumes of runoff outflow, nor their infiltration on site.

9.4.2 A Context-Specific Set of Principles

It is evident from the body of policies and plans of the City of Padua that there is a willingness to consider runoff management not exclusively as flow management, but also to consider other aspects of the hydrologic cycle such as quantity, destination, and quality. These aspects must be taken into account in the design of runoff management when there is a land use transformation and in the implementation of the various planning tools described earlier. This can be enabled by prioritising different runoff management aspects to have a systematic and coordinated approach among plans.

Pursuing these priorities in each land use intervention allows for systemically acting on the urban hydrological cycle and increasing resilience in runoff management by flanking existing infrastructure with diffuse and decentralised solutions. However, given the multitude of tools and contexts of application in the territory of the City of Padua, a more flexible form of regulation is preferable to a rigid regulation approach that allows for the necessary flexibility while maintaining coordination between different interventions. Therefore, the guiding principles for managing runoff are organised in a hierarchy of decreasing priorities (Fig. 9.4), ordered according to the main objective: retaining as much runoff on site as possible.

The hierarchy of priorities has a guiding function in evaluating techniques to be used in the design of a specific intervention. Principles placed in the first levels of the hierarchy generate more significant benefits locally and at the watershed scale and are preferable to those in lower hierarchies. A virtuous intervention applies multiple principles: allocating part of the runoff according to principles of a higher hierarchy, and thus with greater benefits triggered by lower rainfall, and then managing the remaining amount according to principles of a progressively lower hierarchy. Should

Table 9.3 Comparative table between stormwater management policies and regulation framework at the regional (Veneto Region) and local level (Padua Municipality) and selected performance indicators

Category	Objective	Performance indicator	Regional law on hydraulic invariance	Local policies and regulations
Surface water	Reduction of hydraulic stress	Reduction of one-year event peak runoff rate [%]	x	x
		Reduction of one-year event volume runoff [%]	–	–
		Total interception of a 5 mm-event	–	–
	Support flood prevention	Reduction of 100-year event peak runoff rate [%]	x	x
		Reduction of 100-year event volume runoff rate [%]	–	–
	Avoid deterioration of surface water quality	Reduction in pollutants loading of first-flush [%]	x	x
		Reduction in pollutants loading of total runoff [%]	–	–
Urban liveability	Increase in micro-climate	Evaporation rate [%]	–	x
Biodiversity	Support for ecosystem	Increase in vegetated area (m^2)	–	x
		Increase in floristic diversity [number of species]	–	–

(continued)

Table 9.3 (continued)

Category	Objective	Performance indicator	Regional law on hydraulic invariance	Local policies and regulations
Water resources	Local groundwater level	Increase in runoff infiltration [%]	–	–
	Reduction of building freshwater consumption	Rainwater recycling and reuse for service water [%]	–	x

Based on Ferreira Santos et al. (2019)

there be impediments to implementing a higher priority principle, it is possible to scale up to the next level. Impediments arising from peculiar contexts or project conditions can be various. The most recurrent impediments are: water quality and the presence of possible pollutants; insufficient available space; the level of surface or vulnerable aquifer; particular soil composition; excessive cost of construction or maintenance; and possible constraints.

The premise of applying these priority hierarchies for runoff destinations concerns water quality. Stormwater runoff washes polluted surfaces and thus has suspended pollutants that preclude some destinations if not adequately treated. The nature of these pollutants varies; generally, they are present in the first 2.5–5 mm of rainfall called "first rainwater". In designing an intervention, it is necessary to assess the susceptibility of surfaces to be contaminated by pollutants in advance, and, if so, provide appropriate devices that meet the filtering priority. Provision should be made to treat these potentially polluted waters as close as possible to the source of pollution, as it is easier to isolate potentially polluting areas and manage treatment points. There are a variety of devices to treat first rainwater, including some sustainable urban drainage techniques that have a broader spectrum of benefits than the widely used first rain tanks.

The first level in the hierarchy considers water as a resource for local use. This is considered the top priority for a few reasons. The first is that all water that is stored and reused later is a share of precipitation that will not generate runoff and, therefore, is subtracted from the quantities that need to be managed. This principle can generate benefits with both extraordinary and abundant rainfall. Its reuse also makes it possible to reduce the water requirements that the same uses would later require. Finally, the amount that can be allocated for reuse is equal to the volume of reservoir available at that time, so to handle more intense events, it must necessarily be combined with one of the following principles to handle excess volumes. The storage methods and subsequent reuse vary by context and location, most commonly for irrigation, washing, and general non-potable uses, including industrial and private uses.

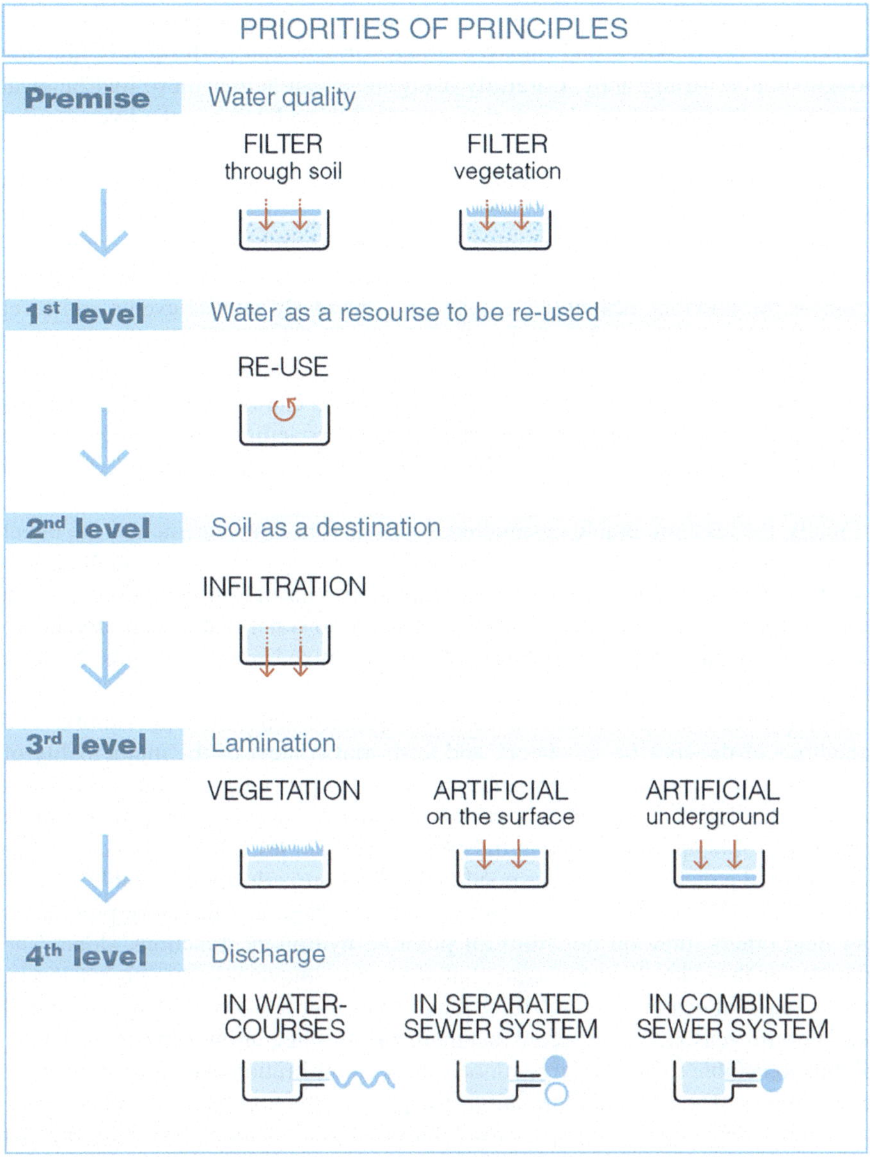

Fig. 9.4 Representation of the hierarchy of principles for the municipality of Padua

The next level of priority concerns considering soil as a destination for runoff. Compatible with hydrogeological characteristics, soil can be considered as a volume into which rainwater can be allocated through slow penetration, emulating the behaviour of undeveloped soil and prioritising permeability. Soil infiltration also supports the processes of evapotranspiration and biological adsorption that allow both to support vegetation and remove significant amounts of runoff to be managed downstream. Many paved and unpaved urban surfaces can perform this function through suitable stratigraphy. Carefully using these can help achieve hydrological invariance, retain runoff in place for minor rainfall, and recover the hydrological functions of soils.

The third preferable level regarding benefits to the hydrological regime concerns runoff lamination, understood as the temporary storage of stormwater in a volume to attenuate the outflow. This principle is of a lower hierarchy than the previous ones because it has a limited capacity to generate benefits. It can be employed downstream of the previous principles to manage particularly intense events and meet the requirements of the hydraulic invariance standard. In general terms, the most sustainable volumes are vegetated, followed by artificial volumes on the surface and, finally, artificial volumes underground. More benefits can be generated by linking to infiltration priority or reuse priority, with the understanding that in order to serve as a storage, a volume must be available and thus at least partially emptied after an event.

Finally, the least preferable destination for urban runoff management is the direct discharge of runoff into the receiving body. This principle should be applied only when the ones mentioned above are incompatible with design conditions and, in any case, for the least amount of runoff. Discharge to water bodies is preferable to discharge to dedicated sewers, while discharge to mixed sewers remains the least preferable destination overall.

The proposed hierarchy of principles was prepared based on the opportunities and criticalities of the specific regulatory and territorial context of the municipality of Padua. Table 9.4 takes the performances listed in the previous tables and compares them with the objectives to which the proposed principles respond. Since there is more than one possible option in some of the proposed principles, where necessary, it has been specified which option can respond to the PIs if properly applied. Table 9.4 also highlights how, although some principles are at a higher, and therefore preferable, level than others, they do not fulfil all possible hydrologic functions alone. The rationale for these levels is consequential since it is imaginable that higher levels, such as reuse allocation, cannot fully handle the runoff from a 100-year rainfall event. For this reason, it is expected that there will be integration between the levels, and thus to be able to handle large runoffs through infiltration and lamination if the context allows. The context's limitations to applying the upper layers can be many. For example, depending on the space available or the composition of the soils, the first level (on-site infiltration of runoff) may not be entirely applicable. For this reason, the principles are organised in descending levels, allowing them to scroll down to the next less preferable level if there are evident obstacles. Thus, the value of the table is not quantitative, as it is not important how many PIs the individual level responds to, but its function is to highlight the values of the individual levels. And thus enables the policymaker and practitioner to know which PIs their chosen combination of levels contributes to reach.

Table 9.4 Comparative table of the proposed principles and selected performance indicators

Performance indicator	Proposed principles				
	Premise	1st level	2nd level	3rd level	4th level
Reduction of one-year event peak runoff rate [%]	–	x	x	–	–
Reduction of one-year event volume runoff [%]	–	x	x	x	–
Total interception of a 5 mm-event	–	x	x	x	–
Reduction of 100-year event peak runoff rate [%]	–	–	x	x	–
Reduction of 100-year event volume runoff rate [%]	–	–	–	x	–
Reduction in pollutants loading of first-flush [%]	x	–	–	–	–
Reduction in pollutants loading of total runoff [%]	x	–	–	–	–
Evaporation rate [%]	–	x	x	x vegetation	x water courses
Increase in vegetated area (m^2)	x	–	x	x vegetation	–
Increase in floristic diversity [number of species]	x	–	x	–	–
Increase in runoff infiltration [%]	–	–	x	x vegetation	–
Rainwater recycling and reuse for service water [%]	–	x	–	–	–

Based on Ferreira Santos et al. (2019)

9.5 Conclusion and Recommendations

The current climate change framework does not allow for defining a safety level within which to consider the Padua area protected from flooding. Coping with this uncertainty requires a flexible and evolutionary approach that can gradually change a territory already challenged by current weather phenomena. Adapting urban spaces to intense rainfall requires taking advantage of every opportunity for action to manage stormwater systemically and diffusely, in parallel to current centralised management. Any amount, however small, managed locally and removed from the current collection infrastructure is, therefore, an adaptive action that contributes to improving the response to intense events and creating widespread resilience. In addition, adopting the principles and techniques of sustainable urban drainage allows both local and watershed benefits to be multiplied, also contributing to adaptation to other climate change impacts and rebalancing the hydrological regime.

Four recommendations can be distilled based on this practical application of supporting policymakers to systematise urban flood risk reduction.

1. **Use best-available data, models, and projections according to the needs of policymakers**

There is a need to connect long-term climatic scenarios, useful to inform National Adaptation Plans, with Disaster Risk Management approaches (UNDRR, 2015) based on the collection of impacts from past climate events, in an effort to sensitise local levels to the ongoing impacts of climate change. The disaster risk management approach to hazard identification is an effective language for informing policies that have a medium to short time horizon, as local policies typically do. This look at past events is not meant to replace future climate forecasts, but to inform, based on science and data, which precise events have caused and will cause impacts in the city, communicating a clear message of "it could happen this afternoon". This mixed approach matches data and models with known events, providing space for policymakers to link current criticalities with long-term scenarios and understand the short-term intervention needs.

This allows for identifying climate phenomena that have caused and will cause impacts in the city, while simultaneously identifying areas that have already experienced impacts. And allows a greater connection to the local knowledge advocated in the Paris Agreement, providing space for policymakers to link current criticalities with long-term scenarios and intervene in the short-term.

2. **Move beyond invariance in stormwater management**

In light of the frequent floods recorded in many cities, it is imperative to move beyond the approach of maintaining the runoff generated by urban soils unvaried with each rainfall event, whether in terms of flow (hydraulic) or quantity (hydrological). To manage runoff challenges generated by frequent intense rainfall events, adopting an approach aiming at the "variance" of the current hydraulic and hydrological imbalance is necessary through widespread and systemic interventions. This involves enhancing the territories' response to the stresses caused by intense rainfall events while simultaneously yielding benefits for ordinary rainfall. Along with flood protection, this can be done by embedding objectives of reduction of hydraulic stress, avoiding deterioration of surface water quality, increase in micro-climate, support for ecosystems, increase recharge of local groundwater level and reduction of building freshwater consumption.

3. **Exploit existing policy opportunities and act systematically**

The different levels of cogency and the different ways local policies and regulations are implemented do not allow for an integrated application of stormwater management principles. This is partially inherent in the sectoral nature of the various documents, where the power to prescribe and implement certain aspects of runoff management ends, where the jurisdiction of another sector and regulatory apparatus begins. For example, the green plan's prescriptions of increased vegetated area do not, rightly, enter into the merits of building codes related to construction, although this is an effective tool to enter into the merits of private space and thus a substantial portion of land. A simultaneous and comprehensive review of all regulations is not easy and quick to implement as it would require continuity of policies and not inconsiderable organisational capacity. In addition, the complexity and diversity of different contexts would also not allow for establishing a single set of regulations, e.g., regional, that could cascade to local contexts facing different challenges of hydraulic risk management and climate change.

Integrating a hierarchy of principles, based on the Anglo-Saxon approaches outlined in this chapter, can coordinate different sectors towards a systematic application of a set of context-specific objectives. These objectives are presented in the form of principles to be followed or qualities that the various systems should have, allowing a more systematic consistency between different planning tools that can be flexibly integrated in different contexts.

4. **Include performance indicators in the mainstreaming and options evaluation**

The compromise for having maximum flexibility in applying the proposed principles approach lies in not having direct normative force and not uniquely identifying those responsible for its implementation. The implementation of the described objectives is then derogated from existing planning instruments, which can refer to the proposed objectives and listed performance indicators in their updating and evaluation of options. Some are purely hydraulic, such as the 100-year event peak runoff rate and volume reduction and have related legislation. Others, however, such as the increase in evaporation rate and of the vegetated area or the reuse of water, can be achieved through a plurality of planning methods. They can become part of the requirements to build, and thus in the Building Code, or to be reward criteria for access to green incentives and certifications. Mainstreaming the principles and related indicators into instruments that give them normative value and accountability for implementation is therefore linked to local opportunities to include them and contribute to the systematic implementation of sustainable urban drainage and subsequent risk reduction.

Expanding on the international importance of the study, it is evident that the challenges faced in the Padua area resonate with urban areas worldwide grappling with the escalating impacts of climate change. The strategies outlined herein for managing stormwater and mitigating flood risk offer valuable insights applicable beyond regional borders. Indeed, as climate change continues to manifest in erratic weather patterns globally, the need for adaptable and proactive approaches to urban planning becomes increasingly urgent. By embracing sustainable urban drainage principles and integrating them into policymaking frameworks, cities worldwide can enhance their resilience to climate-induced hazards while fostering environmental sustainability and community well-being. Furthermore, the emphasis on data-driven decision-making and the integration of diverse stakeholders echoes the principles advocated in international agreements such as the Paris Agreement, underscoring the relevance of this study in the broader context of global efforts towards climate adaptation and disaster risk reduction. As cities worldwide confront the multifaceted challenges posed by climate change, the lessons learned from the Padua case study offer a roadmap for fostering resilience and sustainability in urban environments worldwide.

References

ARPAV. (2022). Intense precipitation events in Padua. Interreg Response.

Ashley, R., Lundy, L., Ward, S., Shaffer, P., Walker, L., Morgan, C., Saul, A., Wong, T., & Moore, S. (2013). Water-sensitive urban design: opportunities for the UK. In *Proceedings of the ICE-Municipal Engineer* (p. 166).https://doi.org/10.1680/muen.12.00046

Brenna, S. (2012). Effetti ambientali del consumo di suolo. ERSAF—Ente Regionale per i Servizi all'Agricoltura e alle Foreste della Lombardia, Milano, Italia.

Böhme, K., & Gløersen, E. (2023). Territorial Agenda 2030: Implementation Review (Brussels: European Parliament, Policy Department for Structural and Cohesion Policies). https://doi.org/10.2861/305300

Braithwaite, J. (2002). Rules and principles: A theory of legal certainty. *Australian Journal of Legal Philosophy, 27.*

Butler, D., & Davies, J. (2000). *Urban drainage.* E & FN Spon, London, UK. ISBN 0-419-22340-1

Carraro, C., & Mazzai, A. (2015). Il clima che cambia. Non solo un problema ambientale, Il Mulino Editore.

Chocat, B., Krebs, P., Marsalek, J., Rauch, W., & Schilling, W. (2001). Urban drainage redefined—From stormwater removal to integrated management. *Water Science and Technology, 43*(5).

Comune di Padova. (2014). *Piano di assetto del territorio (PAT).* Padua, Italy.

Comune di Padova. (2021). *Piano d'azione per l'energia sostenibile ed il clima di Padova (SECAP).* Padua, Italy.

Comune di Padova. (2022). *Linee guida per il drenaggio urbano sostenibile.* Padua, Italy.

Comune di Padova. (2023). *Piano di assetto del territorio, aggiornamento (PAT).* Padua, Italy.

Council of the European Union. (2019). *Horizon Europe package: Framework programme for research and innovation 2021–2027.* Brussels, 14298/19.

DEFRA. (2015). Non-statutory technical standards for sustainable drainage systems. Report PB14308. Dept. Environment & Rural Affairs (DEFRA), London, UK.

EEA. (2021). *Economic losses from climate-related extremes in Europe.* Retrieved from https://www.eea.europa.eu/data-and-maps/indicators/direct-losses-from-weatherdisasters-4/assessment

Ferreira Santos, L., Fonseca Galvãoa, A., & Cardoso, M. A. (2019). Performance indicators for urban storm water systems: A review. *Water Policy, 21*(1), 221–244. https://doi.org/10.2166/wp.2018.042

Fletcher, T. D., Shuster W., Hunt, W. F., Ashley, R., Butler, D., Arthur, S., Trowsdale, S., Barraud, S., Semadeni-Davies, A., Bertrand-Krajewski, J. L., Mikkelsen, P. S., Rivard, G., Uhl, M., Dagenais, D., & Viklander, M. (2015). SuDS, LID, BMPs, WSUD and more—The evolution and application of terminology surrounding urban drainage. *Urban Water Journal, 12*(7). https://doi.org/10.1080/1573062X.2014.916314

Global Covenant of Mayors for Climate and Energy & UN-Habitat. (2022). Discussions and Findings from the 2021 Innovate4Cities conference: New insights for the 2018 Global Research and Action Agenda on Cities and Climate Change Science. https://doi.org/10.26188/62795d16f3bab

Henriques, C. N., Dragovic, S., Auer, C., & Gomes, I. (2020). The territorial agenda 2030: Towards a common language? A review of a conceptual framework. *Europa XXI, 38*, 157–174. https://doi.org/10.7163/Eu21.2020.38.8

IPCC. (2013). Climate Change 2013: The physical science basis. Contribution of Working Group 1 to the Fifth Assessment Report of the Intergovernmental Panel on Climate Change, Cambridge University Press.

IPCC. (2023). Climate Change 2023: Synthesis report. Contribution of Working Groups I, II and III to the Sixth Assessment Report of the Intergovernmental Panel on Climate Change. IPCC, Geneva, Switzerland. https://doi.org/10.59327/IPCC/AR6-9789291691647

Pasi, R., Negretto, V., & Musco, F. (2019). *Diversi approcci al drenaggio urbano sostenibile: un confronto tra il contesto normativo inglese e quello italiano.* Archivio Studi Urbani e Regionali, Franco Angeli.

Provincia Autonoma di Bolzano. (2004). Deliberazione del Consiglio Comunale n. 11 del 10 febbraio 2004.

Regione Lombardia. (2017). Regolamento regionale n. 7 del 23 novembre 2017 recante criteri e metodi per il rispetto del principio dell'invarianza idraulica ed idrologica ai sensi dell'art. 58bis della L.R. 12/2005

Regione del Veneto. (2012). Deliberazione della Giunta Regionale n. 3637 del 13 dicembre 2002.

UN General Assembly, Transforming our world: the 2030 Agenda for Sustainable Development, 21 October 2015, A/RES/70/1, available at: https://www.refworld.org/docid/57b6e3e44.html. Accessed November 13, 2023.

UNDRR. (2015). *Sendai framework for disaster risk reduction 2015–2030.* Sendai, Japan.

UNEP. (2023). Adaptation gap report 2023: Underfinanced. Underprepared. Inadequate investment and planning on climate adaptation leaves world exposed. Nairobi. https://doi.org/10.59117/20.500.11822/43796

Welsh Government. (2016). *Recommended non-statutory standards for Sustainable Drainage (SuDS) in Wales: Designing, constructing, operating and maintaining surface water drainage systems.* Welsh Government, Cardiff, UK.

Wong, T. H. F. (2007). Water sensitive urban design: The journey thus far. *Australian Journal of Water Resources, 110*(3). https://doi.org/10.1080/13241583.2006.11465296

Woods-Ballard, B., Wilson, S., Udele-Clark, H., Illman, S., Ashley, R., & Kellagher, R. (2015). The SuDS manual. Report C753, second ed. Construction Industry Research & Information Association (CIRIA), London, UK.

Chapter 10
Flood and Adaptation Financing: From "Project" Mode to "Living with" Mode at the Service of the EU

Jerome Dendura, Mercedes Sanchez Varela, and Massimiliano Granceri Bradaschia

Abstract Flood management has been approached as infrastructure projects to reduce the impact of floods heights or waves. Experiences have exposed, as the recent floods in Emilia Romagna, the lack of preparedness beyond flood management tools and policy instruments. With climate change, it is no longer just a matter of minimising flood and damages; it is adapting to and living with new conditions, including safeguarding people's lives, health, culture, and the economy. This perspective firmly places flood management into the integrated adaptation policy realm, spanning well over many fields and jurisdictions. Policy dimension is determined both by the fiscal space and the recognition and acceptance that risks call for change. Understanding risks and creating social acceptance are paramount to financing policies that no longer focus on the "building back better" approach. Taking measures now to avoid disaster, damages, and losses to offer a perspective for living and producing with new conditions is a paradigm shift. This begs the question who should pay? This study, which can be considered as a policy paper, addresses this question by criticising the policies and financing approaches employed hitherto, including the "build-back-better", embedding them in the umbrella concept "Project" Mode. The issue is investigated in the EU context, where both adaptation policy and flood risk frameworks are operative, not always in synergy though. By framing this EU framework under the policy integration lens, this study explores the EU adaptation and flood policies through a series of key factors necessary to respond fully to finance and policy effectiveness, which are: money, priorities, proactiveness, and private sector involvement. Insights and recommendations are provided and tailored at the service of the EU context, which is at the beginning of a transition to sustainable financing and policy operativity, going beyond the simplistic "Project" mode.

J. Dendura (✉) · M. Sanchez Varela
Adelante Knowledge and Development, Madrid, Spain
e-mail: jerome@adelante.eu.com

M. Granceri Bradaschia
IUAV University of Venice, Venice, Italy

M. Granceri Bradaschia et al. (eds.), *Climate Change Adaptation, Flood Risk, and Beyond*, Planning for Climate Proof Cities,
https://doi.org/10.1007/978-3-031-65463-3_10

Keywords Climate change adaptation · Adaptation finance · Flood management · European Union · Systemic change

10.1 Introduction

Floods are causing more and more damage to our rapidly urbanising world and rural areas, with disproportionate consequences for the poorest and most vulnerable. In Europe, the European Commission (EC) has reported some 1500 episodes of flooding since 1980, with very high human and economic costs.[1]

Significant damage was reported in May 2023 in Italy's heartland, Emilia Romagna, where the equivalent of six months' rainfall fell in just a few hours, killing 14 people and drowning huge agricultural areas ravaging fields of cereal market gardens and fodder for livestock. Entire villages were washed away by the floodwaters, bridges collapsed, and 400 roads collapsed (Bonnel, 2023). A series of various factors compounded: several months of rain (200 ml/in 24 h) fell over an area already exposed to high hydrogeological risk; smaller basins were put under high stress simultaneously and when the water of the rivers reached the plains, it clogged the drainage network, overflowing and breaking the riverbanks. At the same time, a sea storm was blocking the outflow to the sea.[2] Two weeks earlier the region was already facing other floods whose damage would be around 300 million euros according to the Coldiretti syndicate (ibidem). Following the massive floods of the summer 2021, affecting Germany and Belgium especially, this latter episode should have alerted us to the lack of preparedness of our societies and their socio-economic activities, beyond infrastructures' management. The effects of climate change make this approach ineffective and possibly prone to a higher risk of damages: it is not just a matter of minimising flood and damages to assets and activities; it is about adapting to and living with new conditions, including safeguarding people's lives, health, culture, and the economy.

This shift of perspective firmly places flood vulnerability into the policy arena, not just investments into infrastructure projects, spanning well over multiple fields of work, administrative boundaries, and jurisdictions. The policy space for adaptation to climate change is determined both by recognition and the level of acceptance that risks induced by climate change call for new forms of living, producing and safeguarding people, assets and by the fiscal space for financing this change. This, in turn, impacts how local planning is done.

Climate change and other human-induced changes are accelerating. As Mauro Rossi, a researcher at the Research Institute for Geo-Hydrological Protection recently stated, "Rising temperatures intensify drought episodes, drying up the soil and

---

[1] https://environment.ec.europa.eu/topics/water/floods_en.

[2] https://www.unive.it/pag/16584/?tx_news_pi1%5Bnews%5D=14642&cHash=59b65d26a343c7d2cc9f58d090b0cb76.

changing its permeability in different ways".[3] The infrastructures that were designed in the last few decades, based on return period rather than prospective changes, cannot adapt to the impacts of those events (Granceri, 2020; Serinaldi, 2015). Currently, 3.3 billion people live in areas that are already vulnerable to climate change and the effects of climate change are amplified in cities, where over half the world's population is concentrated (IPCC, 2023).

The devastating consequences of climate change are known, particularly in terms of flooding, and the likelihood of occurrences are highlighted by the 6th IPCC report (2023). However, understanding future flood risks is increasingly complex given the combined effects of changes happening. Simply put, floods risk increase depending on factors such as catchment areas' rainfall, land use, sea level and wave conditions. The compounded effects of overall climate and other key natural mechanisms deregulation and global warming is estimated to worsen in all regions of the world.

According to Garric (2023), a significant number of individuals face multiple simultaneous challenges, including underdevelopment, extreme poverty, inadequate governance, and insufficient funding. Governance and funding need to be adapted to new conditions in all regions and countries, including in Europe. In more vulnerable countries, the concern is that climate change endangers the advancements achieved in development, which is particularly alarming considering that the continent might house 40% of the global population by the century's end (United Nations, 2022).

Understanding risks and communicating clearly to the public is paramount to creating social acceptance for financing adaptation policies beyond the "building back better approach".[4] Taking measures now to avoid disaster, damages, and losses requires rethinking how to live and produce sustainability and safety within our environment. Creating understanding and acceptance of risks requires financing too, prioritising research for understanding the risks and communication for creating trust in science, thereby enabling demand for change to emerge, be voiced and be part of the solution. Social acceptance is a necessary (but not sufficient) condition for financing policies driving resilience to flood (and climate change in general): acting based on research to finance adaptation of our societies and the transition to sustainable practices is new and goes beyond projects. Hence, this article critically addresses the policies and financing approaches employed hitherto embedding them in the umbrella concept "Project" Mode.

European Union (EU) policies are increasingly reflecting this approach and are being translated into legal and regulatory instruments. EU competences on environmental policies have been in place for some decades and the Treaty of Paris in 2015 on climate change triggered the EU to launch the EU Green Deal in 2019 that set the pace and vision for EU ambitious climate targets. New policies and regulations were passed pushing beyond mere initiative, e.g. carbon tax turned into Cross border Adjustment Mechanism. Very rapidly, EU policy makers took advantage of

[3] https://www.euronews.com/green/2023/05/22/were-italys-floods-caused-by-climate-change-experts-analyse-what-happened-as-36000-left-ho.

[4] https://reliefweb.int/report/thailand/lessons-learned-tsunami-recovery-key-propositions-building-back-better.

a stable coalition at the European Parliament to drive a greener policy agenda with climate ambition and dedicated financing instruments to better serve this new climate agenda. Policy making and ambition was clearly driven by the EC while the Member States (MSs) were more concerned with crisis management mode to deal with the increasing effects of climate negatives such as wildfires, floods, and other disasters. For this, EU increased support to make MSs able to call and count upon EU Civil Protection Mechanism to reinforce their national response mechanisms.[5]

In spite of these advances, this begs the questions: who should pay for the cost of floods and adaptation? To what extent can people be asked to face the costs (e.g., through insurance) when conditions and regulations are changing? Should the State finance through its budget (tax and debt)? Should the regulations drive economic agents to take responsibility for some costs? Which level of government should finance this transition? How can market and price mechanisms be harnessed and regulated?

This exploratory essay addresses these questions by proposing an analytical framework on the way to address effective financing (Sect. 10.2), presenting the current EU adaptation policy framework and the local level challenges (Sect. 10.3), going through a series of key factors necessary to respond fully to adaptation needs (Sect. 10.4), which are: money, priorities, proactiveness, and private sector involvement. Section 10.5 wraps up on reconciling how the adaptation strategy at the EU still has a way to go transitioning from the 'Project' mode to the 'Living with' mode.

10.2 Addressing Inaction and Charting a Path Towards Effective Financing Solutions: An Analytical Framework

10.2.1 How to Enhance Adaptation Action and Policies

The 6th IPCC report (2023) explains that the policies in place at the end of 2020 are currently leading to global warming of 2.4–3.5 °C degrees by the end of the century, with a median value of 3.2 °C.

A recent research report published by Basque Centre for Climate change indicates that in half of the 136 coastal cities surveyed, there was no evidence of implementation of adaptation measures, and in almost 85% of cases, planned adaptation measures were not motivated by current or future climate impacts or risks (Olazabal et al., 2023).

To compound matters, the French National Council for Ecological Transition deplores the fact that current trends in global greenhouse gas emissions will lead to global warming of 4 °C by the end of the century, whereas current governments' commitments would result in global warming of around 3 °C. Hans-Otto Pörtner

[5] https://civil-protection-humanitarian-aid.ec.europa.eu/what/civil-protection/resceu_en.

even considers it impossible to adapt to 4 °C global warming path, as France is now planning in accordance with its own projections (Garric, 2023).

In particular, the National Council's report calls to integrate the consequences of global warming beyond the Paris Agreement's target into all top policies and reaffirms the necessity to plan adaptation to climate change at all levels of the territory. This will require support to local authorities, businesses, and citizens in all the actions required for adaptation (technical, financial, educational) and to make much more significant advances in the fields of finance, banking, and insurance (Radulska, 2022). Clearly, such calls impact how governance, policy and financing have to change to focus on what will be done, at which level and who will bear the costs.

Concerning the complexities and uncertainties surrounding the assessment of economic impacts of climate change and the evaluation of adaptation solutions, Olazabal and colleagues asserted that frameworks to monitor and evaluate local adaptation initiatives are still unclear and unshared (2023). These knowledge gaps in planning adaptation and creating effective financing frameworks need to be plugged and reflect both sectoral and regional factors. Research helps exposing risks, for example flood, and exploring a viable adaptation path, particularly in urban settings. Policy needs to be evidence based and open to reviews using facts to sustain adaptation efforts (ibidem). Hence, to effectively mitigate risks and promote long-term adaptation at all levels of government, people, and businesses, it is crucial to gain a comprehensive understanding of the possible risks, like flooding events (factors, frequency, and severity).

As a practical application, Cameron Rye in the Willis Towers Watson's website emphasises that traditional models used to adjust for frequency and severity in catastrophe models are inadequate when it comes to handling fat-tailed events, which have a higher probability of extreme outcomes compared to a normal distribution. The author calls for a careful examination of frequency-severity adjustments when assessing climate change risks, urging risk management professionals to consider not only short-term losses but also tail risks for a more comprehensive analysis.

The translation of research findings into practical models is essential to estimate the cost of adaptation and then source the financing, calling on the public and private sectors including the insurance industry. Knowledge gaps lead to insufficient preparedness (governance, planning, and financing) and to vulnerability. In essence the current state of affairs leaves those facing new vulnerabilities, like those living in flood prone areas, facing misunderstood risks, extreme events, losses and damages, higher insurance premiums or even denied insurance and compensation, creating significant challenges for affected individuals, businesses and communities. Similar cases occur in the North of France since Ciaran in November 2023 or houses damaged by soil structural changes due to climate change. Worse, land use, soil management, and natural resources models increase risks, unabated by regulations and legislative frameworks.

A comprehensive approach resting on an understanding of risks, first at the local level where vulnerability hits, is needed. It should be supported by governance and finance frameworks that truly enable adaptation at local level, including trans-boundary and direct actors (public and private, including household level). Acknowledging the limitations of our current models, conducting extensive research, and acting on evidence is necessary. In that direction, the European Union launched the Global Covenant of Mayors for Climate and Energy in 2008, and the initiative's Mayors Adapt in 2014 adding to other existing cohesion instruments (e.g. LIFE).

10.2.2 Fostering Public Acceptance of Risk to Plan and Finance Policies

While climate change and its consequences were considered a "distant issue" by most in society, there has since been a shift. While increasingly reported in the mainstream media, George Monbiot notes a "climate science denial", with many viewing climate action as hopeless (Monbiot, 2023). The European Bank for Reconstruction and Development's just transition initiative labelled it as the "trust deficit"; this may be evidenced by varying commitments to climate change adaptation across countries, including in Europe (Aguiar et al., 2018).

More fundamentally, Daniel Kahneman's research (2011) reveals that our risk aversion is not consistent: we can be cautious and risk-averse when it comes to gains, but we can be more inclined to take risks when faced with potential losses or unfavourable options, sometimes making irrational decisions as a result. If applicable to decision making in the economics of climate change, research needs to be funded, robust and trusted, and then well communicated to policymakers and the wider public. Policymakers must then consider the asymmetry of human behaviour and what science tells us when communicating about the necessary measures to mitigate the effects of climate change.

According to a recent report from France Stratégie (French Government, 2023), the transformation needed is holistic, should be much faster than currently planned, and primarily driven by public policies rather than technological innovation and markets. It estimates that the high economic cost of the transition (for France) will only be politically and socially accepted if it is equitably distributed (ibidem). However, this equity can only be achieved through the political-policy spheres.

Understanding climate change and the likelihood of scenario compels us to re-think and re-evaluate the policies and projects, for example flood management, based on new policy and financing frameworks. However, current investments may still exacerbate risk-taking and be predominantly assistance projects, leading to what is commonly known as the Moral Hazard (Hsiao, 2023): it is easier to raise finance for a project with a short-term implementation, generating business interests, rather than investing in long term strategies that consists in changing how we live and produce. As highlighted by Tim Smedley,[6] the authorities responsible for averting

[6] https://www.theguardian.com/news/2023/jun/15/drought-is-on-the-verge-of-becoming-the-next-pandemic.

the impending water crisis seem to lack the necessary commitment. Smedley argues, "There is an alarming absence of long-term decision-makers, resulting in an excessively short-sighted approach" (2023). Furthermore, he emphasises, "The inefficiency in our water systems, pollution of our rivers, leakage in underground pipes, and lenient building codes enabling the construction of water-inefficient homes—all of these issues persist even before we turn on a tap" (ibidem). It is no longer sufficient to solely focus on minimising the effects of floods or "Building Back Better" after losses occur. Instead, adaptation should be prioritised, encompassing the preservation of people's lives, health, culture, and economy. This perspective firmly places flood management and risk reduction within the adaptation to climate change policy and political spheres. The possibility to act is however determined by several aspects, which are the willingness to pay as consumers, by the fiscal capacity to release the necessary financial means needed as well as the collective acceptance of risk—reminding the need to act proactively for the sake of both risk management effectiveness and economic efficiency.

10.3 Adaptation Policies at EU Level: A Concise State of the Art

In its climate action approach, the EU first efforts concentrated on climate change mitigation, where the EU raised its ambition to act and to set targets at both EU and MSs' levels. On the adaptation side, it is at the local level where climate change impacts get more visible, concentrate direct impact on populations and raise more concerns and more demands for solutions that will go beyond simple build-back-almost-as-before (Reyner, 2023). It is at the local level that access to finance is often more challenging. While local planning and finance are closely intertwined, the acceleration of natural disasters calls the EU to increase capacity to deliver on all its competences: financing, and policy efforts have to set the pace for adaptation by local authorities more rapidly (Kampa, 2019).

Today, two main drivers are shaping tomorrow's adaptation approach. First, a wider policy approach: EU and MSs have actively been engaged in shaping policies that address mitigation and enhance the continent's resilience to climate change. The EU pushes further embedding climate targets in all policy sectors notably agriculture, infrastructure, urban planning, the so-called adaptation mainstreaming (EU, 2021), although there is a push back from the base. The EU policy approach is different directorates that often follow different logics, have different levels of intervention and varying financing capacities. Secondly, a deepening of the understanding of current risks: without identification of current climate risks, there will be no effective adaptation. Those risks, let it be frequent heatwaves, changing precipitation, rising sea levels are not uniform across Europe and local assessments are essential to tailormade measures specific to local and regional vulnerabilities.

10.3.1 How EU Tackles Floods

In 2007, the EU put forward a directive to prevent floods and manage flood risks at European scale and organise proactively flood prevention. The Floods Directive[7] (2000/60/EC) on the assessment and management of flood risks was clearly developed with a view to prevent physical property damages. As a result, flood hazard and flood risk maps started to be more widely used as tools to anticipate and prevent floods across Europe. Earlier, in 2000, the Water Framework Directive (WFD) offered a resource protection perspective as one of the most influential pieces of European water legislation presenting a general framework for integrated river basin management in Europe to meet environmental objectives.

The link between prevention and protection of people and water quality was concretely made and the management tools and mapping that those directives imposed to MSs to understand where action and implementation of adaptation measures were due to become reality. The reality however has seen a very varied level of implementation, limited local capacity to manage complex adaptation approaches, lacking prioritised financing availability and finally, local political capacity limited in its willingness and capacity to move the adaptation agenda further.

While policy design transposed in the two directives discussed in Table 10.1 was responsive based on the knowledge available at their conception, most MSs have been lagging to implement at scale. For floods, MSs delayed to fully implement the directive, did not fund enough neither the structures put in place by the directive nor the full approach, and the challenge of fully integrating climate change, flood insurance schemes and spatial planning into flood risk management.[8] For Water management, it missed better consideration of climate change particularly on investments, embedding water into other policies. This directive is subject to be revised and the announcement of a Water Resilience Initiative has been postponed to the next Commission to consider, a strong indication of the sensitive importance of the question in regard to agriculture-related discussions, all this despite the urgency alert for the sector on droughts and floods.[9] In this context too, the final approval of the EU Nature Restauration law in February 2024 is a clear contribution to EU climate's mitigation and adaptation objectives with the obligation to restore degraded ecosystems, habitats and species across the EU's land and seas; the law is further promoting nature-based solutions and it clearly targets to improve the ability of soil to retain water.[10] In recent years, numerous[11] Directorate-General (DG) REGIO's projects have been focusing on funding flood related issues at regional and inter regional scale with the use of Cohesion Funds (for e.g. Interreg projects around the Danube

[7] https://eur-lex.europa.eu/legal-content/EN/TXT/?uri=celex:32007L0060.

[8] https://www.eca.europa.eu/lists/ecadocuments/sr18_25/sr_floods_en.pdf.
https://environment.ec.europa.eu/publications/proposal-amending-water-directives_en.

[9] https://www.politico.eu/article/eu-put-water-resilience-plan-hold-drought-grip-spain-italy/.

[10] https://www.europarl.europa.eu/news/en/press-room/20240223IPR18078/nature-restoration-parliament-adopts-law-to-restore-20-of-eu-S-land-and-sea.

[11] https://climate-adapt.eea.europa.eu/en/mission/the-mission/mission-projects.

Table 10.1 How two EU current directives are getting MSs prepared for floods

EU directives	MSs roles	Objective
The objective of the **Floods Directive FD (2007/60/EC)** framework for the assessment and management of flood risks	– Assess all areas where significant floods could take place – Map the flood extent: assets and humans at risk in these areas – Take adequate and coordinated measures to reduce this flood risk	6-yearly cycles currently 3rd finishing in 2027 It also publishes regularly a flood risk areas map presenting the areas in the EU which, according to national authorities, carry a potentially significant flood risk
Water Framework Directive (WFD) integrated approach to water management, respecting the integrity of whole ecosystems, including by regulating individual pollutants and setting corresponding regulatory standards[13]	Use River Basin Management Plans and Programmes of Measures to protect and, where necessary, restore water bodies in order to reach good status, and to prevent deterioration	River Basin Management Plans are drawn up after extensive public consultation and are valid for a six-year period

plains[12] or central Europe water management adaptation). With Covid times, it was the turn to DG REFORM via the Technical support initiative to again support MSs in a more critical strategic level around drafting of national climate plans.

In short, the Flood and water directives highlighted the limitation at a national level to deal with adaptation, and a better response was to finance adaptation projects at local and regional level via cohesion funds.

10.3.2 The EU Adaptation Strategy: The Long Way to a Systemic Approach

The European treaties[14] are the base for the EU to intervene on climate change. They establish clearly that the EU's environmental policy should, apart from protecting the environment, contribute to the 'protection of human health and the prudent and rational use of natural resources. The starting point was that environmental policy is using the precautionary principle and concentrating on preventive action. In recent years, the adaptation to the climate change agenda has been better addressed and is now receiving wider attention and urgent consideration. In the EU set-up, the responsibility for adapting to climate change is shared between MSs and the EU

[12] https://www.interreg-danube.eu/approved-projects/danube-floodplain.

[13] https://environment.ec.europa.eu/topics/water/water-framework-directive_en.

[14] Articles 191 and 192(1) TFEU Union policy on the environment shall contribute to pursuit of the following objectives: (..)—promoting measures at international level to deal with regional or worldwide environmental problems, and in particular combating climate change.

while most climate and adaptation knowledge and research is being done at EU level (e.g. Copernicus and satellite observation), there is insufficient focus on the local level. This is typically a question of scale advantages and cross-border spillovers, as well as of how other EU competences are also involved.[15]

The first adaptation Strategy was developed by the EU in 2013 and concentrated on developing adaptation solutions. Only in 2021, the EU adaptation strategy moved from planning to implementation. Various modelling studies show that southern[16] (and south-eastern) EU countries will be more affected than their northern counterparts, exacerbating existing internal EU tensions notably on the scale of financing to be mobilised. Overall, the EU adaptation strategy was welcomed as it innovates, promotes nature-based solutions, synergies with biodiversity, nature restoration, water management and makes adaptation as a priority for the recovery from the Covid-19 pandemic economic impacts. Its overall vision is to make the EU the first carbon neutral continent by 2050, fully adapted to the unavoidable impacts of climate change and made of resilient societies. This ambition was instrumented in the EU Green Deal that initiates an unprecedented wave of policy started in 2019.[17] The political lead for the EU green deal was given to Frans Timmermans, Commissioner Vice president who faced different institutional hindrances on his mandate, starting from the scope of his own institutional scope for intervention; some DGs were limited by their mandates and their financial means to fully deliver on objectives; and the continuous MSs call for ad hoc support and financial means tapping into Cohesion funds notably.[18] DG ENER, DG AGRI and even DG CLIMA and DG ENV had to share agendas on adaptation with 'richer' DGs such as DG REGIO and DG REFORM that were able to tap more easily their funding more directly to EU green deal objectives. DG REGIO notably had the mandate to play a major role in bridging its support to administrations at regions and local level. In short, 'even though the width of Timmermans' portfolio is designed to give him the tools to actually deliver a Green Deal, institutional governance questions remain'.[19]

The EU adaptation strategy is due to be a decadal document and an "whole-of-government" and "all-of-society" effort (Commission Communication to the European Parliament and the Council, 2023). This buy-in and alignment by Government, local authorities, citizens, and political actors is highly desirable but is also paved with obstacles, like the recent push-back, limiting access to financing means. Notwithstanding, local administrations are also called upon to move towards a more adaptive management in the face of uncertainties and changes and to adjust to new norms.

[15] How can the European Union adapt to climate change while avoiding a new fault line? Bruegel, June 2022.

[16] See the PESETA report (JRC); https://www.intereconomics.eu/contents/year/2022/number/5/article/how-can-the-european-union-adapt-to-climate-change.html.

[17] https://www.bruegel.org/policy-brief/how-make-european-green-deal-work.

[18] https://geopolitique.eu/en/articles/the-green-deal-origins-and-evolution/.

[19] https://www.iddri.org/en/publications-and-events/blog-post/european-union-five-years-deliver-green-deal.

The mandatory effect for MSs to act derives from the 2021 European Climate Law (Regulation (EU) 2021/1119) enshrining the EU's climate neutral objectives through a legally binding target of net zero greenhouse emissions by 2050 at country level. It also called explicitly for the EU and MSs to make progress on adaptation, and finally introduce mandatory adaptation strategies, assessments of progress, consistency of adaptation measures and adaptation mainstreaming such as National Energy and Climate Plans.

10.4 On the Adaptation Public Financing: Money, Priorities, Proactiveness, and Private Sector Involvement

10.4.1 Why Do We Need to Finance More?

As we are already aware of the damages that climate change can trigger, such as pollution, diseases, infections, heatwaves, and more, it is imperative to also consider the material and economic toll of natural disasters, particularly flooding.

Mitigating these damages is no longer an option, as the assessment of risk has shifted, and the cost of inaction in the face of these damages is unfathomable. According to the European Environment Agency (EEA), the annual minimum cost of failing to adapt to climate change would amount to "100 billion EUR in 2020 and 250 billion EUR in 2050 for the entire EU". The damages linked to flooding, for instance, present amounts that are impossible to manage for individuals, insurance companies, and even governments.[20]

The greatest need thus lies with adaptation, as effective adaptive measures can help avoid more significant long-term economic losses, but this requires a systemic transformation with substantial funding. The French Government's report (2023) on the economic implications of climate action, written by Jean Pisany Ferry and Selma Mahfouz, compares it to an 'industrial revolution' and puts the focus on the cost and who should pay. Estimates are complex to make, for new investments to reach carbon neutrality, for expenses, for change in revenue structure (loss of fossil fuel). The figures are high for France only: 2 points of GDP by 2030 for investments, major impact on labour markets, loss of revenue of up to 25 points of GDP, need to finance via tax, debt, and the private sector (regulations impacting enterprises and prices to consumers). The report obviously calls for debate on how to finance and just transition as the cost and benefits will not occur at the same time and to the same stakeholders. The European Environment Agency stated that multi-risk adaptation actions are cost-efficient when the benefit–cost ratio exceeds 1.5—a proxy useful also for the mono-risk case of floods. The call for financing adaptation is estimated today

[20] JRC PESETA IV, EU Science Hub: https://joint-research-centre.ec.europa.eu/peseta-projects/jrc-peseta-iv_en.

to range from 80 bn/year in the EU to stay in a scenario 1.5 °C to reach 175–200 bn/year should we go beyond 3.0 °C scenario. Over this background, the MSs are largely underfunding their own national adaptation plans and estimations. To compare, the cost of climate change adaptation in developing countries could range from 140 to 300 billion dollars annually by 2030 and escalate to 280–500 billion dollars per year by 2050, according to a recent report by the United Nations Environment Programme (UNEP, 2020). In the realm of flood management, it becomes essential to discern what to fund (research), how to fund it, by whom, and what potential impacts might ensue.

The EU Adaptation Strategy missed the opportunity to set binding and measurable targets. This is in contradiction to the clear call on this from the European Parliament for binding and quantifiable targets.[21] The overall goal 'shift the focus from understanding the problem to developing solutions, and to move from planning to implementation' cannot be met without setting legally binding targets and specific measures. Instead, the Commission mostly promises to 'promote', 'support', 'explore', 'continue to incentivize' or 'help' (see for instance the EEB[22] critical assessment).

The EIB acknowledges that as a "Climate bank",[23] there is a dramatic need to do more on adaptation, and to push forward the mainstreaming of adaptation policies. The call on biodiversity notably, because of its nature of 'non bankable' sector, aims at finding ways and instruments that work to help the adaptation agenda to scale.

Already on the table is the current guideline for Members States to adaptation[24] which aims to regularly assess the consistency of relevant national measures with progress on adaptation, and that supports "adaptation and mitigation actions, that prioritise equity, social justice, climate justice, rights-based approaches, and inclusivity". It also issues recommendations when EU MSs' measures are inconsistent with ensuring progress in enhancing adaptive capacity, strengthening resilience, and reducing vulnerability to climate change.

The first ever European Climate Risk Assessment (the future EUCRA[25]) is due to be published by the EEA in Spring 2024 and will be on time to feed into the process of setting priorities for the next Commission. Also, the Commission has published technical guidance on the climate proofing of infrastructure projects in 2021 and on enhancing the climate resilience of buildings in 2023.

---

[21] European Parliament resolution of 17 December 2020 on the EU strategy on adaptation to climate change (2020/2532 RSP).

[22] https://eeb.org/wp-content/uploads/2021/02/EEB_First-Assessment-2021-Adaptation-Strategy_24-Feb-2021.pdf.

[23] Fayolle at the EU Climate Stocktake event on 27 October 2023 https://climate.ec.europa.eu/news-your-voice/events/european-climate-stocktake-eu-and-global-progress-towards-goals-paris-agreement-2023-10-27_en.

[24] https://climate.ec.europa.eu/system/files/2023-07/Guidelines%20on%20MS%20adaptation%20strategies%20and%20plans.pdf.

[25] The first EUCRA will support the identification of adaptation-related policy priorities in Europe and EU policy development in climate-sensitive sectors. It may also provide an EU-wide point of reference for conducting and updating national or subnational climate risk assessments.

Within the frame and ambition of the EU Green Deal, the end of the current Commission (Von Der Leyen, 2019–2024) and 2024 European Parliament elections might be key to how much EC will do on adaptation in the next years to come. The massive needs for finance directed at adaptation might not be met if the binding targets and measures are not and if the Just Transition is not addressed urgently. The Negotiations of the next Multi Annual Financial Framework (MFF, 2027–2034) will definitely be expected to be centred around financing adaptation in Europe, if not overshadowed by other political priorities driven by the outcome of the EP elections in 2024.

10.4.2 What Should We Prioritise for Funding?

Even though we are acknowledging that "the economic cost of climate inaction far exceeds that of action" (French Government, 2023) tensions are emerging between immediate and long-term economic, social, and environmental funding needs. Fiscal constraints, political views (denying climate change, prioritising immigration control) faced by governments and the EU[26] limit action. However, addressing climate change requires sufficient funding and collective action. It is only by understanding the necessary changes, that we can identify how to finance them. This requires 3 fundamental steps:

– **Research**. Understanding the risks[27] we face is crucial in finding effective solutions as well as being able to assess progress and risks frequency and severity, to know how the world is adapting and must adapt.[28] To effectively address climate change, it is important to accurately measure the costs and benefits of our efforts, both in human and material terms. Therefore, when evaluating climate-related actions, a comprehensive range of factors, including material gains, immaterial aspects such as well-being, monetary benefits, and non-monetary advantages need to be considered. Furthermore, analysing the effects of these measures on behaviours is essential for effective decision-making. It is important to establish a precise research mapping to align on the research and findings, particularly by conducting research at a consistent level (in this case, the European level) and then disseminating it globally, especially through informational tools like

[26] France stratégie, *Les incidences économiques de l'action pour le climat*, 2023 (p. 18). «Le coût économique de l'inaction excède de loin celui de l'action».

[27] C-PREE, *Sea Level Rise and Urban Adaptation in Jakarta* (2023). The author highlights how certain state interventions create Moral Hazard and prevent the proper protection of the population.

[28] Basque Centre for Climate change (BC3), *Evaluating Adaptation: An Analysis of Policy Progress in Coastal Cities and Regions* (2023). "Progress made to date to adapt is still poorly understood and tracked due to a lack of a means to evaluate how well the world is adapting".

Climate-Adapt[29] and make it the authoritative European Platform for adaptation knowledge. Also, in March 2024, the first ever EU Climate risk assessment (EUCRA) was published and covered current and future climate change impacts and risks relating to the environment, economy, and wider society in Europe (EEA, 2024).

- **Communication**. As trust in both politics and science varies, and aversion to risk when addressing loss versus gains[30] (Kahneman, 2011) is not consistent with probabilities, it is imperative for policymakers to factor in this asymmetry of human behaviour and the insights provided by science when communicating the necessary steps for mitigating the impacts of climate change. Research becomes compelling as effective communication relies on understanding risks, their severity, and frequency. Effectively conveying the risks associated with climate change, the consequences of inaction, and the opportunities tied to proactive action becomes pivotal in garnering societal acceptance of adaptation. Stakeholders' management is also required.
- **Governance and Resilience to Flood**. Governance and resilience in the face of climate change within the EU are intricately linked to the imperative of acting based on research to finance the adaptation of our societies and transition to sustainable practices. This innovative approach transcends mere projects. However, while research is pivotal, it is undeniable that numerous solutions are already known, quantified, and presented to policymakers, yet struggle to be implemented. Regulatory mechanisms exist but are not always fully operational, and resistance from EU MSs, economic interests and citizens can impede their efficacy. By playing this leading role, the EU helps focusing on all governance levels including regional and local ones. The tangible effects of climate change are most felt at the local level, where financing needs must be flexible and tailored to specific circumstances. To achieve this, tailoring financing to regions and risk assessment are necessary to adapt to the new conditions. In essence, bolstering resilience to climate change effects requires strengthened governance at all levels.

10.4.3 How to Finance Transitioning from a Disaster Management Approach to a Proactive Adaptation Approach to Flood Management

The transition from a disaster management approach to a proactive adaptation strategy requires thoughtful consideration of "costs" allocation to effectively tackle climate change challenges. Financing adaptation and resilience actions involves a

[29] Climate Adapt is an online platform developed by the European Environment Agency (EEA) that serves as a knowledge hub for information related to climate change adaptation in Europe. https://climate-adapt.eea.europa.eu/.

[30] Risk aversion is not constant and homogeneous: we may be cautious and risk-averse when it comes to gains, but we may be more inclined to take risks when faced with losses or unfavourable options, and sometimes make irrational decisions as a result.

Table 10.2 Policy instruments

Policies	Direct	Indirect
Monetary	Financing infrastructures, providing adequate finance pipelines (guarantees), financing adaptation-oriented research	Taxing GES, harmful activities to the seashore, etc.
		Subsidising (households and business) for taking adaptive measures
Non-monetary	Regulating: forbidding (e.g. developing in flood-prone areas), making compulsory (e.g. planning with nature-based solution), insurance	Informing about risks
		Imploring not to take risks

Source Adapted from Flynn and Asquer (2007)

shift from reacting to extreme events to acknowledging the long-term climate changes and their widespread impacts beyond weather phenomena. All policy instruments, as shown in Table 10.2, should be used.

Financing through taxation

The France Stratégie report highlights the crucial role of taxation in financing the transition to a sustainable future, emphasising the limitations of relying solely on debt schemes. This raises new questions: does it become essential to implement a dedicated levy to finance adaptation?

What will be the incidence of such a tax? Recent events show that in France and in Europe at large, political and social acceptance is far from being achieved. Work done by Lucas Chancel already reveals significant inequalities when it comes to carbon emissions. The path to transition remains to be charted so that its incidence on tax, debt and consumer prices reduces inequalities. This can only be achieved through a wide range of public sector programmes.

Despite many preconceived ideas, "delaying the investments needed to achieve climate neutrality in the name of controlling public debt would only improve the situation on the surface, without any substantive benefit"[31] (French Government, 2023). This report establishes that the additional public expenditure induced by the climate transition should be between 25 and 34 billion euros per year by 2030, with the remainder financed by the private sector. Redistributing this additional burden as taxation or through the price mechanism is a societal debate. Who bears the burden and through which mechanism is at the centre of this debate and needs to be addressed openly.

Paying the price

The private sector plays a crucial role in bearing the costs of adaptation. As it transitions towards new regulations, more sustainable and resilient production practices, this process impacts prices. The EU Green Deal and Carbon Border Adjustment

[31] Translated from French to English.

Mechanism intend to put a price tag on carbon to avoid EU imports delocalisation poor practices. So, as confirmed in France Stratégie report, solely relying on the State for financing climate adaptation is likely to prove inadequate (French Government, 2023). The private sector will need to safeguard themselves against escalating natural disasters (insurance) and face the cost of transition. This will also be an engine of growth, but for whom?

As for households, insurance schemes can function only when based on a clear legal and regulatory framework, backed by science; it is only under such conditions that insurance companies will be willing to provide coverage. Applied to flooding this means regulations should protect against flood prone areas being developed. The cost of transition will remain to be addressed, such as relocation and adapting vulnerable regions to avoid recurring damages; so public financing will remain imperative.

10.4.4 From Policy Making to Financing

The local delivery challenge: getting financing at local level

One crucial factor in ensuring sufficient funding revolves around the synergy forged among the EU, its member countries, and local authorities. Historically, this cooperation has stood as a cornerstone within our fiscal management framework. While it is necessary to rethink the articulation between European and national policies to ensure coherence and effective coordination in the pursuit of these objectives, the European Union will be able to initiate adaptation-related efforts through various funds: LIFE program; Horizon Europe research and innovation program; The European Union Solidarity Fund (EUSF); Regional Development Fund (ERDF); Cohesion Fund; Investment Bank (EIB).

Failure to mobilise the private sector

Mobilising the private sector is crucial for effective climate adaptation, addressing challenges that hinder its involvement. One such challenge is the misconception that there is no profit potential in adaptation efforts. Additionally, the private sector is concerned about the extended investment timelines and relatively small-scale projects in adaptation, with their prolonged payback periods (Owen-Burge, 2023). A mere 2% of tracked adaptation finance in 2019 and 2020 came from corporations and institutional investors, with the remaining 98% from public channels. Challenges persist in capturing private expenditures, due to the absence of standardised definitions and monitoring mechanisms. The scarcity of data and gaps in knowledge further obscure the identification of factors influencing private sector adaptation investment (ibidem).

It is evident that although adaptation projects are imperative, they receive significantly less investment compared to mitigation projects, indicating a substantial gap (ibidem). Is this a reflection of a lack of policy focus? Overcoming the private sector's aversion to adaptation investment is pivotal. To this end, public institutions can play a pivotal role by de-risking these ventures. Techniques such as blended finance and risk-tolerant capital structures, exemplified by entities like the Global Fund for Coral Reefs and Light Smith Climate Resilience, absorb initial investment losses, thus making adaptation more attractive to the private sector (ibidem).

A closer analysis on how the existing gap between current volumes of finance and what is required to transform the economy in its adaptation to climate change is needed to design policies and regulations that oblige the private sector to adapt and thus mobilise private funds. Policymakers should then focus on addressing market externalities and lead to increasing financial flows for adaptation.

Insurance gap

Only a quarter of climate-related catastrophe losses are insured.[32] In some countries, the figure is less than 5%.[33] Moreover, the growing effects of climate change mean that coverage is likely to shrink as rising premiums choke demand and insurers withdraw from particularly exposed areas (see also work that EIOPA is developing around insurance protection gap for natural catastrophes[34]).

It is of utmost importance then to establish well-defined standards for insurance companies to ensure protection for individuals against the most severe events. The insurance industry plays a crucial role as a risk manager in enhancing society's and the real economy's resilience against the impacts of climate change. However, a significant challenge lies in the insufficient coverage of extreme weather and climate-related losses. Currently, only 23% of such losses in Europe are insured, creating a substantial protection gap that is projected to widen given the prevailing climate forecasts.[35] Insurers possess a distinctive capacity to contribute by creating innovative insurance products that promote climate-related risk mitigation.[36] They can offer reduced premiums to policyholders implementing such measures, fostering a proactive approach to risk management. This not only safeguards policyholders from losses but also ensures the sustained availability of insurance products and reduces overall insurance costs.

However, achieving effective adaptation to climate change requires extensive research on risk to define clear and practical standards for insurance companies. Insurance companies will seek to maintain profitability to sustain their operations. By conducting thorough investigations and setting transparent guidelines, we can bridge the gap between insurers 'reluctance and individuals' vulnerability. Establishing a balance between regulating activities and protecting individuals from extreme events and ensuring the financial viability of insurance companies becomes

---

[32] https://www.eiopa.europa.eu/system/files/2022-12/dashboard-on-insurance-protection-gap-for-natural-catastrophes-in-A-nutshell.pdf.

[33] https://www.ecb.europa.eu/press/blog/date/2023/html/ecb.blog.230424~4cdc3a38ba.en.html.

[34] https://www.eiopa.europa.eu/tools-and-data/dashboard-insurance-protection-gap-natural-catastrophes_en.

[35] *Impact Underwriting: EIOPA reports on insurers' use of climate-related adaptation measures in non-life underwriting practices.* (2023). EIOPA.

[36] *The role of insurers in tackling climate change: challenges and opportunities.* (2023). EIOPA.

a pivotal challenge that must be overcome through diligent research and policy implementation.

Cost incidence

Additional production costs (linked to new production standards and regulations, green energy, etc.) are likely in addition to taxes linked to carbon. Regarding this cost overrun, for instance in the French case, France Stratégie calls for consideration to be given to the closely related Principles of Equity. Today, France's national budget allocation for helping the transition amounts to around 8.5 billion euros in 2023 and will need to be substantially increased to help households (e.g., aid delocalization of houses, energy renovation). "The economic cost of the transition will only be politically and socially accepted if it is fairly distributed. Public financing is necessary because the transition is spontaneously unequal (even for the middle classes)" (French Government, 2023).

10.5 Final Considerations

The move to a systemic approach for the EU to better operate at local level

Floods know no boundaries: river basins and rivers flow across geographies and countries. Climate change is now adding uncertainty. Prevention and emergency interventions cannot address the need for a wider, more systemic approach, starting in the EU particular settings.

- To best serve a systemic approach, the EU has to deal with its own nature. The EU has boldly intervened into the Climate Change agenda with the EU Green Deal, but in doing so, it did it in its dual capacity as mobilising regulatory and policy action to enable all actors including the private sector to intervene. Also, the EU does also intervene as a public actor enabling central funding to support the policy landscape it is creating. This has to fit into the multi-level governance framework in Europe that allows for both top down and bottom-up planning and implementation of adaptation to climate change demands. Access to open data, mapping, observation programs, scientific studies can best be carried out at European level. This EU knowledge is often open data and can be used as a publicly available source for further research globally. On the other hand, adaptation interventions have also generated ad hoc solutions that are worth knowing, being replicated or put to scale through an accelerated learning process.
- Most of the effects of climate change are concentrated at the local level: but the issues to be addressed (dealing with critical infrastructures, transport and logistics, agriculture, pollution, etc.) and the accessibility of public and private financing have to be addressed at EU level. Financing will come from the public purse and market approaches. However, this obliges the Europeans to debate and agree on "who pays".

- a dedicated set of policy instruments for implementing adaptation to climate change, delivered at the right level for best impact and efficiency is required. Scientific research shows the acceleration of climate change and its effects, both made widely public, but the narrative for adaptation remains not heard and is missing in the current political discussions in Brussels and in MSs.

Transitioning from "Project" mode to "Living with" mode, a long-term vision

Addressing climate adaptation and in the same way addressing flood management requires a multi-faceted approach that considers both short-term and long-term perspectives. The evaluation of adaptation interventions in various sectors, conducted by Deval (2023), emphasises the need for additional measures to tackle current challenges. Nature-based solutions, infrastructure interventions, and education initiatives have proven effective in enhancing adaptive capacities and responding to shocks and stress factors (ibidem). The Adaptation project showcased in Le Monde underlines the importance of transforming through policy alterations and changes in production methods. This emphasises the need to work with nature rather than against it, as eloquently stated by Tim Smedley in The Guardian (Chapuis, 2023a, 2023b; Smedley, 2023).

To effectively live with water and bolster infrastructure, it is essential to rethink our cities and encourage lifestyle and production changes. This involves incentivising sustainable land use and farming practices, for instance, and enhancing comprehensive risk management with a focus on vulnerable and pre-existing residents in flood-prone zones (DEval, 2023).

For this, the shift from the "Project" mode, as discussed in this article, to the "Living with" mode is necessary. As we look towards the future, EU actions must be "lived with" long-term perspectives. Extensive research on flood-related risks is essential, preventing moral hazard and enabling effective communication and fostering societal acceptance of these changes while bolstering trust in scientific advancements. Adaptation involves sustained investment based on individual cases rather than temporary reconstruction projects. Adequate funding for adaptation and resilience hinges on robust governance, well-coordinated international cooperation, localised action initiatives, productive engagement with the private sector, and accurate delineation of protection standards solutions.

Acknowledgements The Authors would like to thank Giulia Dunand, from the International Climate Finance Accelerator, for the support and work done by gathering and systematizing information.

References

Aguiar, F. C., Bentz, J., Da Silva, J. S., Fonseca, A., Swart, R., Santos, F. D., & Penha-Lopes, G. (2018). Adaptation to climate change at local level in Europe: An overview. *Environmental Science & Policy, 86*, 38–63. https://doi.org/10.1016/j.envsci.2018.04.010

Aligishiev, Z., Bellon, M., & Massetti, E. (2022). *Macro-Fiscal implications of adaptation to climate change*. IMF. https://www.imf.org/en/Publications/staff-climate-notes/Issues/2022/03/16/Macro-Fiscal-Implications-of-Adaptation-to-Climate-Change-512769

Barnhoorn, A. (2023). Comparing responses to climate-related security risks among the EU, NATO and the OSCE. https://doi.org/10.55163/rdou1271

Bonnel, O. (2023). *Italy's Emilia-Romagna region, devastated by floods, questions land use*. Le Monde.fr. https://www.lemonde.fr/en/environment/article/2023/05/22/italy-s-emilia-romagna-region-devastated-by-floods-questions-land-use_6027614_114.html

Borunda, A. (2022). *As climate disasters grow more costly, who should pay the bill? Environment*. https://www.nationalgeographic.com/environment/article/as-climate-disasters-grow-more-costly-who-should-pay-the-bill

Chapuis, N. (2023). *Tout savoir sur le projet Adaptation lancé par «Le Monde»*. Le Monde.fr. https://www.lemonde.fr/planete/visuel/2023/06/11/tout-savoir-sur-le-projet-adaptation-lance-par-le-monde_6177170_3244.html

Chapuis, N. (LeMond). (2023b). *La forêt française sous le feu du réchauffement. (2023, 11 juin)*. Le Monde.fr. https://www.lemonde.fr/planete/visuel/2023/06/11/la-foret-francaise-sous-le-feu-du-rechauffement_6177166_3244.html

Colenbrander, S., Pettinotti, L., & Cao, Y. (2022). *A fair share of climate finance? An appraisal of past performance, future pledges and prospective contributors*. ODI Working Paper. ODI. https://odi.org/en/publications/a-fair-share-of-climate-finance-an-appraisal-of-past-performance-future-pledges-and-prospective-contributors/

Commission to the European Parliament and the Council EU. (2023). *Commission staff working document report on the implementation of the EU strategy on adaptation to climate change accompanying the document report from the Commission to the European parliament and the council EU climate action progress report 2023 {COM(2023) 653 final}—{SWD(2023) 339 final}*.

Commonwealth Secretariat. (2023). *Accessing development and climate finance: Issues and challenges in the commonwealth countries*. https://thecommonwealth.org/publications/accessing-development-and-climate-finance-issues-and-challenges-commonwealth-countries

Deval. (2023). *Adaptation: Agriculture, water and coastal protection*. https://www.deval.org/en/evaluations/our-evaluations/evaluation-of-adaptation-interventions-in-agriculture-water-and-coastal-protection

EEA. (2012). *Climate change, impacts and vulnerability in Europe*. https://www.eea.europa.eu/publications/climate-impacts-and-vulnerability-2012

EEA. (2024). *European climate risk assessment executive summary*. EEA Report 01/2024. https://www.eea.europa.eu/publications/european-climate-risk-assessment

EIOPA. (2023a). *Impact underwriting: EIOPA reports on insurers' use of climate-related adaptation measures in non-life underwriting practices. (2023, 6 février)*. European Insurance and Occupational Pensions Authority. https://www.eiopa.europa.eu/impact-underwriting-eiopa-reports-insurers-use-climate-related-adaptation-measures-non-life-2023-02-06_en

EIOPA. (2023b). *The role of insurers in tackling climate change: Challenges and opportunities. (s. d.)*. European Insurance and Occupational Pensions Authority. https://www.eiopa.europa.eu/publications/role-insurers-tackling-climate-change-challenges-and-opportunities_en

EU. (2021). *COM (2021) 82—EU climate adaptation strategy*. https://climate.ec.europa.eu/eu-action/adaptation-climate-change/eu-adaptation-strategy_en

European Commission, Directorate-General for Economic and Financial Affairs. (2023). *Disaster risk financing: Limiting the fiscal cost of climate-related disasters*. Publications Office of the European Union. https://doi.org/10.2765/270746

Ferguson, T., & Storm, S. (2023). *Central banks raising interest rates makes it harder to fight the climate crisis.* The Guardian. https://www.theguardian.com/commentisfree/2023/may/06/central-banks-interest-rate-hike-climate-crisis

Flynn, N., & Asquer A. (2007). *Public sector management.* SAGE Publications Limited.

French Government. (2023). *Les incidences économiques de l'action pour le climat.* France Stratégie. https://www.strategie.gouv.fr/publications/incidences-economiques-de-laction-climat

Garric, A. (2023). *«Nous poussons la planète vers un avenir climatique où seule une partie de sa surface sera habitable».* Le Monde.fr. https://www.lemonde.fr/planete/article/2023/06/12/nous-poussons-la-planete-vers-un-avenir-climatique-ou-seule-une-partie-de-sa-surface-sera-habitable_6177325_3244.html

Granceri, M. (2020, June 9). *Understanding climate change adaptation mainstreaming and planning challenges.* Insights from Barcelona and Turin municipalities, Politecnico di Torino, Turin. Retrieved from https://iris.polito.it/handle/11583/2839863#.YV3getpBxaQ

Hsiao, A., & Princeton University. (2023). *Sea level rise and urban adaptation in Jakarta.* Center for Policy Research on Energy and the Environment (C-PREE). https://cpree.princeton.edu/events/2023/sea-level-rise-and-urban-adaptation-jakarta

IPCC. (2022a). *Summary for policymakers* (pp. 3–36). Cambridge University Press eBooks. https://doi.org/10.1017/9781009157964.001

IPCC. (2022b). Sea level rise and implications for low-lying Islands, coasts and communities. In *The Ocean and Cryosphere in a changing climate: Special report of the intergovernmental panel on climate change.* Cambridge University Press (pp. 321–446).

IPCC. (2023). *Rapport de synthèse: Changements climatiques.*

Jongman, B. (2018). Effective adaptation to rising flood risk. *Nature Communications, 9*(1). https://doi.org/10.1038/s41467-018-04396-1

JRC-Peseta-IV. (2022). *EU Science Hub.* https://joint-research-centre.ec.europa.eu/peseta-projects/jrc-peseta-iv_en

Kahneman, D. (2011). *Thinking, fast and slow.* Farrar, Straus and Giroux.

Kampa, E. (2019). *The Implementation of the Water Framework Directive (2000/60/EC) and the Floods Directive (2007/60/EC) Second River Basin Management Plans | First Flood Risk Management Plans.* https://www.ecologic.eu/16371

Monbiot, G. (2023). *The hard right and climate catastrophe are intimately linked. This is how.* The Guardian. https://www.theguardian.com/commentisfree/2023/jun/15/hard-right-climate-catastrophe-extreme-weather-refugees

Olazabal, M., Lewis, W., Venner, K., de Gopegui Aramburu, M. R., Sainz de Murieta, E., Mahawelatanne Goonesekera, S., & Galarraga, I. (2023). *Evaluating adaptation: An analysis of policy progress in coastal cities and regions.* https://www.bc3research.org/index.php

Owen-Burge, C. (2023). *What it takes to attract private investment to climate adaptation—Climate Champions.* Climate Champions. https://climatechampions.unfccc.int/what-it-takes-to-attract-private-investment-to-climate-adaptation/

Radulska, N. (2022). *How to break down the barriers to transformation.* WiseEuropa. https://wise-europa.eu/en/2022/12/29/report-how-to-break-down-the-barriers-to-transformation/

Reyner, T. (2023) *Adaptation to climate change: EU policy on a mission towards transformation?* The Nature Journal. https://www.nature.com/articles/s44168-023-00068-z

Serinaldi, F. (2015). Dismissing return periods! *Stochastic Environmental Research and Risk Assessment, 29*(4), 1179–1189.

Slovic, P., & Weber, E. U. (2002). *Perception of risk posed by extreme events.* Decision Research and University of Oregon. https://www.researchgate.net/publication/209805350_Perception_of_Risk_Posed_by_Extreme_Events/citations

Smedley, T. (2023). *Drought is on the verge of becoming the next pandemic.* The Guardian. https://www.theguardian.com/news/2023/jun/15/drought-is-on-the-verge-of-becoming-the-next-pandemic

UNDP. (2015). A Methodological Guidebook: Climate Public Expenditure and Institutional Review (CPEIR) | United Nations Development Programme. https://www.undp.org/asia-pacific/publications/methodological-guidebook-climate-public-expenditure-and-institutional-review-cpeir

UNDRR. (2015). *Sendai framework for disaster risk reduction 2015–2030*. https://www.undrr.org/publication/sendai-framework-disaster-risk-reduction-2015-2030

UNEP. (2020). *Adaptation gap report 2023*. https://www.unep.org/resources/adaptation-gap-report-2023

UNFCCC. (2022). *Fifth Biennial assessment and overview of climate finance flows*. Standing Committee on Finance. https://unfccc.int/sites/default/files/resource/J0156_UNFCCC%20BA5_2022_Report_v4%5B52%5D.pdf

United Nations. (2015). The Paris Agreement | United Nations. https://www.un.org/en/climatechange/paris-agreement

United Nations. (2022). *World population prospects*.

WTW. (2023). *Why relying on frequency-severity adjustments could underestimate your tail risk from climate change*. https://www.wtwco.com/en-ca/insights/2023/06/why-relying-on-frequency-severity-adjustments-could-underestimate-your-tail-risk-from-climate-change

Chapter 11
How to Improve M&E in Climate Adaptation Action in the Case of International Projects Dealing with Socio-hydrological Resilience

Geraldo Carreiro

Abstract Climate change adaptation after COP21 (Paris 2015) has started gaining policy and finance attention. Several are still the challenges to make adaptation operative and one of them is the establishment of robust systems for monitoring and evaluating adaptation actions. In fact, there is still no single indicator or methodology agreed at international level to measure adaptation. The evolution of climate change impacts on socio-economic conditions at local level remains mostly uncertain, which hinders the development of capacities needed for effective adaptation. Least Developed Countries and Small Islands Developing States are the strongest hit by climate change yet also characterised by climate data paucity and limited capacity of interpretation on potential impacts. These caveats are barriers to adaptation action at both national and international levels. This paper formulates lessons learned from a review of monitoring and evaluation systems in adaptation projects supported by international funds, i.e. Adaptation Fund, with a focus on socio-hydrological resilience. It concludes on recommendations for establishing results frameworks that improve transparency and accountability on adaptation action.

Keywords Climate change adaptation · Indicators · Monitoring · Evaluation · International cooperation

Abbreviations and acronyms

AF	Adaptation fund
AR6	Assessment report n°6
CCA	Climate change adaptation
CCM	Climate change mitigation

G. Carreiro (✉)
Adelante Knowledge and Development, Madrid, Spain
e-mail: geraldo@adelante.eu.com

M. Granceri Bradaschia et al. (eds.), *Climate Change Adaptation, Flood Risk, and Beyond*, Planning for Climate Proof Cities,
https://doi.org/10.1007/978-3-031-65463-3_11

CMA	Conference of the parties serving as the meeting of the parties to the Paris Agreement
COP	Conference of the parties
EbA	Ecosystem based adaptation
EU	European Union
GGA	Global goal on adaptation
IPCC	Intergovernmental panel on climate change
JRC	Joint Research Centre
M&E	Monitoring and evaluation
NDC	Nationally determined contribution
UK	United Kingdom
UNEP	United Nations Environmental Programme
UNFCCC	United Nations framework convention on climate change
WMO	World Meteorological Organisation

11.1 Introduction

COP26 in Glasgow was determinant for international action on climate change adaptation. Parties were requested to double the provision of climate finance for adaptation to developing countries by 2025, compared to a 2019 baseline. The Glasgow–Sharm el-Sheikh work programme on the global goal on adaptation (GGA) also intended to improve the assessment of progress toward adaptation and establish robust systems for monitoring and evaluating adaptation actions.

Yet, monitoring and evaluating adaptation is no easy task. There is still no single indicator or methodology agreed at international level to measure adaptation. The evolution of climate change impacts on socio-economic conditions at local level remains mostly uncertain, which hinders the development of capacities needed for effective adaptation. Least Developed Countries and Small Islands Developing States are the strongest hit by climate change yet also characterised by climate data paucity and limited capacity of interpretation on potential impacts. These caveats are barriers to adaptation action at both national and international levels.

This paper formulates lessons learned from a review of monitoring and evaluation systems in adaptation projects supported by international funds, with a focus on socio-hydrological resilience. It concludes on drawing recommendations for strengthening adaptation through improved use of available information and internal project monitoring and evaluation systems. It focuses on projects associated with floods, a typical adaptation area which includes environmental, social, and economic dimensions.

To this, it identifies key elements in the basic features of Monitoring and Evaluation (M&E) systems of Adaptation Fund (AF) projects that can contribute to M&E functions including improved future project preparation and contribution to learning processes, should these be at local, national, or international levels. The indicators dataset for AF projects dealing with adaption were analysed to show how indicators are being used in practice.

The analysis underlines the difficulty of the reviewed projects to indicate how they could help the intended beneficiaries to reach a desirable level of adaptation. Ensuring the success of climate adaptation, notably in the case of projects related to floods, requires a comprehensive and multifaceted approach that involves a combination of planning, policy, review of infrastructure capacity, community engagement, and ensuring the availability of relevant information.

Projects may also fail to give robust evidence on their contributions to Nationally Determined Contributions (NDCs) or other adaptation plans, potentially due to a lack of adaptation metrics in these plans. This limits the capacity of projects to feed into future adaptation strategies or plans including at sector level.

The study also formulates recommendations to help organisations and projects in setting up and improving the M&E of hydro-related adaptation programmes. Project rationale for engaging in adaptation action should be based on a strong risk analysis, including a vulnerability analysis of human systems. It should also determine positive and negative pressures that act as incentives or barriers for stakeholders to support the change that is needed. These would allow stakeholder dialogue to be informed on realistic options for planning and agree over a common adaptation path. Further, adaptation monitoring should integrate public management systems including development planning. Examples and characteristics are given, of reference adaptation frameworks that provide to climate change adaptation stakeholders information on the key steps and methods for assessing adaptation progress, which should remain relevant for each stakeholder at its own level.

11.2 Analytical Framework

11.2.1 The Challenge of M&E Climate Change Adaptation

11.2.1.1 The Complexity of Adaptation

Adaptation refers to adjustments in ecological, social, or economic systems in response to actual or expected climatic stimuli and their effects. It refers to changes in processes, practices, and structures to moderate potential damages or to benefit from opportunities associated with climate change.[1] Adaptation is therefore a relative

[1] https://unfccc.int/topics/adaptation-and-resilience/the-big-picture/introduction#:~:text=Adaptation%20refers%20to%20adjustments%20in,opportunities%20associated%20with%20climate%20change.

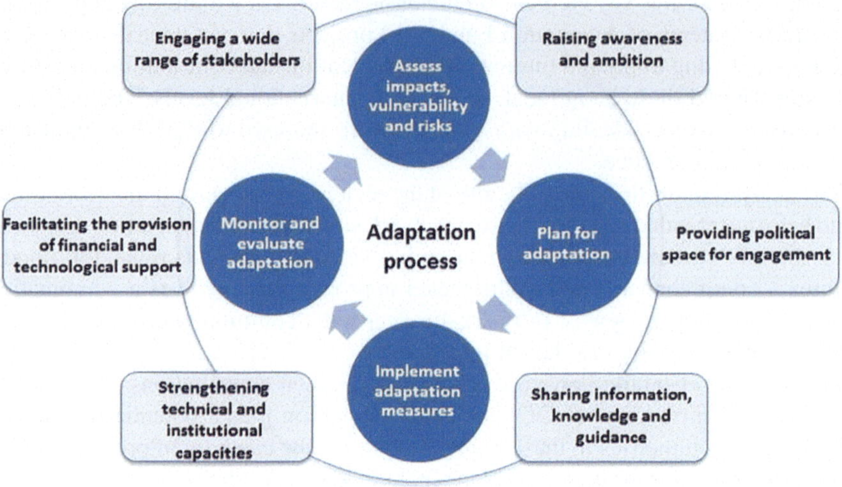

Fig. 11.1 The adaptation process. *Source* UNFCCC

concept that is highly contextual. To adapt to climate change, human societies need to consider interactions with their natural environment in a holistic way, through the prism of vulnerability. Interventions related to climate change adaptation belong to a broad spectrum whose characteristics depend on the scale and scope of the action, the sector considered, and how it fits within social and economic constructs.

While it recognises there is no "one size fits all" solution for adaptation, the UNFCCC recommends using a process in four steps, as shown in Fig. 11.1.

There is no general consensus on what successful adaptation to climate change looks like.[2] Article 7 of the Paris Agreement establishes six adaptation principles.[3] Adaptation should be:

- Country-driven;
- Gender-responsive;
- Participatory and transparent;
- Addressing vulnerabilities;
- Guided by best science and local knowledge; and
- Supportive of national development, integrated into wider development policies and plans.

[2] https://www.weadapt.org/knowledge-base/governance-institutions-and-policy/monitoring-and-evaluation-of-climate-change-adaptation-an-introduction.

[3] Paris Agreement. https://unfccc.int/sites/default/files/english_paris_agreement.pdf.

Furthermore, equity is an important element underpinning UNFCCC recommendations, reiterated in the Paris Agreement. The Global Goal on Adaptation (GGA), launched in 2021 at COP 26,[4] maps out the way for the world to enhance adaptive capacity, strengthen resilience, and reduce vulnerabilities associated with climate change.

11.2.1.2 Adaptation Related Costs, an Underestimated Challenge

In 2020, the annual costs of adaptation in developing countries were estimated at USD 70 billion.[5] In 2022, when part of the deadline had passed, these revised costs were estimated at USD 160–340 billion in 2030 and USD 315–565 billion in 2050.[6]

The financial requirements for adaptation at global level are colossal. Already in 2022, international financial flows for adaptation in developing countries were 5–10 times lower than estimated needs (UNEP).

This situation is not likely to change quickly. In fact, even without taking into account financial flows for adaptation, which tend to evolve slowly, the number of disasters is changing rapidly. The number of disasters has multiplied by a factor of five over the last 50 years. This is due to the effects of climate change, the increase in extreme meteorological phenomena and improved global reporting (WMO).

11.2.1.3 Why Monitor and Evaluate Adaptation Effectiveness

Monitoring and evaluating progress through metrics or unities of measurement are crucial components of climate action planning. Information about the impacts and successes of adaptation actions helps urban and rural communities to improve their decision-making, planning and implementation of adaption solutions.

Yet, decision makers and adaptation stakeholders, i.e. the society at large, still have difficulties in having access to quality information about the effectiveness of adaptation policies, measures and actions. This calls for the strengthening of or establishment of monitoring programmes related to adaptation efforts. Such programmes can contribute to knowledge, learning and data, assessing the successes, difficulties and needs of adaptation initiatives, to support adaptation decision making and governance.

[4] https://unfccc.int/topics/adaptation-and-resilience/workstreams/glasgow-sharm-el-sheikh-WP-GGGA.

[5] UNEP GAP report 2021.

[6] UNEP GAP report 2023.

11.2.1.4 Issues in M&E Adaptation

International cooperation players supporting climate change adaptation action have identified conceptual and operational challenges to measuring, monitoring and evaluation of adaptation, including[7]:

- Developing suitable metrics for adaptation due to the diversity and complexity of potential climate impacts and adaptation responses. There is no standard metric or one single and internationally recognised indicator for adaptation, which makes tracking and aggregating results across different sectors and localities very challenging. Identifying the best possible proxy outcome indicator is therefore a key challenge in designing M&E frameworks. Lack of a well- established standard of 'best practice' M&E methodology and indicators for adaptation interventions, as is generally available for many regular (i.e. non-climate change-focused) development interventions.
- Establishing baselines and comparing progress to a counterfactual. Due to the nature of adaptation as an additional but not easily distinguishable factor in an already dynamic development process, the definition of specific baselines for an isolated adaptation investment and defining what would have happened in the absence of the intervention is difficult;
- Adaptation takes place against shifting climatic, social and environmental baselines, which produce changing risk contexts which can be confounding factors in the assessment;
- The results of an adaptation intervention are sometimes only visible in the long-term. Timeframes for the expected benefits of adaptation interventions are usually much longer than the normal lifetime of standard projects and programmes. This means that, paradoxically, impacts will often need to be documented before they have fully materialised;
- Unclear reporting requirements and a lack of M&E capacities, or even to have access to solid, credible datasets in the countries most affected by adverse impacts of climate change, which often are also among the poorest;
- The problem of attributing outcomes in the form of increased resilience directly to specific adaptation investments, as adaptation is inherently a complex process cutting across sectors and levels of interventions.

[7] Synthesised from Author, UNEP/DTU Working Paper "Monitoring and Evaluation for Climate Change Adaptation—A summary of key challenges and emerging practice", https://backend.orbit.dtu.dk/ws/files/157400902/MandE_challenge_guidance_note_01_07_16.pdf and WeAdapt document "Monitoring and evaluation of climate change adaptation". https://www.weadapt.org/knowledge-base/governance-institutions-and-policy/monitoring-and-evaluation-of-climate-change-adaptation-an-introduction.

11.2.2 Socio-hydrological Resilience: Adapting to Floods

11.2.2.1 Climate Change and Increased Flood Risks

Adaptation concerns all economic and social activities, yet links between hydrometeorological events notably those leading to floods and related adaptation needs are particularly relevant, notably in terms of disaster risk reduction.

There is an important number of weather and human-related factors that influence the occurrence of a flood. Also, it is difficult to identify the causes of past floods, if they can be compared to those happening today and at the same time determine the potential influence of a changing climate.[8] The Intergovernmental Panel on Climate Change (IPCC) noted in its special report on extremes that climate change "has detectably influenced" several of the water-related variables that contribute to floods, such as rainfall and snowmelt. It becomes increasingly clearer that climate change exacerbates the factors that contribute to floods.[9]

In terms of investments, the EU's Joint Research Centre (JRC)[10] relates adaptation in the water related sector to a 1:4 return on investment in Europe. This means that each euro invested in creating water detention areas in Europe can save four euros in avoided damages, in the context of a 3 °C warming scenario.

11.2.2.2 How Climate Change Contributes to Floods

Floods can be defined as "an overflow of a large amount of water beyond its normal limits, or on an area that is usually dry".[11] This definition underlines the notions of "normal" and "usual" states, which are being disturbed by climate change. Floods depend on many factors, such as weather events (e.g., rains, storms, snowmelts, ...), and elements derived from a human construct, including the alterations made to the landscape and to natural waterways; the location, design and management of water related infrastructure; and their relative impacts at the local level.

Climate change can lead to the creation of new conditions for floods, through:

- **Precipitation**. Rains shall intensify, bringing as much as 50% more heavy rain[12] by 2100. Climate change is also increasing extreme precipitation events, which only occurred at rare occasions. This implies events with a long return period, such as 100-year precipitation events, are now happening more frequently. Heavier rainfall does not automatically lead to floods but contribute to the conditions[13] for

---

[8] https://www.nrdc.org/stories/flooding-and-climate-change-everything-you-need-know#causes.

[9] https://www.ipcc.ch/site/assets/uploads/2018/03/SREX-Chap3_FINAL-1.pdf.

[10] https://joint-research-centre.ec.europa.eu/scientific-activities-z/natural-and-man-made-hazards-0/floods_en.

[11] Oxford Dictionary.

[12] https://climate.nasa.gov/news/2740/climate-change-may-lead-to-bigger-atmospheric-rivers/.

[13] https://www.ucsusa.org/sites/default/files/attach/2018/07/gw-fact-sheet-epif.pdf.

them. Even modest amounts of additional rainfall may cause damage, particularly in exposed urban areas.
- **Storms**. Climate change is increasing the frequency and strength of storms, a trend expected to continue[14] in the future. In the US Atlantic basin, category 4 and 5 hurricanes, i.e. the most destructive, shall increase by 80% over the next 80 years.
- **Sea levels**. The IPCC indicates that sea level rise is "extremely likely" to continue across all continents and will increase the risk of coastal flooding under high tides, also increasing the impacts from waves and storm surges. Global projections show that in the business-as-usual scenario, the area of global land inundated by coastal flooding may increase by 15%, leading to a 17% increase in population exposure, and a 16% increase in exposed assets.[15] Globally, around 148 million people in low-lying coastal cities are exposed to episodic coastal flooding[16] while 58 million people are exposed to river flooding globally.

Sea level rise is a slow but continuous and certain process that may lead to 186 cm increase in the Extreme Total Water Level in Europe by the end of the century in a business-as-usual scenario (climate scenario leading to a warming between 3° and 4 °C by 2100). In the absence of better coastal protection, this sea level rise will increase the frequency of extreme coastal floods in Europe up to a factor 1000 depending on the location and climate scenario.

11.2.2.3 Impacts of Floods

IPCC AR6 warns that flooding events will continue to increase across most regions of the world for global warming scenarios superior to 2 °C. At 3 °C warming, minimum exposition risk is 90–186 million people could be affected by river flooding,[17] with maximum exposition risk 113–241 million people.[18] Meanwhile, the global population exposed to extreme river flooding with a 100-year return period is projected to increase 2.5–3% by mid-century for global warming scenarios above 2 °C.[19]

Floods caused about 5600 fatalities and more than €250 billion losses in EU countries between 1980 and 2021.[20] In the aftermath of such flooding, higher water temperatures accelerate the growth of waterborne pathogens, which encourages the

---

[14] https://www.ipcc.ch/site/assets/uploads/2018/03/SREX-Chap3_FINAL-1.pdf.

[15] https://www.ipcc.ch/report/sixth-assessment-report-working-group-i/.

[16] https://eco-act.com/climate-risk/flood-risk-impact-damage-and-recommendations-for-organisations/#_ftn5.

[17] SSP5—fossil fuel development scenario. https://ourworldindata.org/explorers/ipcc-scenarios

[18] SSP3—regional rivalry scenario. https://ourworldindata.org/explorers/ipcc-scenarios.

[19] https://eco-act.com/climate-risk/flood-risk-impact-damage-and-recommendations-for-organisations.

[20] https://climate.ec.europa.eu/news-your-voice/news/how-climate-change-disrupting-rainfall-patterns-and-putting-our-health-risk-2023-08-03_en#:~:text=According%20to%20estimates%2C%20floods%20caused,and%20proliferation%20of%20deadly%20diseases.

outbreak and proliferation of deadly diseases. Further, 75% of those affected struggle with their mental health.[21]

Given that around 170,000 people in Europe are exposed to flood risk annually,[22] the potential impact of flooding across the European continent could be huge:

- With 3 °C global warming, river flood damage in the EU and UK in 2100 would be six times larger than current losses, reaching €48 billion/year, and about 480,000 people would be exposed to flooding annually. Adaptation strategies could further reduce future flood impacts significantly, in a cost-effective manner, and have co-benefits. For example, reducing river flood peaks by installing retention reservoirs would reduce river flood damage in 2100 by nearly 40 €billion/year and also improve ecosystem quality.
- Without mitigation and adaptation measures, annual damage from coastal flooding in the EU and UK could increase sharply from 1.4 €billion nowadays to almost 240 €billion by 2100. Around 95% of these impacts could be avoided through moderate mitigation and by raising dykes where human settlements and economically important areas exist along the coastline. The extent to which adaptation can lessen the effects of coastal flooding and at what cost is sensitive to the investment strategy adopted.

11.3 A Review of Adaptation Indicators, in Practice

11.3.1 The Adaptation Fund

The Adaptation Fund was established in 2001 to finance concrete adaptation projects and programmes in developing country Parties to the Kyoto Protocol that are particularly vulnerable to the adverse effects of climate change. Since 2019, the Adaptation Fund (AF) serves the Paris Agreement under the CMA with respect to all Paris Agreement matters, effective 1 January 2019.

As of 30 June 2022, cumulative receipts into the AF Trust Fund comprised USD 1235.06 million, which makes it the largest international single fund with a focus on climate change adaptation.

The fourth review of the Adaptation Fund[23] recognizes its comparative advantages, including the speed of project approval, strategic engagement of stakeholders at the subnational level, various institutional benefits, efficiency of institutional arrangements, enhancement of country ownership in the funding process, direct and regional access modalities and readiness programmes.

[21] https://adventure.discomap.eea.europa.eu/experience/3/page/Floods/?views=Impacts-from-past-events.

[22] https://climate-adapt.eea.europa.eu/en/observatory/evidence/health-effects/flooding.

[23] https://unfccc.int/documents/624414.

11.3.2 M&E in the Adaptation Fund

The AF determines its needs for M&E and type if information required, including:

- *Type of Indicator*: this refers to whether the indicator is measuring the project/programme outcome, output, etc.
- *Indicators*: provide the indicators as identified through the results framework developed in the project documentation.
- *Baseline*: provide the value of the indicator prior to project start; if this was not available at the project design phase, the value must be reported in the first PPR submitted to the Board.
- *Progress since inception*: provide data on the progression of the indicator value since the start of the project.
- *Targets*: provide the end targets as set out in the project document for each of the identified indicators.

Any project or programme funded through the AF must align with the Fund's results framework and directly contribute to its overall objective and outcomes, as defined in Table 11.1.

AF projects use a results chain that shows the causal relationship between activities, outputs, outcomes, and impact over time. The AF use the following definitions[24]:

- ***Results***: Changes in a state or condition that derive from a cause-and-effect relationship. A development intervention can set three types of change into motion: output, outcome, and impact.
- ***Impact***: Positive and negative long-term effects on identifiable population groups produced by a development intervention. These effects can be economic, socio-cultural, institutional, environmental, technological or of other types.
- ***Outcome***: The intended or achieved short-term and medium- term effects of an intervention's outputs, usually requiring the collective effort of partners. Outcomes represent changes in development conditions that occur between the completion of outputs and the achievement of impact.
- ***Outputs***: The products and services resulting from the completion of activities within a development intervention.

During project/programme design, project-level results framework should reflect outputs/outcomes included in the AF Strategic Results Framework. In particular, at least one outcome and output indicator from the Framework must be included at the project design stage. In addition, project specific indicators would also be selected to reflect country specific objectives and reporting requirements.

AF applicants should also provide indicative target figures for each of the five core indicators that are relevant to the particular project.

[24] "Results Framework and Baseline Guidance", Adaptation Fund. https://www.adaptation-fund.org/wp-content/uploads/2015/01/Results%20Framework%20and%20Baseline%20Guidance%20final%20compressed.pdf.

Table 11.1 List of indicators per AF outcomes and outputs

Expected results	Indicators
Goal: Assist developing country parties to the Kyoto Protocol and the Paris Agreement that are particularly vulnerable to the adverse effects of climate change in meeting the costs of concrete adaptation projects and programmes in order to implement climate-resilient measures	
Impact: Increased resiliency at the community, national, and regional levels to climate variability and change	
Outcome 1: Reduced exposure to climate-related hazards and threats	1. Relevant threat and hazard information generated and disseminated to stakeholders on a timely basis
Output 1.1: Risk and vulnerability assessments conducted and updated	1.1. No. of projects/programmes that conduct and update risk and vulnerability assessments (by sector and scale)
	1.2. No. of early warning systems (by scale) and no. of beneficiaries covered
Output 1.2: Targeted population groups covered by adequate risk reduction systems	1.2.1. Percentage of target population covered by adequate risk-reduction systems
Outcome 2: Strengthened institutional capacity to reduce risks associated with climate-induced socioeconomic and environmental losses	2.1. Capacity of staff to respond to, and mitigate impacts of, climate-related events from targeted institutions increased
Output 2.1: Strengthened capacity of national and sub- national centers and networks to respond rapidly to extreme weather events	2.1.1. No. of staff trained to respond to, and mitigate impacts of, climate-related events (by gender)
	2.1.2. No. of targeted institutions with increased capacity to minimize exposure to climate variability risks (by type, sector and scale)
Output 2.2: Increased readiness and capacity of national and sub-national entities to directly access and program adaptation finance	2.2.1. No. of targeted institutions benefitting from the direct access and enhanced direct access modality
Outcome 3: Strengthened awareness and ownership of adaptation and climate risk reduction processes at local level	3.1. Percentage of targeted population aware of predicted adverse impacts of climate change, and of appropriate responses
	3.2. Percentage of targeted population applying appropriate adaptation responses
Output 3.1: Targeted population groups participating in adaptation and risk reduction awareness activities	3.1.1. No. of news outlets in the local press and media that have covered the topic
Output 3.2: Strengthened capacity of national and subnational stakeholders and entities to capture and disseminate knowledge and learning	3.2.1. No. of technical committees/associations formed to ensure transfer of knowledge

(continued)

Table 11.1 (continued)

Expected results	Indicators
	3.2.2. No. of tools and guidelines developed (thematic, sectoral, institutional) and shared with relevant stakeholders
Outcome 4: Increased adaptive capacity within relevant development sector services and infrastructure assets	4.1. Responsiveness of development sector services to evolving needs from changing and variable climate
	4.2. Physical infrastructure improved to withstand climate change and variability-induced stress
Output 4: Vulnerable development sector services and infrastructure assets strengthened in response to climate change impacts, including variability	4.1.1. No. and type of development sector services modified to respond to new conditions resulting from climate variability and change (by sector and scale)
	4.1.2. No. of physical assets strengthened or constructed to withstand conditions resulting from climate variability and change (by sector and scale)
Outcome 5: Increased ecosystem resilience in response to climate change and variability-induced stress	5. Ecosystem services and natural resource assets maintained or improved under climate change and variability-induced stress
Output 5: Vulnerable ecosystem services and natural resource assets strengthened in response to climate change impacts, including variability	5.1. No. of natural resource assets created, maintained or improved to withstand conditions resulting from climate variability and change (by type and scale)
Outcome 6: Diversified and strengthened livelihoods and sources of income for vulnerable people in targeted areas	6.1. Percentage of households and communities having more secure access to livelihood assets
	6.2. Percentage of targeted population with sustained climate-resilient alternative livelihoods
Output 6: Targeted individual and community livelihood strategies strengthened in relation to climate change impacts, including variability	6.1.1. No. and type of adaptation assets (tangible and intangible) created or strengthened in support of individual or community livelihood strategies
	6.2.1. Type of income sources for households generated under climate change scenario
Outcome 7: Improved policies and regulations that promote and enforce resilience measures	7. Climate change priorities are integrated into national development strategy
Output 7: Improved integration of climate-resilience strategies into country development plans	7.1. No. of policies introduced or adjusted to address climate change risks (by sector)
	7.2. No. of targeted development strategies with incorporated climate change priorities enforced

(continued)

Table 11.1 (continued)

Expected results	Indicators
Outcome 8: Support the development and diffusion of innovative adaptation practices, tools and technologies	8. Innovative adaptation practices are rolled out, scaled up, encouraged and/or accelerated at regional, national and/or subnational level
Output 8: Viable innovations are rolled out, scaled up, encouraged and/or accelerated	8.1. No. of innovative adaptation practices, tools and technologies accelerated, scaled-up and/or replicated
	8.2. No. of key findings on effective, efficient adaptation practices, products and technologies generated

11.3.3 Methodology of the Review

The assessment started with a screening of the full set of projects funded by the Adaptation Fund (145 projects), as available in the AF website in March 2023. Project related basic data (including title, country/ies of implementation, funding amounts implementing partners, date of approval) and information on the available documentation (project document, reports) were downloaded and summarised in a database.

The available data and reports were screened independently by two reviewers, who identified projects addressing flood related risks and impacts. This resulted in the selection of 17 projects, totalling 171 proposed indicators. The list of projects is in Annex 11.1.

There are 14 national projects and 3 multi-country[25] projects, covering 22 countries in total (see Table 11.2). A total of eight projects refer to coastal flooding (the remaining nine referring to river floods) and while most projects considered a mix of semi-urban and rural areas, six had a particular focus on urban areas.

The review comprised an analysis of the description of the context, the justification of the project, intervention logic and focused on the proposed indicators. Projects typically included a description of climate impacts with a focus on floods, with the support of visuals such as infographics, maps (see Fig. 11.2), summaries of socio-economic information on flood impacts, and a general description of climate scenarios with identified climate risks. This comprised expected global impacts in the medium term, notably at national or regional level. Proposals may indicate the likeliest consequences in the thematic field of their proposed activities, notably for settlements and livelihoods (e.g., "coastal climate impacts also threaten the already severely limited provision of basic services such as water supply, both for domestic and agricultural use"). These descriptions contained relevant information but with varying levels of detail, often lacking local and quantitative data.

---

[25] These are called "regional" projects by the AF but their analysis shows they are mostly based on a partnership between countries and lack a true regional nature.

Table 11.2 Geographical repartition of reviewed projects

Region	Reviewed projects	Countries
Africa	1	Ghana
Caribbean	3	Haiti, Dominican Republic, Cuba
Central Asia	4	Tajikistan, Kyrgyzstan, Kazakhstan, Uzbekistan
Europe	1	Georgia
Latin America	4	Chile, Ecuador, Costa Rica, Nicaragua
Oceania	1	Papua New Guinea
South Asia	2	Bangladesh, Pakistan
South-East Asia	5	Indonesia, Lao PDR, Cambodia, Thailand, Vietnam
West Asia	1	Armenia

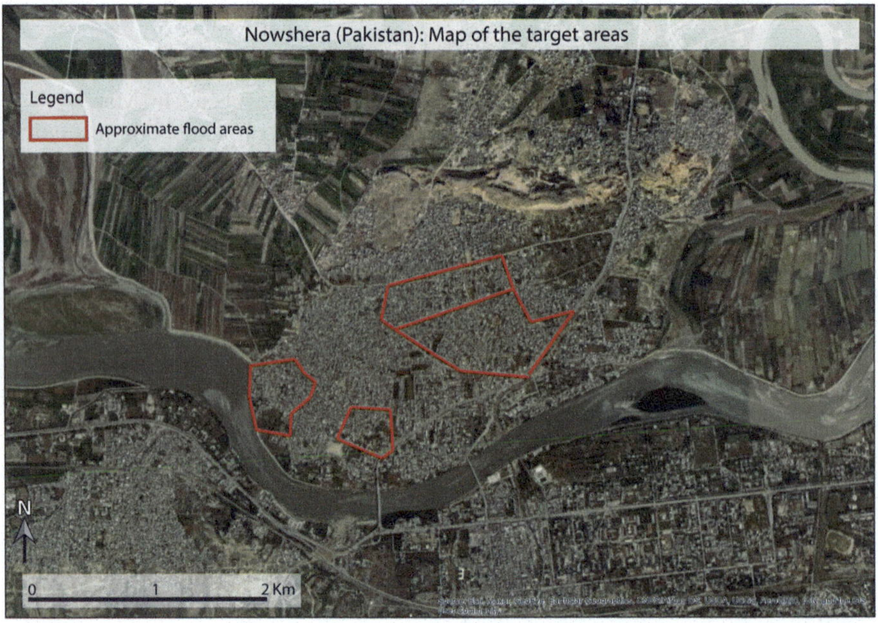

Fig. 11.2 Example of identification of project areas (UN Habitat proposal for Pakistan—made by the Author)

The indicators were categorised in relation to the AF Outcomes (Table 11.3) and their closest AF core indicator (Table 11.4). The same indicators were also categorised as per the assessment of the reviewer, looking at their nature as Output or Outcome indicators.

Attributes related to Baseline, Progress since inception and Targets were not reviewed, mainly due to lack of available information in many projects.

Table 11.3 Thematic repartition of indicators, per AF outcome

Expected results	% of total proposed indicators	% projects including a related indicator
Outcome 1: Reduced exposure to climate-related hazards and threats	13	41
Outcome 2: Strengthened institutional capacity to reduce risks associated with climate-induced socioeconomic and environmental losses	12	59
Outcome 3: Strengthened awareness and ownership of adaptation and climate risk reduction processes at local level	23	88
Outcome 4: Increased adaptive capacity within relevant development sector services and infrastructure assets	12	53
Outcome 5: Increased ecosystem resilience in response to climate change and variability-induced stress	8	29
Outcome 6: Diversified and strengthened livelihoods and sources of income for vulnerable people in targeted areas	18	76
Outcome 7: Improved policies and regulations that promote and enforce resilience measures	12	65
Outcome 8: Support the development and diffusion of innovative adaptation practices, tools and technologies	2	24
Total	**171**	**17**

11.4 Description of the Reviewed Projects

The summary of the review can be done through a categorisation of studied indicators per AF Outcome (Table 11.3) and per Output or Outcome indicators (Table 11.4).

Table 11.4 Number of project indicators related to AF indicators, categorised as outcome or output indicators[26]

Indicators	Number of indicators	Reviewed as Outcome indicators*	Reviewed as Output indicators*
1. Relevant threat and hazard information generated and disseminated to stakeholders on a timely basis	5	0	5
1.1. No. of projects/programmes that conduct and update risk and vulnerability assessments (by sector and scale)	2	0	2
1.2. No. of early warning systems (by scale) and no. of beneficiaries covered	9	0	9
1.2.1. Percentage of target population covered by adequate risk-reduction systems	7	5	2
2.1. Capacity of staff to respond to, and mitigate impacts of, climate-related events from targeted institutions increased	13	2	11
2.1.1. No. of staff trained to respond to, and mitigate impacts of, climate-related events (by gender)	4	1	3
2.1.2. No. of targeted institutions with increased capacity to minimize exposure to climate variability risks (by type, sector and scale)	1	1	0
2.2.1. No. of targeted institutions benefitting from the direct access and enhanced direct access modality	2	0	2
3.1. Percentage of targeted population aware of predicted adverse impacts of climate change, and of appropriate responses	14	1	13
3.2. Percentage of targeted population applying appropriate adaptation responses	16	7	9
3.1.1 No. of news outlets in the local press and media that have covered the topic	0	0	0
3.2.1. No. of technical committees/associations formed to ensure transfer of knowledge	6	0	6
3.2.2. No. of tools and guidelines developed (thematic, sectoral, institutional) and shared with relevant stakeholders	4	1	3

(continued)

[26] * Category as per the reviewer's assessment of each Project indicator.

** AF indicator 4.1.2 was interpreted as being close to AF indicator 4.2. Project indicators mainly comprised targets in numbers of infrastructure, it was then preferred to use AF indicator 4.1.2 rather than 4.2.

*** During the review, project indicators mentioning the integration of climate adaptation in policies and/or strategies were associated with the AF indicator 7.2 category.

Table 11.4 (continued)

Indicators	Number of indicators	Reviewed as Outcome indicators*	Reviewed as Output indicators*
4.1. Responsiveness of development sector services to evolving needs from changing and variable climate	2	0	2
4.2. Physical infrastructure improved to withstand climate change and variability-induced stress	0**	0	0
4.1.1. No. and type of development sector services modified to respond to new conditions resulting from climate variability and change (by sector and scale)	7	0	7
4.1.2. No. of physical assets strengthened or constructed to withstand conditions resulting from climate variability and change (by sector and scale)	12**	2	10
5. Ecosystem services and natural resource assets maintained or improved under climate change and variability-induced stress	3	3	0
5.1. No. of natural resource assets created, maintained or improved to withstand conditions resulting from climate variability and change (by type and scale)	10	0	10
6.1. Percentage of households and communities having more secure access to livelihood assets	0	0	0
6.2. Percentage of targeted population with sustained climate-resilient alternative livelihoods	5	0	5
6.1.1. No. and type of adaptation assets (tangible and intangible) created or strengthened in support of individual or community livelihood strategies	22	1	21
6.2.1. Type of income sources for households generated under climate change scenario	3	3	0
7. Climate change priorities are integrated into national development strategy	2	2	0
7.1. No. of policies introduced or adjusted to address climate change risks (by sector)	0***	0	0
7.2. No. of targeted development strategies with incorporated climate change priorities enforced	18***	0	18
8. Innovative adaptation practices are rolled out, scaled up, encouraged and/or accelerated at regional, national and/or subnational level	1	1	0
8.1. No. of innovative adaptation practices, tools and technologies accelerated, scaled-up and/or replicated	1	0	1

(continued)

Table 11.4 (continued)

Indicators	Number of indicators	Reviewed as Outcome indicators*	Reviewed as Output indicators*
8.2. No. of key findings on effective, efficient adaptation practices, products and technologies generated	2	0	2
Total	**171**	**30**	**141**

11.5 Main Findings

Indicators by generic thematic categories

The thematic review of the indicators (Table 11.3) shows that globally there is a good balance of indicators per AF Outcome. All AF Outcomes are represented in the sample, projects putting comparatively more focus on awareness (Outcome 3) and improved livelihoods (Outcome 6).

A detailed analysis shows some of the AF Outcomes can be used by a limited number of projects, indicating a trend for projects to specialise in a type of AF Outcome (e.g., one project alone gathers 40% of indicators related to Outcome 5. Another example is that AF Outcome 1 and 7 are related to a similar number of indicators, yet there are much less projects involved in Outcome 1 (41% of the sample) than in Outcome 7 (65% of the sample).

Indicator types

The review (see summarised data in Table 11.4) indicates a majority of Output indicators (82.5%) in relation to Outcome indicators (17.5%). Projects tend to report on activities and outputs, rather than outcomes and potentially impacts. These indicators have a stronger tendency to be used as an accountability tool to measure accomplishments and report on them, or as a management tool to track performance during implementation, than as a learning tool by giving evidence on what works and why.

This also is related to the fact that although some indicators are related to the Outcome level in the AF priority Outcomes, in practice they are used by projects for Output monitoring. For instance, Indicator 6.2 "percentage of targeted population with sustained climate-resilient alternative livelihoods" is presented at Outcome level, yet in practice projects measure their direct beneficiaries i.e., at Output level (e.g., "Percentage of farms in each micro-watershed with access to irrigation by means of hydraulic works built with programme funds".)

The impact level is missing, which limits the collection of information of higher-level effects of projects, for instance environmental monitoring and the relation between project activities and the status of the environment. For instance, projects could estimate the reduction of GHG emissions compared to BAU or their contribution to carbon sequestration, allowing them to underline their capacity to provide potential climate change mitigation co-benefits.

We note that the AF table with proposed indicators at Outcome level can be confusing as, outside the AF context, the same indicators would probably be categorised as Output indicators. For instance, the proposed indicator for Outcome 1 "Relevant threat and hazard information generated and disseminated to stakeholders on a timely basis" relates to a typical project Output of information generation and dissemination.

Reaching desired adaptation levels

Adaptation implying a holistic approach, its performance should be covered by indicators related to a diversity of adaptation solutions, as captured through the AF Outcomes and core indicators (including, capacity building, improved livelihoods, improved infrastructure, etc.…).

Addressing climate-related hazards, vulnerabilities and risks should be at the base of any adaptation action. Adaptation performance should also align to local/national objectives and find common metrics. Operating otherwise would imply its monitoring system focuses on demonstrating project outputs and on fulfilling project reporting requirements. Project related adaptation monitoring should contribute to improve knowledge on adaptation practices, notably from the perspective of the beneficiaries.

Most reviewed projects focus on the improvement of one or several of the following points:

- Developing and disseminating knowledge on climate change related risks, and identified measures for disaster risk management;
- Development of decision-making and planning tools (e.g., vulnerability maps, risk studies);
- Specific economic drivers of climate resilience (e.g., assets for climate smart agricultural production, availability of climate proof infrastructure);
- The management of environmental resources (with a focus on water in the reviewed sample) which degradation affects human settlements and livelihoods (e.g., coastal degradation, adverse impacts to agriculture).

As such, the projects lacked to link their proposed activities to an intervention logic from an adaptation point of view, that would indeed connect to local/national adaptation plans. Adaptation capacity as a whole could be addressed through a holistic approach, ensuring a mix of solutions is implemented, ideally using UNFCCC adaptation principles and GGA recommendations. Projects do not need to implement everything by themselves and should count with activities implemented by others. Yet they should explicitly indicate in the project document how they intend to contribute to the overall adaptation plan.

Clarity in the formulation of indicators and potential contribution to adaptation

Many of the proposed project indicators rely on complex concepts, often without a clear indicator definition and robust means for data collection. For instance, how to verify "land rehabilitated with increased adaptation capacity"? Or an "avoided decrease in income"?

In such cases, it remains unclear how the related information will be collected. Validation may also rely on the interpretation of the project itself, rather than of external and independent actors.

Further, the contribution of activities to adaptation attributes is sometimes not captured by the indicators themselves. For instance, the extent to which the "number of (renewable energy) nano-grids installed" can contribute to climate resilience depends largely on the context, including location, climate proofing of the new infrastructure, and climate vulnerability of the intended beneficiaries. Without a clear international and single set of indicators for adaptation, most frequently used types of adaptation indicators do not fundamentally differ from those used in development programming. Yet, they can justify their adaptation nature if they are relevant to the vulnerability context and well combined to measure contribution and impact.

Use of reference indicators

Most proposed project indicators are standalone indicators only relating to the project activities (at Output level). Despite the availability of M&E methodologies for climate change adaptation and for floods, none was mentioned in the reviewed projects. Most rely on self-methods for data collection and seem to lack coordination with national data systems. This is usually accepted in development projects, notably when focusing on Outputs. Yet, this approach doesn't allow to assess potential project contribution to NDCs and national adaptation plans or to sector development plans. Further, this limits the possibility of comparing performance across projects addressing similar issues.

Specific M&E related to floods

The reviewed projects included activities addressing floods, yet most were not focusing on flood issues only. The most common approach was to diversify activities in order to respond to climate resilience through various perspectives, including floods.

Some indicators directly referred to floods (e.g., "Number of houses made resilient against climate disasters including cyclones and floods"; "Number of people participating in concrete on-the ground climate change adaptation interventions (including EbA) to build resilience from drought and flood"). In the project logframes, reference to floods was more frequent at Objective or Outcome level (e.g., "Reduction of the impacts of coastal flooding through the recovery of coastal ecosystems"). However, to reduce flood impacts, it is essential to understand flooding causes. Most of the projects turned to "grey" infrastructure solutions intending to mitigate the visible impacts of floods, mainly on existing settlements.

Disaggregation

Many indicators included a mention of being disaggregated by sex, which is useful in order to collect information on the number of women beneficiaries.

Yet, adaptation is an issue that potentially affects all vulnerable groups. This implies a more detail in the disaggregation of the data, in order to better capture information on who are the beneficiaries, and if those who have the strongest needs are also those who benefit more.

The disaggregation would allow collecting information of outputs or outcomes according to climate vulnerabilities or by vulnerable groups, which are usually the most affected by climate impacts. Types of disaggregation may include e.g.: by locality, income, ethnic group, age, belonging to a minority, class of climate vulnerability.

11.6 Conclusions and Recommendations

The difficulties of building adaptation in practice

While the individual project activities and indicators are relevant to adaptation and show potential to contribute to the reduction of climate vulnerability, the analysis underlines the difficulty of the reviewed projects to indicate how they could help the intended beneficiaries to reach a desirable level of adaptation in a holistic and dynamic approach.

The present analysis shows that the main approach of the reviewed projects is to collect and produce more information related to individual Outputs rather than wider encompassing Outcomes. This implies more limited opportunities for lessons learned, notably on how to obtain results that can be sustained over time and ultimately ensure that beneficiaries attain a tangible change in adaptive status.

Most projects specify that their proposed indicators are well aligned to national sector objectives or local plans (notably, watershed management plans, which is particularly relevant in the context of flood related projects), sometimes mentioning alignment to a climate strategy (e.g., the NDC). Projects also develop evident efforts to tackle adaptation issues including through participatory climate risk and vulnerability assessments and by linking adaptation priorities to the socially and economically most disadvantaged. Yet, they may lack explicit demonstrations or robust evidence of project contributions to NDCs and national adaptation plans or to sector development plans. This is potentially due to the lack of adaptation metrics in these plans. This situation also limits the capacity of projects to directly feed into the development of future adaptation strategies and plans including at sector level.

Projects mainly proposed output-based performance reviews, linked to specific adaptation investments, framed in the time periods of traditional development projects, using static baselines. These determinants are not incompatible with but are probably insufficient to ensure a full implementation of UNFCCC and GGA recommendations in terms adaptation monitoring and evaluation. These include the incorporation of indicators and methods to measure the long-term effect and impact of adaptation, to look ahead and to identify opportunities to better assess transformative adaptation and systems-level resilience over time.

Adaptation: preparing for a "new normal"

Adaptation processes imply constant adjustments and new changes being accepted over business-as-usual practices. Ensuring the success of climate adaptation, notably in the case of projects related to floods, requires a comprehensive and multifaceted approach that involves a combination of planning, policy, review of infrastructure capacity, community engagement, and ensuring the availability of relevant information. In this paper, generic recommendations for strengthening adaptation through improved use of available information and internal project monitoring and evaluation systems are defined, with a focus on response to floods.

Recommendation 1: Build an initial information base and identify information needs for decision-making

Information, data analysis and knowledge products are essential to develop the rationale for engaging in adaptation action, including risk awareness, natural constraints, and vulnerability analysis of human systems. This includes information on climate trends and projections (e.g. changes in temperatures and rainfall patterns), associated climate-related hazards (e.g. floods, droughts, heatwaves, storms and storm surges, sea level rise), exposure and vulnerability of people, livelihoods, infrastructure and ecosystems to such hazards.[27] It also implies determining positive and negative pressures that act as incentives or barriers for stakeholders to support the change that is needed.

More specifically in the case of floods and adaptation, recommended activities include:

- Assess Vulnerability and Risks: assessment of flood vulnerability and risks in the area of concern. Consider past flood events, climate projections, land use, and population density.
- Updating flood models of developing catchments assists stakeholders to understand the impact of new development on existing residents in the catchment.
- Establish effective early warning systems that provide timely and accurate information to communities about flood risks. Ensure that the communication channels are accessible to all, including vulnerable populations.

Recommendation 2: Translate climate data and future projections into options for planning

Sound adaptation calls for a stakeholder dialogue and agreement over a common adaptation path. Adaptation requires all stakeholders to factor in the changing climate characteristics that will affect their usual social and economical patterns, should these be precipitation patterns, sea level rise, and other climate-related factors.

This implies stakeholders are aware not only of the full range of available solutions in order to choose options (technical, nature-based, fiscal, social, ...) that best fit

[27] A description of the status of knowledge on climate trends and projections, climate-related hazards, exposure and vulnerability should be indicated in the documentation of projects focusing on adaptation.

specific contexts, but also of the consequences and implications of these various adaptation solutions. These include societal and organisational choices (e.g., use of taxes, changes in mobility practices); support need from public institutions, private sector and organised citizens (e.g., new services from institutions, development of climate-smart business-to-business solutions); alignment with local, national and regional strategies and plans. Data and data analysis can be used to:

- Make stakeholders understand flood risk, its damage and protection costs. The vulnerability assessments together with stress tests can help communities and organisations to assess their current and future risk to flooding and improve their knowledge on available options (and their costs) on how to adapt to and mitigate risks;
- Identify weaknesses in land use policies and formulate measures to improve them, e.g., by restricting development in flood-prone areas and promoting sustainable land use practices, or by encouraging the use of permeable surfaces and green infrastructure to manage stormwater.
- Identify priorities for adaptation action, with clear references to and aligned with local/national priorities.
- Identify options for flood protection measures. Notably, so-called "grey" measures (e.g. construction works, dikes, levees, channels and reservoirs) can be effective in the short term, yet can be costly, have high maintenance costs and negatively impact biodiversity. "Green" flood protection measures use natural means to slow water flow such as providing space for flood water (e.g. through wetland restoration, reforestation including through mangroves, and natural buffer zones to reduce the impact of floods and enhance resilience by not building in floodplains of rivers) often take more time to implement, but have many co-benefits such as increasing biodiversity, improving water quality and sequestering carbon.
- Options do not need to focus on new structures. Flood proofing usually includes waterproofing existing structures (using waterproof materials), raising of structures to prevent damage, changing building use, or even relocating structures to a non-flood prone zone.
- Ensure to avoid maladaptation.[28]
- Agree on a flood risk management plan that outlines strategies for both mitigation (preventing or reducing flood risks) and adaptation (responding to and recovering from floods).

Recommendation 3: Support the integration of adaptation processes in routine development tools, including monitoring, budgeting and planning

Success in climate adaptation for floods requires a holistic effort among existing development practices, from governments and communities to scientists and NGOs.

[28] Any changes in natural or human systems that inadvertently increase vulnerability to climatic stimuli; an adaptation that does not succeed in reducing vulnerability but increases it instead, UNFCCC glossary of terms.

While it remains essential to approach adaptation, including when addressing floods, as an ongoing process that evolves with changing climate conditions and knowledge, long term perspectives need to be factored in. To adjust strategies based on changing climate conditions and emerging risks, a deeper monitoring of the evolving context is needed. At implementation level, adaptation monitoring implies clearer links with adaptation objectives, and reporting on the effectiveness of the proposed adaptation measures.

This implies climate adaptation stakeholders use a common reference adaptation framework for the design, implementation, monitoring, and evaluation of new adaptation action. Such a framework should provide key steps, questions, and methods for assessing adaptation progress, as well as a common language and structure for communicating results and potential benefits to each stakeholder. The framework can be based on an existing model, to suit the availability of existing data sources and information management capacities and practices.[29]

To be part of a next "new normal" adaptation monitoring should integrate traditional management and development planning, through a reference adaptation framework that:

- Integrates that flood adaptation is a long-term endeavour. The framework should be developed with a multi-decade perspective;
- Provides feedback on whether the proposed innovative adaptation approaches have reached results and compare with updated best practices in adaptation / flood risk management. This can be done through a co-evaluation or a co-monitoring exercise, for a joint assessment and evaluation of project progress and outcomes, led together with regional authorities, national/international peers, notably to address shared flood risks;
- Uses data and lessons learned in capacity building sessions for local governments, emergency responders, and community organisations to effectively respond to floods and implement adaptation measures. Support the establishment of public organisations that specialise in adaptation, supporting communities in their response to changing circumstances and evolving awareness and knowledge about flood risks and adaptation strategies;
- Uses adaptation data and lessons learned to develop laws and policies that support flood risk reduction and adaptation efforts;
- Improves coordination of information systems to ensure alignment with national and regional policies related to climate change and disaster management. Help identifying knowledge and innovation gaps, including in information systems

[29] Examples of existing adaptation frameworks include the Adaptation Monitoring and Evaluation Framework (AMEF) of the Global Environment Facility (GEF), the Climate Resilience Framework (CRF) of the Stockholm Environment Institute (SEI), and the Tracking Adaptation and Measuring Development (TAMD) framework of the International Institute for Environment and Development (IIED). Other useful adaptation tools include, the Climate Change Adaptation Indicator Catalogue (CCAIC) developed by the World Bank, the Climate Change Adaptation Toolkit (CCAT) developed by the United Nations Development Programme (UNDP), and the Climate Vulnerability and Capacity Analysis (CVCA) tools developed by CARE International are all useful adaptation tools.

such as early warning systems, to support investment in research and innovation to develop new technologies and approaches for flood adaptation;

- Uses data on impacts of floods and results of action reducing impacts to improve flood insurance options and financial mechanisms that can help communities recover quickly after a flood event. Consider risk-sharing arrangements and public–private partnerships;
- Raise public awareness about the importance of flood adaptation and the role individuals and communities can play. Promote sustainable behaviours that reduce flood risks;
- Engage and raise community awareness about flood risks and adaptation measures. Involve local residents in the planning and decision-making process to ensure their needs and concerns are addressed.

Acknowledgements The Author would like to thank Giulia Dunand, from the International Climate Finance Accelerator, for the support and work done by gathering and systematizing information.

Annex 11.1—List of Reviewed Projects

Country	Title	Implementing entity	Approved amount in USD
Bangladesh	Adaptation initiative for climate vulnerable offshore small Islands and Riverine Charland in Bangladesh	UN Development Programme	9,995,369
Indonesia	Adapting to climate change through sustainable integrated watershed governance in indigenous people of Ammatoa Kajang Customary Area in Bulukumba Regency, South Sulawesi Province, Indonesia	Partnership for Governance Reform (Kemitraan) of Indonesia	1,125,015
Armenia	Artik city closed stonepit wastes and flood management pilot project	Environmental Project Implementation Unit	1,435,100
Lao People's Democratic Republic	Building climate and disaster resilience capacities of vulnerable small towns in Lao PDR	UN-Habitat	5,500,000
Cambodia	Climate change adaptation through protective small-scale infrastructure interventions in coastal settlements of Cambodia	UN-Habitat	5,000,000

(continued)

(continued)

Country	Title	Implementing entity	Approved amount in USD
Georgia	Developing climate resilient flood and flash flood management practices to protect vulnerable communities of Georgia	UN Development Programme	5,316,500
Pakistan	Enhance community, local and national-level urban climate change resilience to water scarcity, caused by floods and droughts in Rawalpindi and Nowshera, Pakistan	UN-Habitat	6,094,000
Papua New Guinea	Enhancing adaptive capacity of communities to climate change-related floods in the North Coast and Islands Region of Papua New Guinea	UN Development Programme	6,530,373
Dominican Republic	Enhancing climate resilience in San Cristóbal province, Dominican Republic Integrated Water Resources Management and Rural Development Programme	Dominican Institute of Integral Development of Dominican Republic	9,953,692
Haiti	Implementing measures for climate change adaptation and disaster risk reduction mitigation of school facilities in Haiti	United Nations Educational, Scientific and Cultural Organization	9,916,344
Ghana	Increased resilience to climate change in Northern Ghana through the Management of Water Resources and Diversification of Livelihoods	UN Development Programme	8,293,972
Multi-country (South East Asia)	Mekong EbA South: enhancing climate resilience in the Greater Mekong sub-region through ecosystem-based adaptation in the context of South-South Cooperation (Thailand, Viet Nam)	UN Environment Programme	7,000,000
Multi-country (Latin America)	Reducing climate vulnerability and flood risk in coastal urban and semi urban areas in cities in Latin America (Chile, Ecuador)	Development Bank of Latin America	13,910,400
Costa Rica	Reducing the vulnerability by focusing on critical sectors (agriculture, water resources and coastlines) in order to reduce the negative impacts of climate change and improve the resilience of these sectors	Fundecooperacion Para el Desarollo Sostenible	9,970,000

(continued)

(continued)

Country	Title	Implementing entity	Approved amount in USD
Multi-country (Central Asia)	Reducing vulnerabilities of populations in the Central Asia region from glacier lake outburst floods in a changing climate (Kazakhstan, Kyrgyzstan, Tajikistan, Uzbekistan)	United Nations Educational, Scientific and Cultural Organization	6,500,000
Nicaragua	Reduction of risks and vulnerability based on flooding and droughts in the Estero real watershed	UN Development Programme	5,500,950
Cuba	Reduction of vulnerability to coastal flooding through ecosystem-based adaptation in the South of Artemisa and Mayabeque Provinces	UN Development Programme	6,067,320

References

Adaptation Fund. (2010). *Project level results framework and baseline guidance document*. https://www.adaptation-fund.org/document/project-level-results-framework-and-baseline-guidance-document/

Adaptation Fund. (2020). *Guidance document to complete document to complete project performance report (PPR) for projects funded by the Adaptation Fund*.

Christiansen, L., Schaer, C., Larsen, C., & Naswa, P. (2016). *Monitoring & evaluation for climate change adaptation: A summary of key challenges and emerging practice. Understanding, discussing and exemplifying the key challenges of M&E for adaptation*.

FAO and UNDP. (2019). *Strengthening monitoring and evaluation for adaptation planning in the agricultural sectors*. Rome.

GIZ. (2021). *Indicators matter to LNOB—An indicator toolbox to leave no one behind in fighting poverty and inequality. A practical guide for project designers and implementers*.

IPCC. (2021). *Sixth assessment report, climate change 2021: The physical science basis, the working group, 14th session of working group I and 54th session of the IPCC*.

Riahi, et al. (2017). *The shared socioeconomic pathways and their energy, land use, and greenhouse gas emissions implications: An overview*. Global Environmental Change.

Seneviratne, S. I., Nicholls, N., Easterling, D., Goodess, C. M., Kanae, S., Kossin, J., Luo, Y., Marengo, J., McInnes, K., Rahimi, M., Reichstein, M., Sorteberg, A., Vera, C., & Zhang, X. (2012). Changes in climate extremes and their impacts on the natural physical environment. In C. B. Field, V. Barros, T. F. Stocker, D. Qin, D. J. Dokken, K. L. Ebi, M. D. Mastrandrea, K. J. Mach, G.-K. Plattner, S. K. Allen, M. Tignor, & P. M. Midgley (Eds.), *Managing the risks of extreme events and disasters to advance climate change adaptation. A special report of working groups I and II of the intergovernmental panel on climate change (IPCC)* (pp. 109–230). Cambridge University Press.

UNEP. (2021). *Emissions gap report 2021: The heat is on—A world of climate promises not yet delivered*. Nairobi.

UNEP 2023 GAP report 2023, United Nations Environment Programme. (2023). *Adaptation gap report 2023: Underfinanced, underprepared. Inadequate investment and planning on climate adaptation leaves the world exposed*. Nairobi. https://doi.org/10.59117/20.500.11822/43796

UNFCCC United Nations Framework Convention on Climate Change. (2023). *Monitoring and evaluation of adaptation at the national and subnational levels: Technical paper by the Adaptation Committee, the Kyoto Protocol and the Paris Agreement*.

Part III
From Action to Policy

Chapter 12
Rediscovering Rivers in the Brazilian Megacity of San Paolo: Leveraging Water Sensitivity in the Urban Governance

José Bueno, Luiz de Campos Jr., Alessio Mazzaro, and Massimiliano Granceri Bradaschia

Abstract Located on the edge of the Brazilian Atlantic Plateau, the city of São Paulo has its territory (1500 km^2 with over 11.4 million inhabitants) predominantly drained by the Tietê River and its tributaries. From the late nineteenth century to the end of the twentieth century, the city experienced exponential growth, with environmentally impactful transformations such as radical interventions in the river system, a high rate of urban soil sealing, avenues implemented in valley bottoms, and disorderly territorial occupation—especially around water sources. This process led to a growing alienation of the population regarding urban life issues and a distancing from the systemic dimension of nature present in the urban environment. It is in this context that the project Rios e Ruas was created in 2010: a citizen initiative aimed at encouraging public engagement in the sustainable development of São Paulo by promoting a rediscovery of the water bodies that the city has made invisible as it developed. Rios e Ruas methodology is based on historical-cartographic research, field explorations, interaction, collaboration, and participatory interventions. This chapter, involving the two founding activists of the initiative in its writing, presents part of the practical results achieved by Rios e Ruas such as the drafting of two laws, a sticker intervention on street signals to reveal the presence of water under the urban fabric, and the creative and sensitive use of cartography in exhibitions as a way of inspiring care for water bodies. In particular, reflecting on how these experiences influenced the perception of citizens and politicians, the chapter highlights the importance of Rios e Ruas in motivating the citizen engagement in the protection of urban nature and in inspiring innovative initiatives for São Paulo and other Brazilian cities.

J. Bueno · L. de Campos Jr.
Rios e Ruas, São Paulo, Brazil

A. Mazzaro
Politecnico di Torino, Torino, Italy

M. Granceri Bradaschia (✉)
IUAV Università di Venezia, Venice, Italy
e-mail: mgranceri@iuav.it

M. Granceri Bradaschia et al. (eds.), *Climate Change Adaptation, Flood Risk, and Beyond*, Planning for Climate Proof Cities,
https://doi.org/10.1007/978-3-031-65463-3_12

Keywords Urban planning · Urban watersheds · Walking · Brazil · Water governance

12.1 Introduction

Rios e Ruas emerged from the collaboration between the architect and social urbanist José Bueno and the educator and geographer Luiz de Campos Jr., who, for over 15 years, had been researching, exploring, and disseminating the existence of hundreds of invisible watercourses in the city of São Paulo in the field of formal education and teacher training courses. The duo's first exploratory experience took place in 2010. Using old maps that included part of the city's hydrography, they identified the probable location of the springs of a small stream near their homes in the Butantã neighbourhood, west of São Paulo. They decided to embark on a cycling expedition, following the course of these waters that only became visible at the springs, and then flowed through streets and avenues of the neighbourhood for several kilometres until reaching their mouth. This field exploration revealed the forgotten presence of a stream and served as the matrix for the future actions that Rios e Ruas offered to the population to awaken a new perception of the rivers forgotten beneath the city's asphalt. They initiated a growing process of non-formal education beyond classrooms through workshops and expeditions that included conversations, group dynamics, audiovisual presentations, and explorations of the city—on foot or by bicycle—that brought an experiential dimension to the reflection on the environment, sustainability, and citizenship. By encouraging sensitivity to what is alive in the city streets, the reconnaissance expeditions of the relief forms, watercourses, and vegetation lead participants to slow down their minds, appreciate the richness of the environment they live in, and reconsider the quality of coexistence with other citizens. José and Luiz believe that beyond theoretical or formal teaching, collective experiences that stimulate knowledge of the nature of the city where we live create the necessary sense of belonging and ownership that strengthens the emotional bond with space. The goal of Rios e Ruas is to promote and inspire multiple actions so that the 20 million people inhabiting São Paulo can discover, see, and desire their rivers to be clean and free. The article is organised as follows: a first part, Sect. 12.2, which presents the geographical, historical, and urban context of the city of São Paulo where Rios e Ruas's activities take place, followed by Sect. 12.3, which presents the theoretical themes of the research such as Water Sensitivity and the involvement of the population in environmental governance and water policies. Section 12.4 illustrates the methodology developed over the years by Rios e Ruas, Sect. 12.5 discusses the results of three collective experiences, and, finally, Sect. 12.6 provides the key final considerations and potential follow ups of the Rios e Ruas activities.

12.2 São Paulo's Context

The city of São Paulo, the capital of the state of São Paulo, was established in a small sedimentary basin, located on a compartment at the edge of the Brazilian Atlantic Plateau. Its historical centre is approximately 60 km in a straight line from the coastline.

The Tietê is the main river in the hydrographic basin that drains the city, originating 90 km to the east in the municipality of Salesópolis. Despite its sources being just over 20 km from the coast, the river flows westward, penetrating the plateau in a course of over 1100 km until it reaches the mouth at the Paraná River. Therefore, the municipality of São Paulo is in the sub-basin called Alto Tietê, near the region of its headwaters. Even though it is located in the initial stretch of the river, São Paulo has a dense hydrographic network with hundreds of watercourses. Official studies on the extent of this network are unknown, but the most reliable estimates revolve around 4000 km of watercourses.

The city of São Paulo was founded in 1554, and its population in 1872—the date of the first national census—was just over 30,000 people. This slow growth changed dramatically from the last decades of the nineteenth century to the mid-twentieth century when it became exponential, increasing from 65,000 to over 2.1 million inhabitants between 1890 and 1950. The city's population continued to grow at very high rates until the 1980s, and today it is over 11.4 million inhabitants[1] (Fig. 12.1).

This large population, driven and attracted by the economic development of the region[2]—coupled with the urbanisation of extensive areas, the establishment of numerous industrial plants, and the consequent demand for water and energy, produced many environmentally impactful transformations in the territory within a short period.

The urban sprawl of São Paulo grew in tandem with the population increase. Vast peripheral areas, once dominated by rural properties, were rapidly urbanised. Affluent regions closer to the centre and on flat terrain were used for the establishment of planned neighbourhoods catering to the wealthier strata of the population. The three main rivers—the Tietê and its tributaries, Pinheiros and Tamanduateí—were straightened, canalised, and their floodplains were drained, filled, and mainly occupied by industrial plants and working-class neighbourhoods.

Areas farther from the centre were generally subject to uncontrolled real estate speculation, leading to the development of poorly planned neighbourhoods with low-quality urban infrastructure, such as roads, transportation, public lighting, garbage

---

[1] The most recent official census (2022) indicates that the municipality of São Paulo has 11,451,245 inhabitants in a territory of 1521 km^2, resulting in a population density of 7528 inhabitants per square kilometre. Available at: https://censo2022.ibge.gov.br/panorama/index.html.

[2] The introduction of coffee cultivation in the state of São Paulo in the second half of the nineteenth century required a large number of workers. The increasing restrictions on the slave system that were being implemented at the time compelled large landowners to encourage the massive arrival of immigrants to work on the plantations. The wealth generated by the economic cycle of coffee cultivation provided the capital that was invested in the early stages of the industrialization process in the twentieth century.

Fig. 12.1 Localisation of São Paulo state within Brazil (bottom left), localisation of São Paulo municipality within its state (bottom right), and the Tietê River—the main river and its catchment area for the São Paulo urban area (top)

collection, and sewage treatment. The lower-income strata of the population often occupied these neighbourhoods, and areas closer to watercourses prone to floods or in more rugged terrains susceptible to landslides received even more precarious settlements, often built by the residents themselves.

In the central region, the last areas to be occupied were the valleys of the tributary streams of the main rivers. Municipal administrations successively favoured automotive transportation, implementing major avenues in the valleys over the canalised watercourses. The floodplains of the main rivers—channelled in the open—received expressways running parallel to their channels.[3]

The Pinheiros River underwent a monumental engineering intervention: its course was reversed. The purpose of this intervention was to redirect the flow of water from the Tietê and Pinheiros Rivers to a reservoir high up in the Serra do Mar, and then channel it to a hydroelectric power plant located in the coastal plain, taking advantage of a 700-m elevation difference between the top and the base of the mountain range.[4]

As a result of São Paulo's rapid and chaotic growth and the urban planning policies implemented by its leaders, the city witnessed a pronounced degradation in the quality of its watercourses, especially the Tietê River and its main tributaries. Almost all the streams in the central region of the city were culverted (channelled and buried), flowing invisibly beneath the eyes of the residents of São Paulo.

Today, most of the water that supplies the residents of São Paulo comes from eight large production systems, with the main ones located at a distance from the city. The Billings Reservoir, the largest in the metropolitan region, was built to store water for energy production, with no concern for maintaining the quality of this water for

[3] The Tietê and Pinheiros rivers are flanked by expressways known as "Marginais," and the Tamanduateí River canal is bordered by Avenida do Estado. Among the main avenues in the city implemented in the valleys of the streams are:

Nove de Julho (Saracura e Iguatemi), Vinte e Três de Maio (Itororó e Caaguaçu), Pacaembu (Pacaembu), Sumaré (Sumaré), Pres. Juscelino Kubitschek (Sapateiro), Bandeirantes (Traição), Hélio Pellegrino (Uberaba), Jorn. Roberto Marinho (Águas Espraiadas), Prof. Vicente Rao (Cordeiro), Eng. Alberto de Zagottis (Zavuvus), Eliseu de Almeida (Pirajussara), Jorge João Saad (Antonico), Escola Politécnica (Jaguaré), Corifeu de Azevedo Marques (San Remo e Pirajussara-Mirim), Interlagos (Olaria), Salim Farah Maluf (Tatuapé), Abel Ferreira (Capão do Embira), Prof. Luiz Ignácio de Anhaia Mello (Capão do Embira), Aricanduva (Aricanduva), Carlos Caldeira Filho (Morro do S), Dr. Bernardino Brito de Carvalho (Gamelinhas), Gov. Carvalho Pinto (Tiquatira), Jacu-Pêssego (Jacu), João Relicário Eleutério (Itaquera), Nadir Dias de Figueiredo (Divisa), Luiz Dumont Villares (Carandiru), Eng. Caetano Álvares (Mandaqui), Inajar de Souza (Cabuçu de Baixo), Miguel Conejo (Rio da Pedra), Gal. Edgar Facó (Ribeirão Verde).

[4] The so-called "Projeto Serra" was conceived by the engineer Asa White Kenney Billings, an employee of the Canadian company São Paulo Tramway, Light and Power Company, which provided electricity and tram transportation to the city of São Paulo since the early twentieth century. The Serra Project was implemented in the 1920s when the Henry Borden Plant began operating with water drawn from the headwaters of the Pinheiros River. In the 1940s, the construction works were completed, enabling the reversal of the river's flow from its mouth to the Tietê River, a system that operated until 1992 when it was prohibited by law. Today, such an operation only occurs during periods of heavy rainfall, in case of a risk of major flooding downstream.

potable use. After decades of pumping water from the Tietê and Pinheiros Rivers, which received untreated domestic and industrial effluents from a large part of the region, almost all of the water in the reservoir is highly polluted, making it infeasible for potable use.

It was in this environmental, historical, and urban context that it became apparent that the type of development the city experienced in recent decades led the population to accept that the rivers of São Paulo no longer exist. The association of rivers and streams with bad odours, the transmission of diseases, and the risks of flooding may explain the public support for public policies to channel and cover watercourses. Even today, a significant portion of the population approves of stream channelling, unaware that there is sufficient technology to keep them clean and open (Fig. 12.2).

It was observed that people, in general, perceive rivers as "problems" to be solved by someone else. They do not see themselves as part of the solution, believing that the resolution of problems is the exclusive responsibility of public management and sanitation agencies. A commonly heard challenging phrase is: "It's not my problem, it has always been this way, and someone should fix it." This community demobilisation contributes to a low level of self-accountability, a sense of community collaboration, and political agency.

The practitioners concluded that the major challenge to be faced is the alienation of the city's inhabitants from the natural environment in which they live. This alienation results in a low awareness of the impact of water nature on urban life, such as: where does the water we consume come from? Where do the waste products we discard go? Is the poor state of the rivers natural, or is it produced by us? Why can't we enjoy rivers and parks more in the city landscape?

12.3 Research Context

12.3.1 Water Sensitive Approach

Over the years, urban and industrial development has increased the area of impermeable surfaces in our cities, reducing permeable green areas, and consequently, increasing stormwater runoff. For a long time, these waters, which temporarily flooded inhabited areas, carrying pollutants and street sediments, were considered a problem to be removed as quickly as possible meaning, discharged into rivers without treatment.

In the 1980s, a different approach began to emerge, and in arid regions like Australia, stormwater runoff, instead of being seen as a problem, started to be perceived as a water resource that could be stored, filtered, and reused (Radcliffe, 2018). In 1994, Australian states began to release guidelines for an approach to planning that integrates urban design, landscape architecture, and the urban water cycle: "Water Sensitive Urban Design" (WSUD). This holistic planning approach has the primary goals of protecting watercourses and wetlands in the urban fabric,

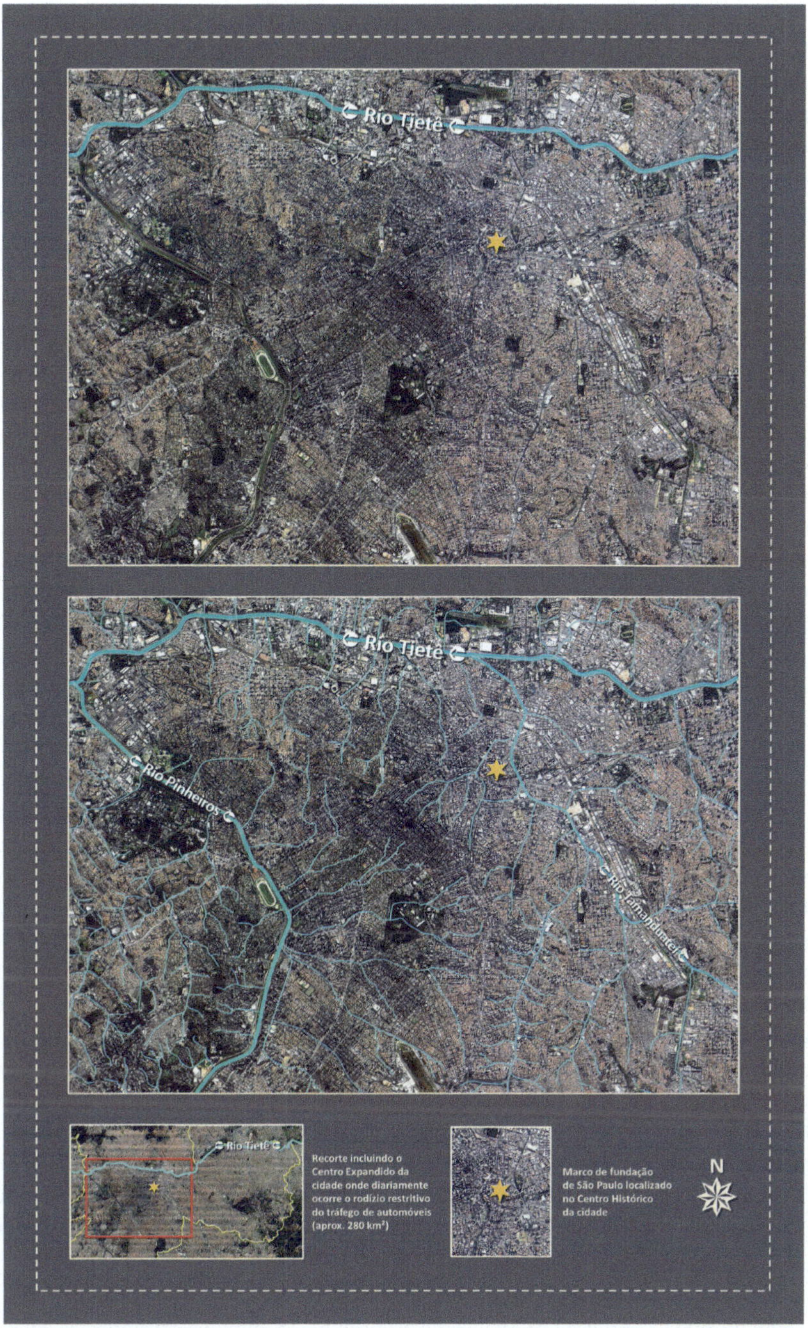

Fig. 12.2 The "Expanded Center" of São Paulo, the rivers Tietê, Pinheiros, Tamanduateí and the simplified hydrography

improving their quality, and using them as social spaces; reducing surface runoff by increasing permeable surfaces and creating tanks to partially contain it; using vegetation as a filter for pollutants and artificial swamps and ponds to allow sedimentation; maximising groundwater protection; reducing the use of potable water by storing rainwater and surface runoff, reusing this water for domestic purposes, and releasing it slowly into the aquifer. WSUD considers all water inputs as a resource and proposes relying more on the water that can be captured within the city than on that brought from external basins, thereby managing the city as a watershed (cit. "*managing the city as a catchment*"[5]).

With the increase in urban flooding due to climate change, the application of WSUD has gained importance as an approach to promote the sustainability and resilience of water resources. However, WSUD was developed and primarily applied in cities in the Global North, and only recently has there been a revision for its application in the Global South—where part of the urban space and water supply are characterised by informality (Kumar et al., 2023). From this review, a list of "New global principles of sensitive planning", was proposed, including some key points such as providing legal protection after the renaturalization of water bodies, building or redesigning roads and surfaces to reduce flooding, linking planning to population growth prediction models, and creating strong connections between territorial planning and water planning in the management of Southern cities (Kumar et al., 2023).

Another theoretical advancement, derived from studying the history and future of Australian cities and presented as a tool for water urban managers and policymakers, is a set of six evolutionary stages through which a city can pass on its way to achieving sustainable water conditions (Brown et al., 2009). In the first three phases, the city provides primary services—drinking water, sewage, and flood protection. In the next three phases, it advances to the Water-Sensitive City (WSC) stage, where the city begins to exhibit green infrastructure, climate change resilience, intergenerational equity, a community of citizens supporting a sustainable lifestyle, and flexible urban infrastructure and forms.

In conclusion, in the context of rethinking our relationship with water (floods) in the urban fabric, new sensitivities are emerging, such as the acceptance that certain designated parts of the city can be flooded and should not be conceived as carved-out areas but as spaces for social public life—an example is the Benthemplein square in Rotterdam, designed by Urbanisten.[6] There is also a recent interest in "Wetness" (Mathur & Da Cunha, 2020) and in the practices and existing knowledge of life in wet landscapes—such as in monsoon areas, as seen in the research project "Inhabited Sea".[7] There is also an exhortation not to plan the city as if it were dry land—considering that a fixed and stable separation between land and water, and the idea

[5] Diretrizes do WSUD da cidade de Melbourne, páginas 12–15. http://observatoriaigua.uib.es/repositori/suds_australia_melbourne.pdf.

[6] Projeto praça Benthemplein. https://www.urbanisten.nl/work/benthemplein.

[7] Projeto de pesquisa Inhabited Sea. https://www.inhabitedsea.org.

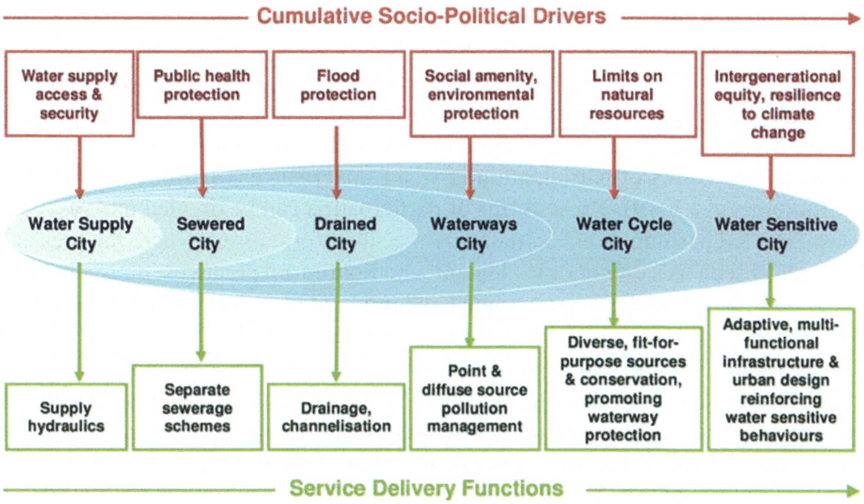

Fig. 12.3 Urban water transitions framework (Brown et al., 2009)

of a river as a fixed body of water confined to the path traced on maps, is culturally constructed (Da Cunha, 2019) (Fig. 12.3).

12.3.2 Citizen Engagement in Environmental Governance and Water Politics

Water in many contemporary cities can be invisible—channelled and hidden—except when it rains, and urban floods occur. Invisibility also manifests as our disconnection from the processes that enable drinking water to reach our homes and wastewater to be removed. Reconnecting people with water therefore means giving inhabitants the opportunity to understand the social context in which water interacts with the city, linking the hydrological cycle based on physical processes -of precipitation, infiltration, subsurface and ground flow, and evaporation- with their human alteration through infrastructures and politics. The resulting hydr-osocial cycle[8] (Linton 2014; Swyngedouw, 2015) allows to portray water territories not only as "mere biophysical nature" (Boelens et al., 2016) and so to consider how water problems and their solutions aren't politically neutral, nor solvable only through technical and managerial knowledge. A second reason to reconnect the inhabitants of a territory with its

[8] The hydrosocial cycle of water considers the physical and social movements of water contextualising them in the political, social, economic and cultural sphere. In addition, this concept proposes to consider that water and society are mutually influenced (Linton and Budds 2014; Pinheiro Imbelloci and Fernandes Fellipe, 2020).

water becomes so the importance of adding the contextual knowledge of residents to technical expertise in water management (Sharp, 2017).

In line with this sensibility, the emerging attention to the role that local communities can play in adapting to climate change and water resilience, as well as emerging forms of activism and the willingness to consult residents in decisions, has led to a proliferation of forms of participation in water governance (Mauroner et al., 2021; Van Buuren et al., 2019). These forms of participation can be divided into two categories: "Invited participation," initiated by the state, formalised, where the presence of citizens is seen as a democratic element; and bottom-up practices, more informal, where participatory processes are guided and initiated by communities (Van Buuren et al., 2019).

Invited participation may aim to empower local stakeholders, including their perspectives in decision-making, but often serves as a way to garner support for government policies, anticipating controversies and assessing citizen satisfaction (Enqvist et al., 2020; Tsouvalis & Waterton, 2012), resulting in limited citizen influence in decision-making (Van Buuren et al., 2019). Moreover, invited participation, especially in the Global South, if not supported by parallel community development actions, risks reproducing existing social inequalities—not every citizen will have the financial means and time to attend public meetings or the technical knowledge to contribute at the same level as experts (Van den Brandeler, 2014). Perkins (2004) highlighted this in the study of water resource management experiences in Brazil—particularly in the state of São Paulo—where since 1988 a decentralised water governance system was developed with comitês de bacias hidrográficas (CBH)[9] composed of representatives from public authorities, water users, and civil society organisations. Furthermore, Perkins notes how in these committees' horizontal relationships can be challenging due to the language used and the differentiated background of the participants. Additionally, as Sharp (2017) writes, for stakeholder engagement in hydro-social water management to work, administrations must have experts in communication and participatory processes.

Bottom-up initiatives, on the other hand, may arise from dissatisfaction with government policies or emerge in spaces where the government has little presence or power has been decentralised. Stakeholders involved in community-based water governance initiatives (CBWG) can develop alternative proposals to official policies to address local practical needs, initiate specific actions, or mobilise as activists to try to change existing policies (Van Buuren et al., 2019).

Van Buuren et al. (2019) defines the real challenge for both types of participation as achieving the co-creation of public value between water users and governments, a challenge because, for example, agendas may not align, or an invited participation

[9] Comitês de Bacia Hidrográfica, Agência Nacional de Águas e Saneamento Básico (ANA).
Comitês de Bacia no Estado de São Paulo. https://www.gov.br/ana/pt-br/assuntos/gestao-das-aguas/fortalecimento-dos-entes-do-singreh/comites-de-bacia-hidrografica.
http://www.comitepp.sp.gov.br/pp/comites.htm.

initiative may occur more as a simple consultation required by law. It should also be considered that another obstacle to co-creation may be the fact that water authorities are bound by legal regulations, budgets, and deadlines that do not always align with user proposals.

An important example of a bottom-up initiative that led to the co-production of public policies is the experience of the "Comités de Agua Potable y Saneamiento" (CAPS) in Nicaragua—committees that manage water resources in small communities. Through the creation of a national network linking different CAPS and with the economic support and knowledge (information, experience, and contacts with government officials) from some NGOs and multilateral agencies, the community initiative gained political and legal recognition in the 2000s (Romano, 2017). Furthermore, the NGOs that are part of the CODA coalition (Coalition for Water Rights) involved the CAPS network in the co-creation of Law 620 (Ley General de Aguas Nacionales). Once again, knowledge transfer and capacity development were crucial for the effective participation of citizens in the design of public policies.

12.3.3 Re-naturalization of Water Bodies and Reconnection with Nature

Born within the ecological movement to respond to the deterioration of watercourses and now seen as a step toward the green economy (Benjamin et al., 2014), River Restoration is a series of activities aimed at improving the physical and ecological characteristics of a river, restoring biodiversity, and where urban space allows, returning a channelized river to a more natural and meandering course.[10] In cities, this can be challenging, but even if only partially achieved, river re-naturalization can have a considerable positive impact on the urban environment, reintegrating the river into the landscape (Guimarães et al., 2021) and addressing a disconnection between citizens and the river landscape due to technocratic water management (Sharp, 2017).

Wantzen et al. (2023) speak of a loss of biological and cultural diversity linked to rivers, a reduction in River Culture,[11] due to the transition from a society where life was determined by the flow of rivers to modernism, where nature was considered something to be controlled, and people—outside indigenous communities—stopped considering the river as a living entity.

Over the years, community participation in river re-naturalization processes has become increasingly important to align ecological goals with the possibility of involving inhabitants with the landscape, improving aesthetic and recreational aspects, and also providing the opportunity to establish more sustainable and long-term management of the re-naturalization process (Benjamin et al., 2014).

[10] Definition of river restoration given by European Center for River Restoration. https://www.ecrr.org/River-Restoration/What-is-river-restoration.

[11] "The sum of biological and cultural adaptation, and cultural connection to nature, developed by organisms (including humans) living in riparian landscapes".

12.4 Methodological Approach of Rios e Ruas

Until 2010, in Brazil the relation between rivers and the city was a matter restricted to an infrastructure of professionals, public management bodies, and a few academic researchers. "Rios e Ruas" instead made it accessible to the average citizen through a simple and inviting language. Furthermore, for five years they worked as archaeologists, seeking maps and information in the archives of the Empresa Paulista de Planejamento Metropolitano—Emplasa and the Instituto Brasileiro de Geografia e Estatística, IBGE- an effort that brought to the publication in December 2015, through the GeoSampa platform,[12] a map showing the city of São Paulo with all its rivers, channelized or not.

As their work developed, José and Luiz identified three general principles that guide all activities, something they called "the Rios & Ruas way": the foreign gaze, being in the place, and connecting with others.

1. Foreign gaze.

> É preciso não se contentar com a primeira camada, é preciso sempre levar o olhar um pouco além.[13]

Every Rios & Ruas project begins with approaching to the space—known or unknown—where an expedition will take place with a foreign gaze: involving estrangement, questioning, and openness to surprises.

2. Being and visiting the place

> O lugar NUNCA é exatamente o que você espera.[14]

The approach to the place can take various forms, such as online research, bibliographic research, maps, field visits, or conversations with residents. Usually, all these forms are adopted, but each specific project may propose a different sequence. However, Rios & Ruas understand presence in the place and experience it through the body as essential: it is during the field survey that they will have more opportunities to be surprised. Being in the place involves talking to people, observing the landscape, smelling the scents, and listening to the sounds. Allowing oneself to be moved. This is the way they discover the senses that will guide the work in that specific place and how they will work with the people who live there (Fig. 12.4).

3. Connecting with others.

> Sonhar grande, agir pequeno e começar agora.[15]

[12] GeoSampa—Digital Map of the City of São Paulo, City Hall, Municipal Department of Urban Planning and Licensing. Accessible at: https://geosampa.prefeitura.sp.gov.br/PaginasPublicas/_SBC.aspx#.

[13] trsl. Do not be satisfied with the first layer; always look a little further.

[14] trsl. The place is NEVER exactly what you expect.

[15] trsl. Dream big, act small, and start now.

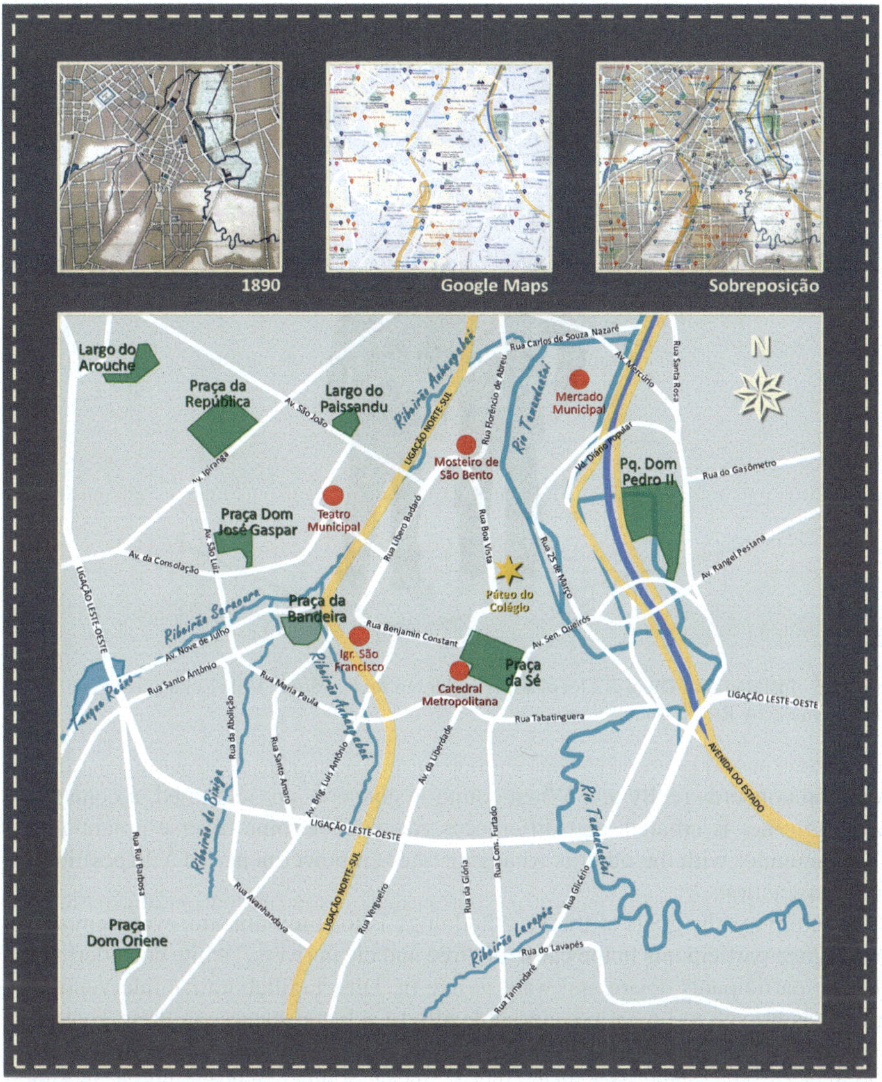

Fig. 12.4 Example of a map of downtown São Paulo used in the Rios e Ruas expeditions, produced by overlaying historical maps on the current street layout (made by the Authors)

The essence of Rios e Ruas' work lies in people, their relationships with each other and the environment. They consider that the work has a political character, much broader than simply "hydrographic" or "environmental." In this sense, they are more interested in differences (a wealth of possibilities) than unanimous (common sense). The first challenge is to connect people and create a predisposition to openness, to differences. Once this first challenge is overcome, the possibility of generating a legitimate interaction between participants is created, identifying what their

Fig. 12.5 Installation "There is a lot of water under our feet", Parque Villa Lobos, São Paulo, 2015 (picture made by Rios e Ruas)[16]

common concerns really are. When common concerns are identified, a connection with the other is created, and differences can then become diverse potentialities. This encounter with the other is what generates empowerment and a disposition for collective action.

In summary, the goals of their methodology is to communicate, evoke emotions, and engage participants in the cause of "free and clean rivers," while considering the city the participants desire and want to live in. Direct collaborative intervention in space has proven to be an indispensable ally to achieve success in these goals. For example, organising collective efforts to restore springs and abandoned public spaces, which are often located near watercourses, or when this is not possible as in highly urbanised regions, promoting artistic and graphic interventions, such as signalling the presence of hidden rivers with street sign stickers and graffiti (Fig. 12.5).

[16] The installation was part of the event "Mostra Rios e Ruas—Planeta no Parque" (translated: Rivers and Streets Show—Planet in the Park). It was on display for two months and consisted of a map of approximately 300 m^2. The cut-out included a representation of the invisible hydrography, important landmarks in the city and 20 "giant pins" with texts explaining short stories, relating that point to curiosities that would arouse an emotional connection between the visitor and the history, culture and hydrography of São Paulo. For example: "The biggest escape in the history of Carandiru Prison was via the stream of the same name. In 2001, 106 prisoners escaped through a tunnel dug in the direction of Av. General Ataliba Leonel, reaching the gallery through which the Carandiru

The scientific and technological foundations that support their approach and interventions in the territory include techniques from Permaculture (Bill Mollison and David Holmgren), Nature-Based Solutions, concepts from Cities for People (Jan Gehl), Transition Towns (Rob Hopkins), and Biophilic Cities (Timothy Beatley). Regarding pedagogy, their knowledge construction strategy aligns closely with the concepts of Paulo Freire: "Knowledge is a social process created through transformative action-reflection by humans on reality." The focus on learning, rather than teaching, through collective, social, transdisciplinary, and integral experiences, reflects the pedagogical fundamentals associated with "interactive learning" and "learning communities" by Augusto de Franco. These concepts derive from research on social networks and the work of authors such as Jiddu Krishnamurti, Carl Rogers, John Holt, José Pacheco, Tião Rocha, and others. In the interaction process with participants, principles of Nonviolent Communication (Marshall Bertram Rosenberg) are included to build a fertile learning environment based on empathic listening and freedom of expression (Fig. 12.6).

Applying their methodology, over the years Rios & Ruas brought centrality to the theme of urban rivers, becoming a reference in schools and universities, as well as inspiring many news articles, films, documentaries, songs, and theatre performances. Furthermore, in 2017 they were invited by Waterlution to conduct two awareness-raising activities for young people at the Water Innovation Lab in São Paulo Rios e Ruas and in 2018, they played a key role in moderating the Urban Rivers panel at the 8th World Water Forum in Brasília.

12.5 Experiences

12.5.1 Project "Bacias Hidrográficas nas placas de rua de São Paulo" (PL 858–2013 SP)

In September 2012, Ricardo Young, a candidate for City Councilor in São Paulo—*Vereador*[17]—approached numerous socio-environmental initiatives in the city to listen to their needs with the intention of exploring ways to contribute, if elected, through his mandate in the Municipal Chamber. Rios e Ruas was one of the chosen initiatives and had the opportunity to convey its work to Young's team through an experiential exploration in the Botanical Garden, where the Pirarungáua stream originates and flows. For over 70 years, this stream had been channelized and covered under the main avenue of the location.

Rios e Ruas proposed to Young that he could contribute by presenting a project that would include information about the micro-basins of unnoticed streams on the

---

flows. Today the stream runs through Parque da Juventude (where the prison used to be) in the open air."

[17] The councillor is a political agent who represents the people in the municipal legislature.

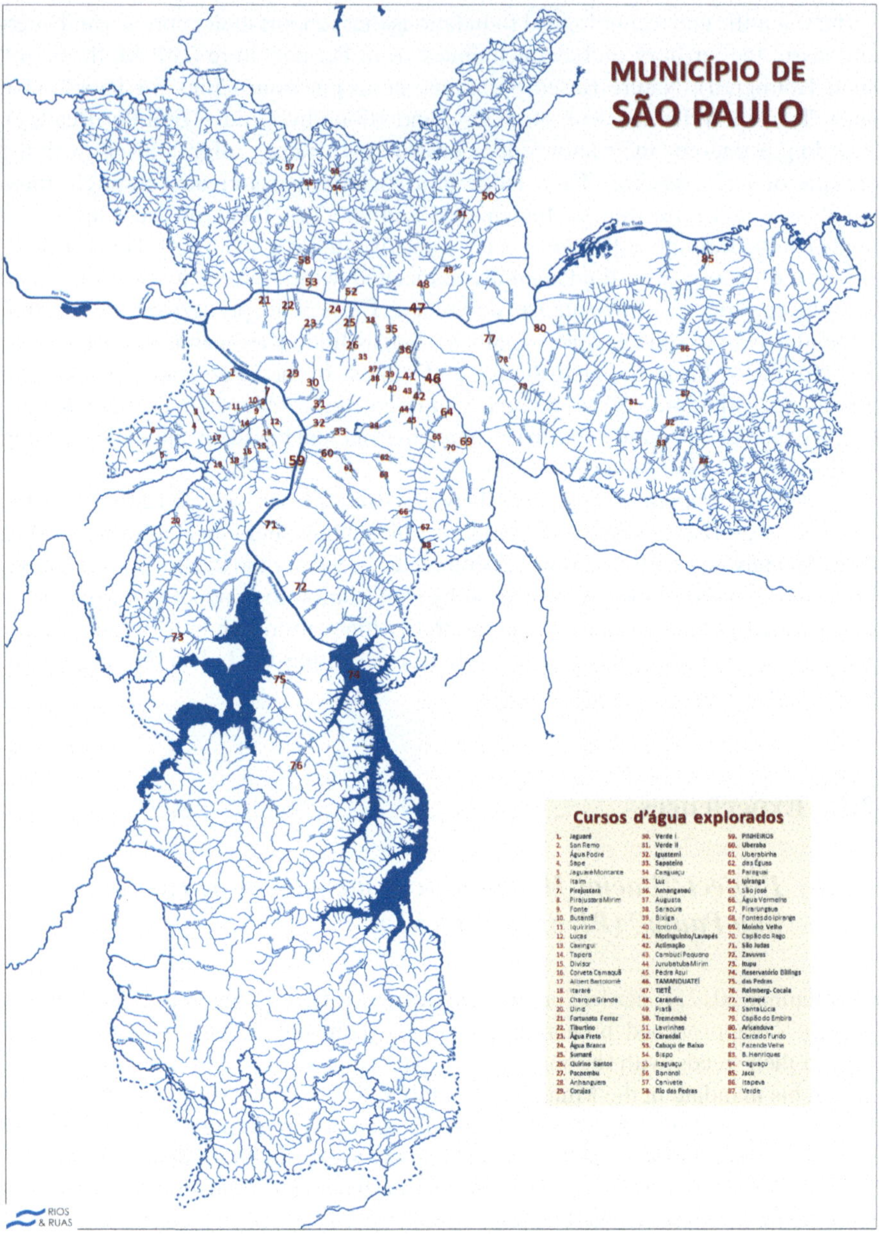

Fig. 12.6 Rivers of São Paulo where Rios e Ruas performed activities

Fig. 12.7 Workshop with the politician and his team, September 2012 (Picture by Rios e Ruas)

city's street signs, encouraging citizens to realise that the city's streets still belong to a territory that was originally water-based. Just as streets and avenues are considered as part of a neighbourhood, they would be perceived as part of a watershed. In this way, the population would understand that distant addresses, even if far from a river, belong to the same watershed. A proposal that would aim to generate a map of São Paulo where the regions are divided by river basin: a management map (Fig. 12.7).

Ricardo Young was elected as a councillor for a four-year term (2013–2017) and proposed transforming this idea into Bill 01-00858/2013, which provided for the amendment of Article 10 of Law 14.454 of 27 June 2007 to oblige the inclusion of the name of the river basins on the nameplates of the roads and public places on which they are situated.[18]

The bill was approved on 12 July 2014 after two votes in the São Paulo City Council and was taken to the then mayor Fernando Haddad for sanction. To the surprise and frustration of Rios and Ruas, the bill was rejected on 13 August with the following justification: "Including the name of the catchment area in which the street is located would result in an excessive amount of data"[19] compromising good visibility and prompt and immediate guidance for citizens. In addition, the mayor used as second motivation the fact that in order to make the action of raising awareness

[18] PROJETO DE LEI 01-00858/2013 do Vereador Ricardo Young (PPS).

[19] RAZÕES DE VETO, Projeto de Lei no 858/13, Ofício ATL no 130, de 13 de agosto de 2014.

of the city's natural resources effective, "all street signs should be replaced, at a significant cost to the municipal treasury".[20]

In response to these complaints, Rios e Ruas created a sticker with a transparent background that used the same font of the street signs to report information about the basin. The sticker could be applied directly to the signs, thus avoiding replacement, and demonstrating that there would be no need to spend public money on replacing the signs. That's how, in November 2015, while participating in an episode of "Causando na Rua", a tv series by the Brazilian movies' director Tata Amaral, Rios e Ruas carried out a first activist intervention by stickering some street signs in the micro-basin of the Saracura River, which rises near Avenida Paulista and runs through the Cerqueira César and Bexiga neighbourhoods to its mouth on the Tamanduateí River.

From that moment on, applying stickers to the signs became a common practice in all the Rios e Ruas expeditions, in which each time first the participants cleaned a sign, then applied the sticker, photographed it and published it on social media to demonstrate the educational impact of signposting the micro-basins present in the city. Among the dozens of signs that received the informal application of the stickers with the respective basins, we highlight the signage of the Sapateiro River Basin in the streets of Vila Mariana, the Saracura River Basin in the streets of Cerqueira César, the Anhangabaú, Itororó and Saracura Basins in the Centre, the Verde River Basin in Itaquera and the Iquiririm Basin in Butantã.

Carrying on these actions, Rios e Ruas learnt that although simple and feasible, the signposting of river micro-basins on street signs reveal to ordinary citizens the presence of water under the urban fabric of streets, avenues, and buildings. Furthermore, it demonstrated the resistance of public authorities in São Paulo to contributing to citizen education about the hundreds of invisible rivers: Rios e Ruas even received a technical report from CETESB stating that ghost rivers flowing underground have no environmental function for the city. Despite this, they strongly disagreed, and they continued to believe that as long as São Paulo rivers are ignored, urban development will continue to serve only the interests of real estate companies with public authorities approving master plans that disregard the presence of water in São Paulo.

Rios e Ruas stickering actions in São Paulo, later inspired similar initiatives by other independent groups: street signs showing the names of the river basins to which they belong started to appear in the cities of Buenos Aires and Belo Horizonte. Showing the potential, for more and more cities in Brazil, to realise in the near future, that street signs can be an excellent way of raising citizen awareness about the territory in which they live (Fig. 12.8).

[20] RAZÕES DE VETO, Projeto de Lei no 858/13, Ofício ATL no 130, de 13 de agosto de 2014.

Fig. 12.8 Intervention on a street sign (pictures from Rios e Ruas)

12.5.2 Project "Mapeamento, Proteção e Recuperação Das Nascentes Urbanas" (PL 719–2021 SP)

In March 2018, people from different collectives and the SOS Mata Atlântica NGO, together with Councillor Toninho Vespoli, joined forces to create the PL das Nascentes group on WhatsApp. Thanks to the knowledge of municipal management brought by the Councillor and civil servants, and the technical expertise brought by academic researchers, the group begun to exchange on the creation of a draft law for the Atlantic Forest (PL 0719/2021) aiming for the "registration, monitoring, protection, conservation and recovery of existing springs in the municipality of São Paulo".[21] This proposal was based on the fact that in São Paulo it is not enough for a spring to exist, it is necessary to legally prove its existence. Furthermore, municipal authorities tend to make this recognition as difficult as possible, demanding documentary evidence and technical reports, while on the other hand, when real estate developers' want to prove that a body of water present in a construction site is not a spring, they often accept their arguments.

The WhatsApp group worked on the proposal for PL 0719/2021 applying the knowledge gained from the previous legislative proposal processes that didn't work, so as to avoid a rejection from the mayor. In particular, Rios e Ruas built on what they learned during the 2013 experience with Ricardo Young. Understanding that in that previous case, they lacked consistent political support -having just one councillor on their side- with the PL 0719/2021, from the start they paid a visit to all the councillors

---

[21] JUSTIFICATIVA—PL 0719/2021. https://www.radarmunicipal.com.br/proposicoes/projeto-de-lei-719-2021.

who they knew wouldn't have anything a priori against the proposal—for example, without involving those most closely linked to the interests of real estate developers. In addition, in November 2021, they launched an Instagram page, "SOS Nascentes Urbanas", with the aim of publicising the law draft to the population. The social media action had also as second objective: to start a collaborative mapping of springs in Sao Paulo[22] Involving citizens in the mapping was a further way to make them understand the importance of the PL 0719/2021 and the number of urban springs present in São Paulo.

While the first objective of PL 0719/2021, was to demonstrate that there are springs in São Paulo, the second one was to propose a public campaign to recover water springs through the creation of a Programme called "*Adote uma Nascente*" (trsnl. Adopt a Water Spring).[23] This program covering the entire territory of the municipality of São Paulo, aims to promote community participation in the recovery of springs in degraded areas while preserving those that remain intact.

Finally, in December 2021, the bill was voted on and approved in a first discussion on the 14th in Extraordinary Session 84, Legislature 18. Although the bill is still awaiting a second vote in the City Council, in 2022 began a strong campaign of visits, this time to all the councillors and secretaries of the municipal administration in search of support for the bill so that the mayor would approve it.

12.5.3 *Rios e Ruas Exhibition at Centro Cultural FIESP 2023/2024*

The exhibition MOSTRA RIOS E RUAS that opened the 12th of October 2023 at the FIESP Cultural Centre, at the headquarters of the Federation of Industries of São Paulo, was another example of Rios e Ruas efforts to improve São Paulo residents' low systemic perception of nature in the urban environment. The show was planned as a stimulus to engage the population in the sustainable development of the city by making the visitors understand the consequences on the rivers of how São Paulo was planned.

---

[22] Mapa. https://bit.ly/47gNDFG.

[23] The "Adopt One Spring" program covers the entire territory of the municipality of Sao Paulo and aims to promote community participation in the recovery of springs in degraded areas and preserving those that remain intact. Among the actions foreseen in the program are: physical delimitation and characterization of the source area, signalling of the area, recovery of degraded areas, prevention against erosions, periodic cleaning for the removal of solid waste and surveillance to prevent environmental degradation actions.

JUSTIFICATIVA—PL 0719/2021. https://www.radarmunicipal.com.br/proposicoes/projeto-de-lei-719-2021.

The FIESP building is located in the region of Avenida Paulista—São Paulo Financial Center—a nursery of dozens of rivers and the largest divisor of the two main watersheds in the city, the ones of the Tietê and Pinheiros rivers. This location was therefore an ideal site to attempt an expansion of the socionatural perception of the city.

In the free exhibition, a 120 m^2 walkable map of part of São Paulo—"*Cartografia dos Sentidos-São Paulo*" —surprised the visitors by revealing how the rivers that have been covered up by the city's growth, are still alive beneath so many buildings, vehicles, and human beings. Conceived by multimedia artist and visual anthropologist Maurício Panella, founder of Casadágua Instituto e Estúdio de Criação, in partnership with Instituto Harmonia Educação e Sustentabilidade and Rios e Ruas, the map invited the public to an analytical, affective, and sensory immersion into São Paulo cartography. Walking slowly through the city from above, following the blue lines revealing the course of rivers under some of the city's important avenues, aroused the desirable curiosity of São Paulo residents to know what São Paulo would be like if it had opted for another development model, a more sustainable one. Furthermore, the invitation to walk "on" the city's territory prompted a reflection on what São Paulo could be like if some of its water bodies were reintegrated into the urban landscape.

The feeling of being giants over a "small city", while instead Sao Paulo with its scale is always overwhelming the citizens, allows the visitors to reach the heights of birds and clouds, and to feel finally a possible agency on this megalopolis. In addition, this potentially joined with the discovery of the invisible water in Sao Paulo, re-signified the senses of belonging, care and responsibility of the citizens. In fact, for Rios e Ruas, knowing the presence of water could favour in the citizens a sense of caring and responsibility towards this water, and consequently for the city that contains it. For example, visitors in the exhibition had the opportunity to realise with just a few steps that the Ibirapuera lake is supplied by streams that originate close to the green line of the Metro, that the Allianz Parque—Palmeiras club's stadium— is located in the floodplains of the brook Córrego Agua Preta, and that the sources of the Pacaembú river are in the Consolação Cemetery region. Furthermore, they noticed the direct connection between the construction of public roads to transport automobile traffic in the city and the channelling of rivers and streams in the valley bottoms.

Despite the possible agency gained from the reversal of the observation point, Sao Paulo remains also a cause that required actions from us (Bucci, 2010). In Cartografia dos Sentidos-São Paulo, looking down from above, the living organism that is the city of São Paulo summons us to plant our feet on the ground and assume our roles as dreamers and builders of the future; a future that has long revealed the environmental and social challenges we must face. In order to make these challenges manifest clearer, Rios e Ruas invited cultural workers from two institutions linked to water and memory in São Paulo, Instituto Casa D'Água and Museu da Memória Afetiva, to present and discuss the map to the audience. In this same line, in order to further activate the exhibition, a fortnightly Cycle of Conversations about the City was held, inviting selected guests to share their emotional memories (Fig. 12.9).

Fig. 12.9 Rios e Ruas showcase, Centro Cultural da FIESP, October 2023 (pictures of Rios e Ruas)

Among the guests were João Prata, author of the book Rios e Ruas published in 2022; Mauro Calliari, business administrator from Fundação Getulio Vargas, and Ph.D. in urbanism from FAU-USP, whose research focused on mobility and walkability, columnist for the newspaper Folha de São Paulo, author of the book "Espaço Público e Urbanidade em São Paulo" and member of the Municipal Transport Council; Carla Ferro, columnist for the newspaper Folha de São Paulo, author of the book "Espaço Público e Urbanidade em São Paulo" and member of the Municipal Transport Council; Denise Baen, psychoanalyst dedicated to philosophy, literature and translation, a, Sociologist and Master in Education from USP, Education Manager for Sustainability and Citizenship at Sesc SP; Natália Garcia, creator of the New City Academy, with which she leads people, projects and organisations in the creation of a new culture of habits for a happier and more regenerative urban life and Hélio da Silva, a "Tree Planter", graduated in Business Administration until recently executive and entrepreneur, creator of the Tiquatira Linear Park in the East Zone, currently with 39. 800 trees of native Atlantic Forest species have been planted since 2003.

Finally, the conversation cycle that brought together visitors of the exhibition with activists and experts from different areas interested in the role of rivers in city life, can be seen as an effort to recover urban River Culture in São Paulo.

12.6 Final Considerations and Future Perspectives

The chapter presents some projects of the activist group "Rios e Ruas" (Rivers and Streets) and how they sensitise citizens to the presence and history of water in São Paulo, helping to enhance 'water sensitivity' in the local contexts, with a particular interest in the regeneration of buried rivers, the perception of the city as a collection of river basins, and the engagement of the population in sustainable development. Focusing in particular on a recent exhibition at the FIESP Cultural Centre and two examples in which the group's activities led to the co-production of legislative proposals, the chapter shows how Rios e Ruas practices innovation in the way of doing political activism through knowledge co-production and the citizen empowerment so that anyone can be an agent of transformation. Moreover, it shows how an activist initiative can dialogue with politicians to obtain some changes, moving from creating awareness to drafting laws.

In fact, Rios e Ruas, with their activities are not only influencing the perception of citizens, but also that of politicians who started to understand the importance of water in the city, both for the future of Sao Paulo and for their personal political ambitions. Consequently, the activist group has shown the potential to prompt the creation of laws favouring the sustainable development of Sao Paulo and the improving and possible recovery of its water infrastructure. In addition, their experiences of PL 858-2013 and PL Nascentes highlighted the power dynamics influencing and dominating the governance of water and urban development of Sao Paulo.

One of the most innovative aspects of Rios e Ruas is its discussion of the theme of urban rivers through a non-technical language and on-field experiences that go beyond the formality of technical and academic education. The transdisciplinary vision, which includes the socio-environmental, urban planning, legal, political, ancestral, poetic, and spiritual dimensions, has transformed the way São Paulo's citizens look at rivers and cities and has led them, regardless of age, gender, social class, or level of education, to understand and to feel part of their urbanities.

As well as informing, Rios e Ruas believes that there is innovation in emotion—and this emotion arises from the encounter of knowledge with a living, sensory and collective experience. For example, the experience at the Rios e Ruas exhibition, of walking through the urban fabric of the municipality and sensorially realising the presence of countless rivers covered by streets and avenues, opened up the possibility of reproducing the Cartography of the Senses for other territories and municipalities.

Finally, the work of Rios e Ruas showed that making citizens aware of the presence of water that previous urban plans had made invisible, has the potentiality to activate citizens' desire to contribute in the writing of water policies and in demanding a more sustainable development for the future of a city. In the case of Sao Paulo, the progressive maturity in the perception of its residents regarding the decisions of the public power, a maturity to which Rios e Ruas actions highly contributed, is recently producing a massive mobilisation and engagement in defence of the green areas along the Jurubatuba River and Pinheiros River that the forthcoming extension of the road Marginal Pinheiros will destroy. In conclusion, the city of São Paulo, where Rios e Ruas actions began, has been and continues to be a nerve centre for continuing to promote socio-environmental actions that can inspire innovative initiatives in other Brazilian cities as it happened for example in Belo Horizonte.

References

Benjamin, S., et al. (2014). The changing nature of river restoration. *Wires Water, 1*, 249–261.

Boelens, R., Hoogesteger, J., Swyngedouw, E., et al. (2016). Hydrosocial territories: A political ecology perspective. *Water International, 41*(1), 1–14.

Brown, R., et al. (2009). Urban water management in cities: Historical, current and future regimes. *Water Science and Technology, 59*, 5.

Bucci, A. (2010). *São Paulo, razões de arquitetura: Da dissolução dos edifícios e de como atravessar paredes*. Romano Guerra Editora.

Da Cunha, D. (2019). *The invention of rivers*. University of Pennsylvania Press.

Enqvist, J., et al. (2020). Informality and water justice: Community perspectives on water issues in Cape Town's low-income neighbourhoods. *International Journal of water resources development*.

Guimarães, L. F., et al. (2021). The challenges of urban river restoration and the proposition of a framework towards river restoration goals. *Journal of Cleaner Production, 316*.

Kumar, A., et al. (2023). Water Sensitive planning for cities in the global South. *Water, 15* (MDPI).

Linton, J. (2014). Modern water and its discontents: A history of hydrosocial renewal. *Wires Water, 1*, 111–120.

Linton, J., & Budds, J. (2014). The hydrosocial cycle: Defining and mobilizing a relational-dialectical approach to water. *Geoforum, 57*, 170–180.

Mathur, A., & Da Cunha, D. (2020). Wetness is everywhere. *Journal of Architectural Education, 74*(1), 139–140.

Mauroner, A., et al. (2021). *Planning water resilience from bottom-up to meet climate and development goals*. Unesco and AGWA.

Perkins, P. (2004). "Participation and watershed management: Experiences from Brazil". Roundtable "Ideology, Democracy and Governance for Sustainable Development". In *Conference of the International Society for Ecological Economics, Montreal, Canada July 10–14*.

Pinheiro Imbelloni, A., & Fernandes Fellipe, M. (2020). Ciclo Hidrossocial e o Reabastecimento Social das Águas: Uma experiência na Comunidade Quilombola da Tapera (RJ). *Geographia, 22*(48), 260–270.

Prata, J. (2022). *Rios e Ruas*. Prata Editorial.

Radcliffe, J. (2018). History of water sensitive urban design/low impact development adoption in Australia and internationally. In A. Sharma, T. Gardner, & D. Begbie (Eds.), *Approaches to water sensitive urban design*. Elsevier.

Romano, S. T. (2017). Transforming participation in water governance: The multisectoral alliances of rural water committees and NGOs in Nicaragua. *International Journal of Water Resources Development, 35*(3), 430–445. https://doi.org/10.1080/07900627.2017.1363722

Sharp, L. (2017). *Reconnecting people and water*. Routledge.

Swyngedouw, E. (2015). *Liquid power: Contested hydro-modernities in twentieth century Spain*. The MIT Press.

Tsouvalis, J., & Waterton, C. (2012). Building 'participation' upon critique: The Loweswater care project, Cumbria, UK. *Environmental Modelling and Software, 36*, 111–121.

Van Buuren, A., et al. (2019). Understanding emergent participation practices in water governance. *International Journal of Water Resources Development, 35*(3), 367–382.

Van den Brandeler, F. (2014). Decentralization, participation and deliberation in water governance: A case study of the implication for Guarulhos, Brazil. *Environmental and Urbanization, 26*, 489–504.

Wantzen, K. M., Tharme, R. E., & Pypaert, P. (2023). Life as a dance to the rhythm of the waters: An analysis of current trends and suggestions to restore river culture. In K. M. Wantzen (Ed.), *River culture—Life as a dance to the rhythm of the waters* (pp. 1–22). UNESCO Publishing. https://doi.org/10.54677/NIGG2511

Chapter 13
The City of Waters: Art, Perceptions, and Historical–Geographical Studies on the Hydrographic Network of Belo Horizonte (Brazil)

Alessandro Borsagli, Isabela Prado, and Massimiliano Granceri Bradaschia

Abstract Belo Horizonte is currently the sixth most populous city in Brazil, with approximately 2,315,560 inhabitants. Planned and built between 1894 and 1897 to be the capital of the State of Minas Gerais, occupying the territory of thirteen hydrographic sub-basins, all belonging to the sub-basin of the Arrudas stream, the city has been plagued by frequent floods since the 1920s. This was when the intervention process in the hydrographic network began, aiming to incorporate streams into the city's layout. Due to the significant environmental impact and the processes that led to regular overflow occurrences in recent decades, without the full knowledge of a portion of the population that still supports channelization as the sole solution to the urban water issue, it became necessary to encourage a paradigm shift on the subject. In this context, an artist, who has been developing numerous artistic proposals since 2006 that highlight the existence of the hydrographic network in the city of Belo Horizonte, and a geographer, who has been conducting geographical-historical studies on the urban evolution process of Belo Horizonte since 2010, seek through their projects to contribute to breaking the paradigm of "controlling and hiding waters" in the urban environment. Among the results obtained, the works carried out by "*Curral del Rey*" and the projects "*Sobre o Rio*" and "*Entre rios e ruas*" influenced the emergence of various actions related to the watercourses of the capital. They also led to the publication of numerous books, exhibitions, interviews, and scientific publications that significantly reached a portion of the population regarding the need to consider possible alternatives for the urban rivers of Belo Horizonte in the face of the current climate changes.

A. Borsagli (✉)
PPGG-TIE Pontifícia Universidade Católica de Minas Gerais, Belo Horizonte, Brazil
e-mail: borsagli@gmail.com

I. Prado
Universidade Federal de Minas Gerais, Belo Horizonte, Brazil

M. Granceri Bradaschia
IUAV University of Venice, Venice, Italy

M. Granceri Bradaschia et al. (eds.), *Climate Change Adaptation, Flood Risk, and Beyond*, Planning for Climate Proof Cities,
https://doi.org/10.1007/978-3-031-65463-3_13

Keywords Belo Horizonte · Urban rivers · Urban landscape · Science and art · Urban geography

13.1 Introduction

The city of Belo Horizonte was planned and built between 1894 and 1897 to serve as the capital of the state of Minas Gerais.[1] The choice of the location for the construction of the new capital had the hydrographic network as a determining factor, considering that the region encompasses the territory of thirteen hydrographic sub-basins, all belonging to the sub-basin of the Arrudas stream, a tributary of the Rio das Velhas.

Except for the Arrudas stream—the only watercourse integrated into the urban fabric of the city planned by the *Comissão Construtora da nova Capital* (CCNC) as a reference element in the landscape—the other watercourses that traverse the chosen site, all tributaries of the Arrudas, were disregarded.

The impacts on the hydrographic network resulting from the urban development process of Belo Horizonte have been felt since the early decades of the twentieth century when the straightening and channeling of watercourses within the planned zone began (FJP, 1997). This was due to the continuous growth of the urban fabric of the capital and the need for the construction of valley-bottomed avenues and widening of existing roads, causing environmental, social, and sanitary problems.

Channeling, seen as the sole solution for sanitary and aesthetic issues and flood control, has been systematically applied by the municipality for about a century. However, there is still not full awareness among a portion of the population regarding the impacts caused by the use of this technique and the impermeabilization of slopes, contributing to the increase in flow and surface runoff. Consequently, this leads to regular valley-bottom overflows in Belo Horizonte since the 1930s.

All these river-related issues, intrinsically linked to the processes resulting from rapid and intense urbanization and soil impermeabilization, where watercourses began to be seen as disease and unsanitary spreaders, led to a rupture between the population and natural elements. These elements should be controlled according to the interests and needs of society, with the starting point being the *Zona Urbana Planejada* of Belo Horizonte.

Several reasons led the authors to choose the hydrographic basin of the Arrudas stream as the starting point for the projects, including:

[1] Between 1720 and 1897, the city of Ouro Preto was the capital of Minas Gerais. In the last quarter of the nineteenth century, faced with the impossibility of modernization of the city, which was located on a steep slope, the government of the state of Minas Gerais chose five alternative locations with the aim of building a new city to house the state administration. After technical studies and political discussions, the village of Belo Horizonte was chosen to host the new capital, which was planned, financed and built by the state between 1894 and 1897, in accordance with positivist, hygienist and sanitary precepts. Belo Horizonte and its technical rationalism, geometry, hierarchy and well-defined spaces definitively broke with the colonial urban morphology of Brazilian cities.

Fig. 13.1 Planta Topográfica da Cidade de Minas, 1895. Detail of the hydrographic network that crosses the planned urban zone and the incompatibility between the watercourse paths and the geometry of the city. *Source* Modified from the Public Archive of the City of Belo Horizonte

- The planned urban zone of Belo Horizonte (Fig. 13.1) is inserted in the sub-basins of the Leitão, Acaba Mundo, and Serra streams and their tributaries, all part of the Arrudas stream basin. This area is still considered a priority by the government for the installation and maintenance of urban facilities necessary for the city's operation and public spaces.
- The interventions performed attest to how technical rationality was imposed on the site, where the city's orthogonal grid can be considered as a plan that did not harmonize with the topography of the region.
- The three watercourses are featured in all plans drawn up between 1895 and 1970, with strong memories, both positive and negative, still existing in relation to them.

In this context, the analysis of the processes that led to the channeling and covering of the streams has allowed us to understand not only the different contexts in which the works occurred but also the consequences of river interventions and the historical repetitions and standardizations that characterize interventions in the valley bottoms of Belo Horizonte. The projects presented in this article aim to bring to light the issue of channelized streams in Belo Horizonte, preserving their memory and inserting their presence into the public debate on land use in the city and the most appropriate policies for the management and preservation of watercourses.

The article is structured as follows: in the first part, we chose to provide a geohistorical contextualization, correlating the process of urban evolution in Belo Horizonte and the interventions carried out in its extensive hydrographic network (Fig. 13.2).

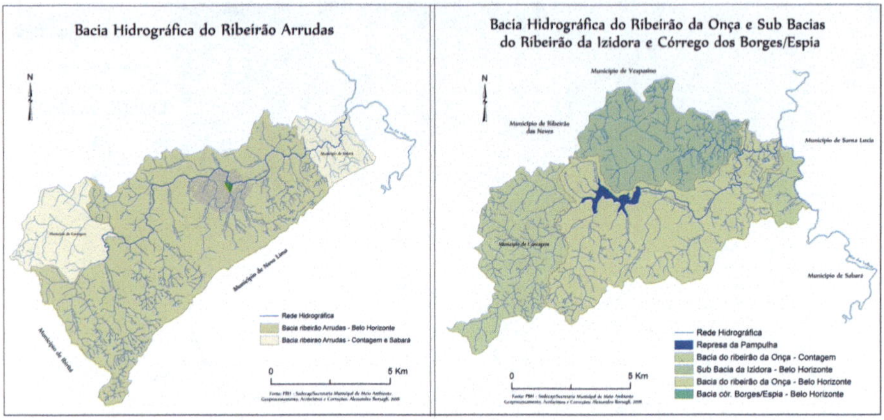

Fig. 13.2 Maps of the watersheds Arrudas e Onça. *Source* PBH/Alessandro Borsagli

This section also includes analyses and reflections on the presence of water in the urban environment today and the political and social nuances surrounding the river issue in the city. The second part discusses the experiences of the Curral del Rey, Over the River, and Between Rivers and Streets projects, bringing to the forefront the relationship between the city and the rivers and the means used to achieve the proposed objectives. The third part addresses the results and conclusions achieved by the projects.

13.2 Research Context

13.2.1 The Pathways to Understanding the Overflows in Belo Horizonte

The relationship between the city and the river is a theme that has been addressed by various researchers in recent years across different fields of knowledge. In this context, many works focus on the rehabilitation of watercourses in cities or the urban-landscape integration of watercourses that flow through urban areas. These works highlight the importance of the hydrographic network for the city. Many of these studies approach the topic from geographical, historical, environmental, cultural, social, and economic perspectives, considering the hydrographic network as a significant reference element in the landscape (Hoyuela Jayo, 2013).

However, little is said about the aesthetic, artistic, and geographical-historical relationship between urban growth and interventions carried out in the hydrographic network. These interventions aim to adapt it according to the interests and needs of urban societies. The reasons and justifications for the straightening and channeling of watercourses are often not fully explored. Ancient Western urban centers often have a notable connection with rivers, with rivers being responsible for shaping the city and determining the territory's occupation (Castonguay e Evenden, 2010).

In this sense, it is important to note that in the case of Belo Horizonte, a city conceptualized, planned, and built on headwaters and valleys, the relationship is not direct. This might be one of the factors that contributed to the consolidation of negative values and perceptions about the presence of watercourses in the city.

To understand why valley-bottom overflows are frequent, even in areas not crossed by watercourses, a quick journey into the past of the capital is necessary. Knowledge and understanding of the past enable an understanding of the phenomena occurring in the present, allowing for the establishment of more coherent plans with urban reality.

In this regard, to understand not only the observations mentioned earlier but also the entire problem and the reasons behind the current river chaos in Belo Horizonte, Borsagli (2016, 2018, 2019, 2020) divided the studies into four historical periods:

1. **First period: 1734–1893**—This period represents the first representations of the hydrographic network of the Curral del Rey region.
2. **Second period: 1894–1920**—This period includes the detailed cartographic representation of the hydrographic network of the new capital site, the first specific interventions, and news about watercourses in Belo Horizonte.
3. **Third period: 1924–1962**—This period saw major rectification and channeling works on part of the watercourses of the Arrudas stream basin, leading to an increase in valley-bottom overflows.
4. **Fourth period: 1963 to the present**—This period witnessed fluvial interventions extended to the entire municipality of Belo Horizonte (basins of the Arrudas and Onça streams), exacerbating valley-bottom overflows.

The first overflows were recorded after the capital's inauguration, including a notable overflow in January 1915 in the Arrudas stream, the mouth of the Leitão stream, and the Serra stream. These caused material damage and disruptions in areas of the city closest to watercourses, such as the lower center and the portion of the Lagoinha neighborhood located in the vast floodplain triangle formed by the Arrudas stream and the mouths of the Pastinho and Lagoinha streams, now occupied by a road complex.

After the completion of the initial interventions in the hydrographic network crossing the planned urban zone of Belo Horizonte in 1929/1930 (Fig. 13.3), the incidence of valley-bottom overflows increased over the years. The interventions, coupled with urban fabric growth and slope impermeabilization through road openings and neighborhood expansion, contributed to altering the hydrological regime of the micro-basins within the Arrudas stream basin and the stream itself.

In other words, the technique of adjusting the course of water according to the geometric layout or the planned neighborhoods of the basins adjacent to the planned urban zone of Belo Horizonte has proven inadequate for addressing a phenomenon over which there is no control, especially considering the hydrological changes that canalizations and urbanization of floodplains bring in urban environments.

Fig. 13.3 Channelling of river Arrudas within the Parque Municipal de Belo Horizonte, 1928. *Source* Museu Histórico Abílio Barreto

In the 1950s, overflows of the Arrudas became more frequent, and from the year 1962, amid the process of metropolitanization of Belo Horizonte, they started to occur almost annually, and in some years, more than once, usually in the same locations where the hydrographic network was canalized, with the exception of the eastern region of the capital. In this region, there was the discharge of the hydraulic whirlwind drained by artificial channels (Fig. 13.4), and the Arrudas, suffocated by sewage directly or indirectly discharged into its waters, remained in its natural bed, flowing squeezed amid the pronounced informal occupation of its floodplains.

The urban pressure exerted on the hydrographic network was one of the main reasons for its denial in the urban environment. At the same time, after the interventions and the disappearance of the watercourse from the urban landscape or its isolation, there was (or is) a considerable gain in land that becomes highly valuable in the market.

The urban infrastructure works carried out between the 1960s and 1970s were also responsible for the marginalization of water in urban areas, as watercourses were canalized and covered to widen public roads (Mesquita, 2013). In the case of Belo Horizonte, these projects were financed with foreign capital, as during that period, the automotive industries were establishing themselves in the country, and major Brazilian urban centers lacked the capacity to accommodate intense motor vehicle traffic (Borsagli, 2016). In other words, rivers were forced to yield their territory to

Fig. 13.4 Floods of creek Leitão in 1966 in the Rua Tupis. *Source* Acervo Arquiteto João Lucas Ribeiro Borges

automobiles, which became the protagonists of urban policies in the capital of Minas Gerais.

While some interventions had local effects in certain sections, the rains of January 1983 showed the city and its administrators that it is not possible to control the uncontrollable—a phenomenon that must be understood and respected, never disregarded. This had already been noted by Saturnino de Brito (1944), who concluded, among other things, that the problem of floods is established by the whims of human activity. The impossibility of solving the problem must be accepted to avoid unnecessary work, the waste of substantial amounts of money, and disappointments (Fig. 13.5).

In this context, it is important to emphasize that the presence of water in urban areas, for over a century, has not been fully debated in political circles. Instead, it has been used as a political tool through fluvial interventions. In the early years of the twentieth century, the discussion was limited to the municipal technical body, which repeatedly observed the problem of water pollution and the lack of interception and treatment of effluents discharged directly into watercourses. Between the 1920s and 1990s, discussions were centered on valley-bottom interventions and the covering of watercourses for road widening—measures widely used for propaganda by municipal administrations and the legislature, responsible for proposing and approving structural works, constructing a negative perception of the presence of river waters in urban areas without addressing the treatment of polluted waters.

Two actions were crucial in the quest for the rehabilitation of Belo Horizonte's hydrographic network: the Environmental Sanitation Program of the Arrudas and

Fig. 13.5 Perspective of the covering works of the Acaba Mundo stream in the 1960s: **a** view of the upstream channel on Professor Morais Street, 1963; **b** widening and deepening works of the bed, October 1963; **c** the covered canal at the intersection with Afonso Pena Avenue, 1965; **d** night view of Professor Morais Street after the completion of the works, 1965. *Source* Arquivo Público da Cidade de Belo Horizonte

Onça Streams Basins (PROSAM), and the Environmental Recovery Program of Belo Horizonte (DRENURBS). The latter aimed to propose measures for the recovery and preservation of watercourses in their natural beds, implemented in five micro-basins of the municipality. Simultaneously, various sectors of society and the academic community began to question the interventionist techniques applied for a century in the valley bottoms of the capital of Minas Gerais, which will be discussed further.

13.2.2 Water Sensitive Approach in Brazil

In Brazil, historically, the presence of water courses in urban areas tends to be denied or ignored. Since the colonial period, urban areas have not been connected to water courses, seen as an element responsible for receiving and conveying sewage and waste. At the same time, many cities that emerged during the colonial period (1500–1822) do not have an integration with the layout of rivers, generally located on the margins of the urban fabric.

One of the first structural interventions on a watercourse in Brazil dates back to 1849, carried out on the Tamanduateí River, a tributary of the Tietê River in the city of São Paulo, with the purpose of eradicating floods in the basin (Botelho, 2011). In this context, similar works of straightening and canalization of watercourses were carried out in Rio de Janeiro in the second half of the nineteenth century, not only for urban drainage improvement but also for sanitary, aesthetic, and road purposes (ibidem). The Mangue Canal stands out in this regard, which can be considered a prelude to the profound urban reforms of the following century.

In the twentieth century, fluvial interventions spread to the main Brazilian capitals, causing a transformation in the city/river relationship. These interventions led to changes related to sediment deposition, alterations in the hydrological regime and fluvial ecology, pollution, canal sealing, reduction of permeable areas, changes in urban climate, and vegetation suppression, as noted by Tucci (1997). According to Tucci, the disregard for environmental issues in urban planning caused significant harm to society, and the hydrographic network was marginalized by urban planning dominated by railway and especially road axes, as observed by Gorski (2010).

In the 1960s, discussions and projects proposing the recovery of watercourses emerged in various European countries, with the recovery of the Thames River initiated during this period being one of the pioneers. This exposed the need for city replanning, recognizing water as a reference element in the urban landscape and of paramount importance to society. The work of the German architect Herbert Dreiseitl, responsible for numerous projects related to urban waters since the early 1980s, is noteworthy. According to the architect, water, despite always being present in the urban fabric, is increasingly invisible to the population (Borsagli, 2017).

In the United States, changes in relation to water also began in the 1960s, with the rehabilitation of the old trade centers in the cities of San Francisco, San Antonio, and New Bedford, abandoned after the consolidation of railway and road modes. The revitalizations in the cities of Baltimore and Chicago, the latter related to the recovery and rehabilitation of the Chicago River, are highlighted by the authors Otto et al. (2004).

Gorski (2010) observes the search for alternative solutions to watercourse canalization, noting that a new attitude is emerging in academic and societal circles due to the exhaustion of previously adopted measures. Costa (2006) also points out the circulation of models and ideas regarding the conflict between fluvial processes and urbanization processes, leading to the disappearance of numerous watercourses from the urban landscape.

Analyzing trends and challenges related to the rehabilitation of urban rivers, Baptista and Cardoso (2013) highlight the complexity of fluvial reintegration into the urban landscape. Urban, social, and environmental demands force the public authorities and society to adopt a new stance on the issue.

In a time of questioning and searching for alternatives that surpass structural models, the city of Belo Horizonte created the Environmental Recovery and Sanitation Program for Valley Bottoms and Natural-Bed Streams of Belo Horizonte (DRENURBS) in 2004. This program sought to establish a new relationship between the city and the watercourses that still flow in their natural beds.

According to Medeiros (2009), the project, in its general conception, aimed not only to understand the factors that led to the crisis in Belo Horizonte's urban drainage system but also to reintegrate watercourses into the urban landscape. This involved adopting a new perspective regarding the integration of natural elements into the urban environment through interventions in intra-urban basins.

Among other objectives, DRENURBS aimed to reduce the risks of urban flooding, enable the recovery of watercourse quality, and ensure the sustainability of urban interventions with the consolidation of a drainage and urban environmental management system.

It should be noted that the urban water requalification policy was partially influenced by changes in attitudes toward urban rivers around the world. In the same year that DRENURBS was established, the American Planning Association published a document titled "Ecological Riverfront Design: Restoring Rivers, Connecting Communities," which recognizes the value of watercourses for urban communities. It emphasizes the need to design future actions based on ecological principles.

Bontempo et al. (2012) point out the technical predominance when considering the river space in Belo Horizonte's urban planning, addressing contradictions between DRENURBS and the covering of the Arrudas stream, initiated in 2005 through a partnership between municipal and state administrations. This divergence prevents its reintroduction into the urban landscape of the capital of Minas Gerais.

In the same context, Baptista and Cardoso (2013) highlight a paradox between the program and the continuous use of structural models in urban rivers in Belo Horizonte. The Arrudas stream covering is presented as an antithesis to the maintenance of rivers in their natural state. The authors note that, conceptually opposed, these measures are contemporary.

Therefore, Belo Horizonte is embroiled in disputes between the reintroduction of rivers into the urban landscape and the continuous adoption of structural models for the straightening and canalization of urban rivers. This is a recurring issue in Brazilian urban centers (Almeida, 2010; Cardoso, 2017; Monteiro Junior, 2011). It is essential to understand that the problems resulting from fluvial interventions persist due to human insistence on inhabiting areas that should not be settled and altered, such as floodplains.

In this sense, it is important to understand that a sensitive approach to water is of great importance for Brazilian cities, where planning and understanding the dynamics that characterize the formation and transformation of urban space is the first way to seek alternatives to achieve satisfactory water management in urban areas based on rehabilitation, urban-environmental integration, quality of life in cities and sustainability.

13.2.3 Public Participation and Civic Engagement in Urban Policies Related to Watercourses

The urban rivers of Belo Horizonte were part of society's daily life in the first decades of the city's existence, using their waters for human and animal consumption, work, and even for leisure. In other words, the waters and their floodplains were territories of shared sociability and reference for the riverside populations.

The techniques imposed on watercourses and rapid environmental degradation led not only to their invisibility in the city but also to an inevitable rupture due to the high degree of pollution and the problems resulting from valley bottom overflow that has been occurring for a century. In this sense, if we consider that a considerable portion of the water network has been invisible for two generations, it is natural that the population's relationship with its rivers is very small today, often only remembered in rainy periods when the waters reclaim their territory usurped by the city.

However, it is important to note that in the last three decades, the academic community and a portion of the population have come to understand that a river is not an obstacle to mobility or the city, and there has been a gradual shift away from the paradigm of canalization and the way natural elements are treated in the urban environment. In Minas Gerais, from the promulgation of the Citizen Federal Constitution in 1988, the decentralization of water governance has become notably evident with the creation of Basin Committees (CBH), constituting a "Water Parliament," an important space where representatives of the community in a watershed discuss and deliberate on water resource management, sharing management responsibilities with the government (Brasil, 2023).

In this context, civil society plays a crucial role in decision-making regarding water basin management, not only having decision-making power but also a fundamental role in formulating policies that aim for better watershed water management, seeking to ensure multiple water uses in a rational and sustainable manner.

Other initiatives, such as neighborhood associations or "collectives," arising from dissatisfaction with government policies or in areas where the government does not directly act or has no interest, are also of great importance. An example is the association of Vila Acaba Mundo in Belo Horizonte, a population enclave between two affluent neighborhoods that has an intrinsic relationship with the stream that gives the village its name.

Through the mobilization of riverside residents that took place throughout the 2010s, the residents not only managed to reverse the stream's canalization process, which was in the project, but also the removal of domestic sewage discharge directly into the stream, integrated into the village and an important part of the population's daily life.

The village and its relationship with the waters are examples of resistance to the imposition of techniques that contributed to the distancing of the relationship between society and the river, a relationship that Lara (2016) describes in a unique way when analyzing human relations with river waters in the urban environment:

> Our watercourses are polluted and need to be restored. The way this can happen is not complex either. We already have the technology. We already have the methodologies for creating sewage treatment plants. The obstacles preventing necessary works like these are purely economic and political. Therefore, the issue takes on its sociopolitical dimension and ceases to be environmental. The environmental aspect is already resolved. What remains unresolved are human relations. These conflicting relationships reproduce our predatory and illogical civilizational model.

The city of Belo Horizonte is on the path towards a shared and effective management of urban waters between the government and civil society. In the context of bottom-up initiatives, programs like DRENURBS, the Drainage Master Plan, and the CBHs (Basin Committees) of municipal microbasins are examples of how the integration and transfer of knowledge, expertise, and experiences can contribute to a new perspective on urban rivers and their relationship with the city, seeking solutions beyond canalization techniques.

However, it is important to emphasize that, as noted by Lara (2016) and Borsagli (2017), human relations, political and personal interests, along with the lack of full integration across knowledge areas, still hinder the establishment of a complete and genuine discussion on the rehabilitation of water in urban areas.

13.3 Experiences

13.3.1 Curral del Rey and the Water and River Memories in Belo Horizonte

Amidst this intricate and meandering plot, discussions and reflections on the process of covering the watercourses in the capital have become more constant in recent years within the academic community. Simultaneously, there has been an increasing awareness among technicians and the population of Belo Horizonte, prompted by the tragedies caused by floods, leading to an understanding of the need to reconsider urban occupation. The current urban occupation conceived for the capital has not been successful in terms of urban drainage and the layout of watercourses, indicating the exhaustion of the development model adopted by Brazilian urban centers post-1930.

Due to the century-long river interventions, a considerable part of the city was unaware of the extensive hydrographic network beneath its feet. This led to the creation of the Curral del Rey project[2] in 2010, aiming to promote not only the

[2] The toponym "Curral del Rey," of Galician-Portuguese origin, was the name of the village demolished at the end of the nineteenth century for the construction of the capital of Minas Gerais.

recovery of Belo Horizonte's urban memory but also to disseminate research initiated in 2008. The project addresses the presence of natural elements in the urban environment, including the issue of tree cover, considered a collective heritage, and the mountains that compose the urban landscape of the capital, highlighting the Serra do Curral del Rey, an important mountain hosting the watercourses discussed in this chapter, and part of a mountain range threatened by iron ore mining.

The Curral del Rey project, through a simple and direct language, innovatively addresses the geographical and historical processes characterizing the rapid urban evolution of Belo Horizonte. It seeks to fill historical gaps forgotten and even ignored by traditional historiography in the capital, proposing a broader view within the landscape studies. The project aims to awaken a commitment to social, cultural, scientific, and environmental aspects within the population.

The methodology adopted in the Curral del Rey project is based on geohistorical contextualization. In the first stage, an understanding of the social, cultural, political, and urban conditions influencing the process of Belo Horizonte's urban evolution and all landscape interventions was sought, establishing a direct relationship between the subject and the object.

The second stage involves the analysis of documents held by various institutions (archives, museums), including photographs, official and unofficial documents, aerial images, maps, and periodicals. These provide insights into how the government and the city conceived and acted during different periods of urban development in Belo Horizonte, helping to understand how the hydrographic network was made invisible according to the needs/interests of the analyzed periods.

The third stage encompasses the culmination of the project, presenting the results obtained in the previous stages. Since 2010 and 2011, numerous works have been published on Belo Horizonte's watercourses from a geographical-historical perspective. This previously unexplored methodology not only attracted attention from the academic community but also from the population of the capital. A part of the population was unaware of the existence of watercourses in the more densely populated regions, or had only heard about rivers that officially no longer existed. It emphasized the need for a swift recovery of water memory, given that the generation that had lived with watercourses in their natural state was disappearing.

It was as if a switch had been triggered in the city's collective memory, leading not only to a search for more information about Belo Horizonte's rivers but also to an understanding of the processes and results of adopting techniques and policies that contributed to the worsening of floods. This necessary and indispensable recognition is essential to assimilate the phenomena caused by human activities on the planet and the consequences of climate change directly affecting the hydrological regime in Belo Horizonte. Heavy rains and extreme heat are already concrete facts in the daily life of the capital.

In the meantime, it is crucial to emphasize the need for interdisciplinary approaches to discuss urban water issues, encompassing not only Geography, History, Urbanism, Biology, or Engineering but all areas of knowledge without exception. This includes health, arts, and other disciplines, where artistic-fluvial works are being developed, crucial for understanding and appreciating Belo Horizonte's hydrographic

Fig. 13.6 Covering works on the river Arrudas in the Avenida do Contorno, 2010. *Source* photo by Alessandro Borsagli

network, profoundly degraded and marginalized but still beautiful and essential for future generations (Figs. 13.6, 13.7, and 13.8).

13.3.2 Art as a Game Changer in the Cities: The Project "Entre Rios e Ruas" and the Urban Intervention "Sobre o Rio"

Creeks are entities of nature present in urban environments. However, they are often erased, marginalized, and stifled by urbanization processes that disregard the specificities of the territory, particularly the pre-existing hydrographic network. They thus become perceived as foreign elements within the city itself.

In the case of Belo Horizonte's hydrographic network, watercourses like the Serra, Acaba Mundo, and Leitão creeks have been channelized beneath urban roads for decades. Over time, an increasing portion of the population has not experienced the uncovered creeks and, therefore, does not recognize their presence in the landscape. Urban-river memories have dissipated to the point where a considerable part of the population is unaware of the existence of creeks in the city, ancient urban landmarks erased from the cityscape and remembered only during rainy periods.

Fig. 13.7 Upstream perspectives from Rua Marília de Dirceu on 2nd January 2020, after the flooding of river Leitão. *Source* Felipe de Ávila Chaves Borges

Fig. 13.8 Alessandro Borsagli speaking about urban river geographies in Belo Horizonte at III Encontro Internacional de Revitalização de Rios (2017). *Source* Wesley Ogando Viana

In this sense, urban interventions are crucial for rescuing memory and sparking reflection on urban space, including the contradictions, conflicts, and problems that constitute the city's landscape. Art has the power to provide the population with a reflection on societal issues and all the necessary public demands for problem-solving.

More broadly, art can fundamentally contribute to addressing issues of the Anthropocene in a sensitive manner. It not only documents and criticizes human impacts on the planet but also promotes awareness, critical reflection, and action. It challenges norms, questions established values, and inspires changes in relationships with others, the environment, and nature.

The discussed approach aligns with contemporary art production operating in the field of subjectivity, emphasizing social interactions and public participation as central components of the artwork. In this sense, a dialogue with the notion of relational aesthetics (Borriaud, 2002) can be pointed out, blurring the boundaries between everyday life and art, creating a space where meaning emerges from shared experiences and social interactions.

This aspect can be better qualified by considering the importance of the concepts of space and territory for this research. Deutsche (1998) argues that space is not a given, pre-established entity but is discursively constructed, based on definitions of its borders—definitions that result from political, social, and cultural relations: "Space is not an entity but a relationship" (Deutsche, 1998, p. 6).

In this regard, the perspective of Milton Santos is important and necessary. For him, space is not just a physical entity but a complex social construction shaped by human interactions and social, economic, and cultural relations (Santos, 2008). As a stage for political struggle, space is not neutral, being an essential element in forming individual and collective identity.

Thus, the concept of territory is defined, in its most symbolic dimension, based on aspects of its history and culture, associated with the construction of identities, elements of common identification, and a collective memory. In this sense, the continuous creation and destruction of territories can be conceived, from the moment their borders are subjectively (re)defined.

It is also worth mentioning the concept of landscape as an element associated with the discussion of space and territory as subjective constructions. Cauquelin (2007) explores how the concept of landscape has evolved throughout history, transitioning from a primarily pictorial term to a broader cultural and perceptual category. Thus, it examines how the notion of landscape has transformed as society and the relationship with nature have changed, influencing how we perceive and represent it. This reinforces the idea that landscape is not just a visual phenomenon but also involves philosophical, social, and political aspects. Once again, Milton Santos can be referred to, defining landscape as the set of forms that, at a given moment, reveal the legacies representing the successive relationships between humans and nature (Santos, 2008).

The intersection between urban art and science gives rise to understandings, (re)recognitions, and analyses, from an incorporation between artistic processes and objective communication, an interaction essential for knowledge and for social, urban, and cultural transformations. In this context, the project "Entre Rios e Ruas," a collection of original and pioneering works that make up Isabela Prado's trajectory, stands out. She is an artist, teacher, and researcher who, since 2006, has developed numerous artistic proposals that highlight the existence of the hydrographic network in the city of Belo Horizonte (Prado, 2013).

Comprising drawings, photographs, objects, videos, installations, and performances, "*Entre Rios e Ruas*" reflects on the relationships between the city, the environment, and the individual, starting from the relationship that Belo Horizonte established since its foundation and still establishes today with the rivers and creeks present in its territory. Through intuition, poetry, and imagination, this project seeks to suggest other possibilities of conceiving, perceiving, and experiencing the world.

With a keen eye on space, "*Entre Rios e Ruas*" brings neglected aspects of life in the city to light, approaching them sensibly, politically, and critically. The project believes in the possibility, through art, of pointing out and bringing to the surface issues about our relationships with the surroundings, fundamentally altering our perception of reality.

"*Entre Rios e Ruas*" had its first public presentation at the seminar "*Diálogos: estratégias do fracasso*" at the Museu de Arte da Pampulha (Belo Horizonte) in 2009. On that occasion, Isabela Prado performed "*Ou Vir o Rio*", in which she distributed water cups to the audience with information printed on the extension of channelized and natural bed creeks in Belo Horizonte, accompanied by the audio of some submerged creeks in the city. The proposal for the audience to "drink" this information reflects the spirit of the project, which often starts from objective, technical information on the topic and proposes a poetic approach, often relying on the viewer's participation.

"*Joia*" is another of the works executed within the scope of the "*Entre Rios e Ruas*" project. It is a gold brooch, 9.5 cm long, replicating on a 1:10,000 scale the course of the last 950 m of the natural bed of the Ribeirão Arrudas within the territorial limits of Belo Horizonte (Fig. 13.9). It explores the relationship between body, space, and scale. The choice of material reaffirms the preciousness of the river, and the use of the "Joia" represents a flag in favor of reflecting on the preservation of rivers, creeks, and springs.

The installation "*Repaisagem*", in turn, proposes the viewer's participation in the construction of new landscapes by manipulating blue magnetic sheets, corresponding to all sections of streams in their natural beds in the municipality of Belo Horizonte (Fig. 13.10). The installation also includes an audio element that corresponds to the sound of these same streams but in canalized sections running beneath the streets of the city center. The interactive nature of the work reinforces the power of public participation as an element of transformation, reconstruction, of reality. The drawing resulting from this interaction is always in formation and reaffirms the idea of impermanence and renewal, so fundamental to watercourses.

Entre Rios e Ruas was awarded the Funarte Contemporary Art Prize in 2011. Among the achievements of this research, the permanent urban intervention "*Sobre o rio*" (Over the River) stands out, carried out by the artist in the watersheds of the Serra, Acaba Mundo, and Leitão streams within the perimeter of Avenida do Contorno, a

Fig. 13.9 *Joia*. Gold brooch, 18k, 9.5 cm, 2010

Fig. 13.10 *Repaisagem.* Installation (wooden boxes, speakers, zinc sheets, magnetic sheets), 2.00 × 6.00 × 3.00 m, 2010

planned urban area in Belo Horizonte. This work consisted of the permanent installation of approximately 230 plaques (Figs. 13.11, 13.12, and 13.13) indicating the existence of the watercourses that cross the central region of the city.

"*Sobre o rio*" arises from a perception, which became increasingly clear with the development of the "*Entre Rios e Ruas*" research, that the work of rescuing memory and raising awareness about the importance of water in the city should occupy the streets permanently. Not only should it move out of the environment of museums and art galleries to reach a broader audience, but it should also persistently fulfill the role of bringing visibility to the canalized streams and promoting greater awareness among citizens on the issue. The underlying understanding was that a cultural change is necessary, and that a fruitful starting point would be the public and collective recognition of the existence of these invisible watercourses in the city.

"*Sobre o rio*" was initially submitted to the Belo Horizonte Cultural Secretariat as a project in the visual arts area. Later, as it was a permanent intervention in the city, it obtained approval from the Subsecretariat of Urban Planning for its execution. Isabela Prado's choice was that the plaques with the names of the streams should not be signed or contain elements that immediately identified them as a work of art. On the contrary, the idea was for the plaques to have characteristics similar to existing street signs, so they would integrate more naturally into the city's landscape.

Thus, the plaques signaling the streams followed the same technical specifications as the street signs in terms of dimensions, material, colors, fonts, and finish. Only a

Fig. 13.11 *Sobre o Rio*. Urban intervention of 230 steel signposts (2018–2020). *Source* Photos by Isabela Prado

color inversion was made on the plaque—white background and blue text—to create differentiation and contrast with the existing signs (Fig. 13.14).

"*Sobre o rio*" blends with an action of the municipal government, especially due to the permanence of the gesture. Like a monument, dispersed and integrated into the city's landscape, it pays homage and rescues the memory of the streams that were rendered invisible during the process of urbanization. This work recreates the urban landscape by revealing the water body that exists beneath the city. Landscape is understood here not as an objective reality or nature but as an entity constructed from social, political, cultural, and aesthetic relations.

The goal of "*Sobre o rio*", as well as the other works that preceded it within the Entre Rios e Ruas research, was to poetically bring to light the presence of the invisible streams in Belo Horizonte. The aim is to motivate a cultural change in the city's relationship with its watercourses by reintegrating the streams into the urban landscape.

Regarding the work, which is notably avant-garde, Isabela Prado notes:

> "*Sobre o rio*" has a public character as it takes place in the streets. It brings visibility to something that is invisible and reintegrates the streams as an element of the city's landscape. Once the presence of the streams is identified and remembered through the signage, their existence becomes something natural, recognized by the citizens moving through Belo Horizonte. The work promotes a new relationship between the city and its inhabitants, engaging society in the (re)discovery of this water body that inhabits the city. (Equipe ArchDaily Brasil, 2021)

Fig. 13.12 *Sobre o Rio*. Urban intervention of 230 steel signposts (2018–2020). *Source* Photos by Isabela Prado

Regardless of the results already observed, "*Sobre o rio*" is a work in progress. The next goal is its transformation into a *patrimonio* (a gift inherited) of the city of Belo Horizonte, so that the public authorities take responsibility for the maintenance and replacement of the signs, just as they do for street signs. In addition, there is an aim to expand it to other regions of Belo Horizonte where the streams are not yet signaled, as well as to other cities in Brazil, such as São Paulo, Rio de Janeiro, Recife, where the presence of invisible streams is also a characteristic of the urban environment.

The process of urbanization in Belo Horizonte led to the erasure of streams from the urban landscape. Signaling the watercourses, now unknown to the majority of the population, is of great importance so that the inhabitants of the Minas Gerais metropolis understand what exists in the city and beneath the city. Promoting the concrete recognition of the existence of these streams to municipal administration and the common citizen represents the first step in ensuring the preservation of these streams and incorporating their presence into urban policies.

Fig. 13.13 Watersheds signposts of *Sobre o rio*. *Source* Made by authors based on Google Maps

Fig. 13.14 Signpost models of *Sobre o Rio* (2018–2020). *Fonte* Images by Isabela Prado

13.4 Final Considerations

The research conducted over the years has gained consistency and resulted in the publication of three books on the urban waters of Belo Horizonte (Rios invisíveis da metrópole mineira, Horizontes Fluviais and Rios urbanos de Belo Horizonte), along with various articles, the impacts of public expressions on the site (www.curraldelrey.com) and Instagram profile (@curraldelrey), public exhibition of the research at CREA (Conselho Regional de Engenharia e Agronomia) and material in major Brazilian media. The growing awareness of the city's waterways in Belo Horizonte led to field practices such as "*Cicloexpedições*", cycling events that traversed the main hydrographic basins of the municipality. Additionally, social media played a crucial role in reaching a considerable number of readers and citizens interested in the subject.

The Cicloexpedições were initially conceived by Curral del Rey and various collectives in the capital. Their purpose was to explore the watercourses of the basin firsthand, understanding their natural aspects and the impacts of intense urban land use that altered the urban fluvial dynamics. This aimed to recognize the territory of waters and the presence of nature in the city, often unnoticed by the population in their daily lives. The goal was to learn how to see, coexist with, and acknowledge urban rivers instead of denying their presence or even their existence in the city.

On the technical side, numerous lectures were given to the technical staff of the municipality (SMMA, SMPU, CMMCE, GEIAT). These aimed to discuss the relationship between urban rivers and the geographical expansion of Belo Horizonte from its inauguration to the present day. The reflection focused on how adopting a single technique for water issues in urban areas led to the current situation, where regular flooding and human and material losses are part of riverside and non-riverside areas.

In the context of the urban-fluvial issue in Belo Horizonte, it's essential to highlight that many toponyms of watercourses (such as the Mendonça stream, the Zoo stream) and others forgotten during the urban evolution of Belo Horizonte "emerged" during the research over the last decade and a half. One example is the *Ribeirão da Izidora*, a watercourse named after an old landowner in the Onça region (Izidora da Costa), which, starting in 1936, became known as "*Ribeirão do Izidoro*". Forgotten amid fluvial invisibility and constant updates of municipal maps, the rescue conducted by Curral del Rey and extensive media coverage in the following years allowed society to understand that under their homes, streets, or neighborhoods, there is a watercourse waiting to be recognized and restored.

Regarding the projects "*Entre Rios e Ruas*" and "*Sobre o rio*", it can be said that their most fundamental and immediate impact lies in changing individuals' perception of their surroundings and the urban spaces they frequent daily. Urban streams, once brought to light by the identification and signaling process promoted by these projects, become geographic references in the city and influence our spatial perception, location strategies, and overall environmental awareness.

A recent episode illustrating this change was a news report in a widely circulated local newspaper about heavy rains causing flooding in some city regions. Unlike previous reports, this one included not only the names of streets or avenues but also the names of the streams passing through those areas.

One characteristic of "*Sobre o rio*" is its ability to foster dialogue across different fields of knowledge concerning the relationship between the city and the environment, with an emphasis on urban streams. This urban intervention gave rise to the eponymous book, bringing together texts from researchers in various areas such as arts, urbanism, geography, and history (Diniz et al., 2022). The book was published in Portuguese and English to broaden its reach, given the uniqueness of the work and the potential interest it may generate for implementation in other cities and countries.

Another indication of the impacts of "*Entre rios e ruas*" and "*Sobre o rio*" is the spontaneous public expressions on the project's Instagram profile (@entrerioseruas). People commented on the presence of canalized streams in their neighborhoods, demanding that they be signaled. Numerous interview requests from various media outlets, including radio, TV, portals, newspapers, podcasts, etc., also highlight the ability of the work to "burst the bubble" and reach a broader audience. Interestingly, most of these outlets are not specialized in art but rather in general city issues.

Finally, "*Entre Rios e Ruas*" and "*Sobre o rio*" have managed to reach the knowledge of the municipal government. Specifically, the projects have the support of the Municipal Secretary of Culture of Belo Horizonte, which has expressed interest in collaborating on the maintenance and expansion of signaling, both in Belo Horizonte and other cities.

Similar to other major Brazilian urban centers, the relationship between the city of Belo Horizonte and its hydrographic network involves coexistence and rupture. The reappropriation of fluvial spaces occupied by the city is necessary and unquestionable, as interventions over a century have not yielded the expected results, particularly concerning bottom-of-valley floods. Societal awareness of the need for coexistence with the environment will lead to changes by the government, which continues to insist on techniques that should have been overcome.

Furthermore, societal awareness of the need to reestablish harmony with the environment is crucial for facing the significant climate changes occurring today. The city must adapt quickly to rising temperatures in urban areas, while the ecosystemic vulnerability of the municipality's two main watersheds approaches a nearly irreversible state due to increasingly frequent extreme hydrological events.

It's also crucial to note that the projects have brought attention to the presence of rivers in the urban environment, evident in the increase in journalistic articles, films (including the short film "*Olhos de Inaiá*"), documentaries, and academic works. Urban rivers, even if covered and hidden beneath the city, have gained centrality in recent years, contributing to the city and society's closer connection with the liquid element.

At its core, the city is a living, multifunctional organism with multiple landscapes of varied forms shaped over time. Despite the impression one might get when observing the city from the summit of the Serra do Curral—a mass of vertical concrete with roads cutting through microbasins deeply altered over the urban evolution process—the work conducted for almost two decades has significantly contributed to recognizing the importance of rehabilitating watercourses in the city. Simultaneously, the awareness that alternatives exist beyond river channeling and covering allows envisioning new horizons beyond the century-old technique responsible for the systematic increase in urban flooding.

Initiatives described in this chapter notably contributed to making the "invisible" rivers "visible" to a significant portion of the population that did not realize that beneath their feet lies an immense body of water waiting for recognition and reintegration into a city born under the auspices of progress nurtured by rivers still flowing from the Serra do Curral and the sporadic greenery that still exists in the municipality. This became evident during the rains of January 2020 and January 2024, creating a significant societal impact due to how the capital treated its waters throughout history.

For decades, the hydrographic network of Belo Horizonte has been signaling that interventions should cease, just as climate and soil indicate that greed and urban and social values need reconsideration. Climate changes have already begun to show that society must adapt to the new reality triggered by the Anthropocene. Faced with present difficulties, it is our task to envision the possibility of a better future. This is the moment when society as a whole must unite to set goals and carry out projects that will undoubtedly benefit not only our brief existence on this planet but also that of future generations.

It is in this sense that the projects described in this chapter contribute to ensure that a considerable part of the population takes a stand in favor of recognizing the environmental and social importance of river rehabilitation. In addition, by throwing lights on the invisible water courses of Belo Horizonte, the projects aim to argue for the development of policies and the adoption of techniques that seek to reduce the impacts of torrential rains in the city.

References

Almeida, L. Q. (2010). *Vulnerabilidades socioambientais de rios urbanos. Tese (Doutorado em Geografia)* (p. 278). Universidade Estadual Paulista.

Baptista, M., & Cardoso, A. (2013). *Rios e cidades: Uma longa e sinuosa história* (vol. 20, no. 2, pp. 124–153). Revista UFMG. jul./dez.

Bontempo, V., Wstane, C., Olivier, Ci., Oliveira, G. (2012) Gestão de águas urbanas em Belo Horizonte: avanços e retrocessos. *REGA - Revista de Gestão de Água da América Latina. Associação Brasileira de Recursos Hídricos (ABRH), 9*(1).

Borsagli, A. (2016). *Rios Invisíveis da Metrópole Mineira*. Clube de Autores.

Borsagli, A. (2017). *Sob a sombra do Curral del Rey: Contribuições para a história de Belo Horizonte*. Clube de Autores.

Borsagli, A. (2019). *Do convívio a ruptura: A cartografia na análise histórico-fluvial de Belo Horizonte (1894–1977)* (p. 224). Dissertação (Mestrado em Geografia)—Pontifícia Universidade Católica de Minas Gerais.

Borsagli, A. (2020). *Rios Urbanos de Belo Horizonte: Geografia histórica da rede hidrográfica da nova capital (1734/2020)*. Clube de Autores.

Borsagli, A., & Ferreira, R. G. B. (2018). *Horizontes Fluviais*. Clube de Autores.

Borriaud, N. (2002). Relational Aesthetics. Les Presses du Réel.

Botelho, R. G. M. (2011). Bacias Hidrográficas Urbanas. In A. J. T. Guerra (Ed.), *Geomorfologia Urbana*. Bertrand Brasil.

Brasil. A. N. Á. (2023). Comitês de Bacia Hidrográfica. Disponível em https://www.gov.br/ana/ptbr/assuntos/gestao-das-aguas/fortalecimento-dos-entes-do-singreh/comites-de-bacia-hidrografica. Acesso em 18 de setembro de.

Brito, F. S. R. (1944). Obras completas de Saturnino de Brito. *Rio de Janeiro: Imprensa Nacional*, v. XIX.

Cardoso, F. J. (2017). Ambientes fluviais urbanos: novos paradigmas de projeto. Tese (Doutorado em Urbanismo), Pontifícia Universidade Católica de Campinas, p. 365.
Castonguay, S., & Evenden, M. (2010). *Urban rivers: Remaking rivers, cities and space in Europe and North America*. University of Pittsburgh Press.
Cauquelin, A. (2007). A Invenção da Paisagem. Martins.
Costa, L. M. S. A. (org.). (2006). *Rios e paisagens urbanas em cidades brasileiras*. Viana & Mosley, Ed. PROURB.
de Lara, D. C. (2016). *O território das águas. Anais do XIV Seminário Cláudio Peres de práticas de ensino e geografia aplicada*. Pontifícia Universidade Católica de Minas Gerais.
Deutsche, R. (1998). The question of "Public Space". Lecture at the American Photography Institute National Graduate Seminar.
Diniz, C., Prado, I., & Mattos, J. (orgs.). (2022). *Isabela Prado: Sobre o rio*. Ed. da Autora.
Equipe ArchDaily Brasil. (2021). *Sobre o Rio: intervenção artística revela rios ocultos de Belo Horizonte*. ArchDaily. Disponível em https://www.archdaily.com.br/br/972769/sobre-o-rio-intervencao-artistica-revela-rios-ocultos-de-belo-horizonte
Fundação João Pinheiro/Centro de Estudos Históricos e Culturais. (1997). *Panorama de Belo Horizonte: Atlas Histórico*. Editora FJP.
Gorski, M. C. B. (2010). *Rios e Cidade. Ruptura e Reconciliação*. Senac.
Hoyuela Jayo, J. A. (2013). *Mapa Histórico Digital de Belo Horizonte*. Disponível em: http://www.mapahistoricodigitalbh.com.br/. Acesso em: 21 setembro 2023.
Medeiros, I. H. (2009). *Programa Drenurbs/Nascentes e fundos de vale: potencialidades e desafios da gestão socioambiental do território de Belo Horizonte a partir de suas águas*. Dissertação (Mestrado em Geografia.), Universidade Federal de Minas Gerais.
Mesquita, Y. M. (2013). *Jardim de asfalto: água, meio ambiente, canalização e as políticas públicas de saneamento básico em Belo Horizonte, 1948–1973*. Dissertação (Mestrado em História), Universidade Federal de Minas Gerais.
Monteiro Junior, L. (2011). *Infraestruturas urbanas: Uma contribuição ao estudo da drenagem em São Paulo*. Dissertação (Mestrado em Arquitetura e Urbanismo), Universidade de São Paulo.
Otto, B., Mc Cormick, K., & Leccese, M. (2004). *Ecological riverfront design: Restoring rivers, connecting communities* (pp. 518–519). American Planning Association, Advisory Service Report.
Prado, I. (2013). *(In)Visível sob a Cidade: o projeto Entre Rios e Ruas* (Vol. 20, No. 1, pp. 298–305) (jan./jun). Revista da UFMG.
Santos, M. (2008). A Natureza do Espaço: Técnica e Tempo, Razão e Emoção. Edusp.
Tucci, C. E. M. (1997). Água no meio urbano. In *Água Doce*. Universidade.

Afterwords

During the last two decades, climate change adaptation narrative has evolved significantly shifting from being considered a taboo, as if human beings would have been able to effectively address climate change through mitigation, to a clear necessity to be addressed in an **integrated** and **holistic** way. However, current climate adaptation action is still inadequate.

Within the climate adaptation narrative several concepts and keywords have been defined, forged, and used in order to make these endeavors long-lasting. For instance, the Intergovernmental Panel on Climate Change in its last report stated clearly that **residual climate risks**—that is risks remaining after ambitious adaptation efforts—will persist even if the Paris Agreement goals won't be reached. As a **cascade effect**, these risks will inevitably lead to both economic and non-economic **losses and damages**. Hence, climate change adaptation is considered nowadays as an urgent problem, which a part of the scholarship labels as a "super wicked problem", to be addressed by scientists and practitioners without any naiveté.

Concerning the information and knowledge chain flow of climate change adaptation, scholars have recently started analyzing the entire narrative through the **Science-Policy-Action** nexus, sometime framed in the dualistic **Theory–Practice** dichotomy. Aware that it is still an unresolved dilemma, nowadays scholars, from different disciplines, and practitioners have tentatively started working together to enhance the decision-making processes on the ground. The climate change adaptation Science-Policy-Action nexus carries gaps that often implies the underestimation of risks, undermines adaptation actions, and triggers maladaptation.

This book delved into this nexus and brought novelty into the climate change adaptation scientific production, addressing especially the water issue in urban context with a preponderant attention to **flood-related risks**.

The first innovation is the **bi-directional attention** in the two interfaces, i.e. Science-Policy and Policy-Action. In this perspective, the contributors of this book demonstrated the sensitivity to transcend their own fields of discipline and knowledge biases in favor of an efficient and effective climate adaptation knowledge production.

M. Granceri Bradaschia et al. (eds.), *Climate Change Adaptation, Flood Risk, and Beyond*, Planning for Climate Proof Cities, https://doi.org/10.1007/978-3-031-65463-3

The second innovation lies in addressing floods in urban context within a **multi-risk** frame. The complexity of pluvial, fluvial, and coastal floods, especially in urban settings, obliges the adaptation actors in widening the analytical perspective and, hence, including into the diagnosis and assessments at least another hazard or impact, e.g. droughts, landslides.

The third innovative insights come from the scale of investigation and action. The book presents climate risk analysis and adaptation research at the **regional scale**, which is still a field of research and practice yet to be properly explored.

Further novelties brought in this book are the inclusion of the **ecosystem service approach** into the climate risks assessments and the **bottom-up initiatives** in leveraging water-sensitivity.

All these lenses have demonstrated to be key and crucial in addressing three gaps, i.e. the **knowledge production gap**, the **implementation gap**, and the **learning gap**.

In Part I, the transition from scientific endeavors to policy formulation was addressed by:

- Gianfranco Pozzer, whose investigation delved into **hydrogeological vulnerability** in the Northern Apennines of Italy that provided nuanced insights into environmental susceptibilities.
- Davide Longato whose examination of flood-prone areas in the Friuli Venezia Giulia Region highlighted dynamics pertinent to **anticipatory measures**.
- Filippo Fraschini, Chiara Franciosi, and Marina Morando's study on climate change adaptation and **regional planning** in Liguria Region offered substantive perspectives on policy integration.
- Alessandra Longo, Linda Zardo, and Davide Longato's research on **ecosystem service mapping** emphasized the necessity of local readiness in flood risk mitigation.

Part II outlined the transition from policy formulation to actionable initiatives.

- Filippo Magni and Katia Federico's work on the Venice Lagoon outlined methodological frameworks for **multiscalar adaptation**, emphasizing adaptive management strategies.
- Jurgi Areizaga and Oliver Canosa's analysis of **coastal infrastructure** and **touristic assets** adaptation in the western Mediterranean exemplified resilience-building efforts.
- Matteo Marchesi, Massimiliano Granceri Bradaschia, Roberto Menghi, Matteo Riccardi, and Claudia Verna's case studies on sustainable energy and climate action plans emphasized **proactive planning's pivotal role** in climate change adaptation.
- Vittore Negretto's examination of systemic approaches to urban stormwater management advocated holistic solutions to multifaceted environmental challenges, providing practical solutions and **planning instruments** at the service of pluvial flood risk reduction.

- Jerome Dendura, Mercedes Sanchez Varela, and Massimiliano Granceri Bradaschia's discourse on **flood and adaptation financing** highlighted the transition towards comprehensive, enduring strategies from the international policy narrative to the local action, arguing about the build-back-better concept.
- Geraldo Carreiro's insights on **monitoring and evaluation** in climate adaptation underlined the significance of adaptive learning and the definition of **key indicators and metrics**.

In the concluding section, Part III, the transformative impact of enacted bottom-up actions on urban water policy paradigms was addressed.

- José Bueno, Luiz de Campos Jr., Alessio Mazzaro, and Massimiliano Granceri Bradaschia's examination of water sensitivity in São Paulo's urban governance exemplified **grassroots initiatives' influence** on policy discourse.
- Alessandro Borsagli, Isabela Prado, and Massimiliano Granceri Bradaschia's inquiry into Belo Horizonte's hydrographic network emphasized the confluence of **art**, **walking's perception**, and **geography** in informing water management policies.

Throughout this journey, a prominent leit motif arises: the necessity of interdisciplinary collaboration spanning various levels of analysis and the need to enhance permanent bi-directionality between science and policy as well as between policy and action.

In conclusion, this volume presents itself, within the technical-educational panorama, as a daring guide to understanding and addressing adaptation to climate change through the science-policy-action nexus. The Editors' hope is that the insights and knowledge shared in these pages will inspire further research and action in this critical field.

Francesco Musco
IUAV University of Venice

SPRINGER NATURE

GPSR Compliance

The European Union's (EU) General Product Safety Regulation (GPSR) is a set of rules that requires consumer products to be safe and our obligations to ensure this.

If you have any concerns about our products, you can contact us on ProductSafety@springernature.com

In case Publisher is established outside the EU, the EU authorized representative is:

Springer Nature Customer Service Center GmbH
Europaplatz 3
69115 Heidelberg, Germany

The manufacturer's authorised representative in the EU is Springer Nature Customer Service Centre GmbH, Europaplatz 3, 69115 Heidelberg, Germany. If you have any concerns regarding our products, please contact ProductSafety@springernature.com

Printed and bound by CPI Group (UK) Ltd, Croydon, CR0 4YY
01/10/2026
02225859-0009